Wireless Mesh Networks for IoT and Smart Cities

Technologies and applications

Edited by
Luca Davoli and Gianluigi Ferrari

The Institution of Engineering and Technology

Published by The Institution of Engineering and Technology, London, United Kingdom

The Institution of Engineering and Technology is registered as a Charity in England & Wales (no. 211014) and Scotland (no. SC038698).

First published 2022

The Institution of Engineering and Technology
Futures Place
Kings Way, Stevenage
Hertfordshire, SG1 2UA, United Kingdom

www.theiet.org

British Library Cataloguing in Publication Data
A catalogue record for this product is available from the British Library

ISBN 978-1-83953-282-5 (Hardback)
ISBN 978-1-83953-283-2 (PDF)

Typeset in India by Exeter Premedia Services Private Limited

Contents

About the Editors

Luca Davoli is a fixed-term assistant professor at the University of Parma, Italy. Since January 2014, he has been a member of the Internet of Things (IoT) Laboratory (https://iotlab.unipr.it) at the Department of Engineering and Architecture of the University of Parma, Italy. His main research interests include Internet of Things, software defined networking, big stream and peer-to-peer networks. He is an IEEE and a GTTI member. He received his PhD in information technologies from the Department of Information Engineering of the same university with a thesis entitled "Architecture and Technologies for the Internet of Things".

Gianluigi Ferrari is an associate professor of telecommunications at the University of Parma, Italy. Since September 2006 he has been the Coordinator of the Internet of Things (IoT) Laboratory (http://iotlab.unipr.it/) and, since 2016, the co-founder, president and CEO of things2i s.r.l. (http://www.things2i.com/), a spin-off company of the University of Parma dedicated to IoT and smart systems. His research activities revolve around signal processing, communication/networking, and IoT. He has published extensively in these areas and coordinated several technical projects, including EU-funded competitive projects. He is a senior member of the IEEE. He received his PhD in information technologies from the University of Parma, Italy.

Chapter 1

Wireless mesh network emulation

Ramon dos Reis Fontes[1], Augusto José Venâncio Neto[2], and Christian Esteve Rothenberg[3]

Experimental research on wireless networks is arduous and usually involves high costs. Hence, means for rapid prototyping along high-fidelity evaluation are highly desirable. This chapter presents Mininet-WiFi as a tool to emulate WMNs allowing high-fidelity experiments that replicate real networking stacks, protocols, and more. In order to bring the most complete experience possible, this chapter also provides practical guidelines on how to emulate WMNs. First, we introduce wireless interface modes tailored to wireless mesh supported by the Linux-based systems in addition to the most common wireless mesh routing protocols. Second, we showcase the emulation of IEEE 802.11p-based networks in vehicle and drone communication scenarios.

1.1 Introduction

The rapid progress in research and development of wireless communication technologies has created different system types, such as Bluetooth, WiFi, and 3/4/5G. These systems are designed to cater to specific service needs and differ in terms of bandwidth, latency, coverage area, cost, and quality-of-service (QoS) requirements. A challenge in this scenario comes up by the need for enabling users to move through the different networks while maintaining their connectivity with a high level of QoS and security. In this context, wireless mesh networks (WMNs) emerge as an alternative to guarantee interoperability among the different existing communication systems. There are a plethora of application scenarios tailored to WMNs, some of which will be explored throughout this chapter. For instance, WMNs can be exploited in projecting home networks, networks in the neighborhood, corporate networks, and metropolitan networks. The ability to self-organize a mesh network reduces the complexity of deploying

[1]IMD/UFRN, Brazil
[2]DIMAP/UFRN, Brazil
[3]FEEC/UNICAMP, Brazil

and maintaining the network itself. Moreover, the mesh network backbone provides a viable solution for users to keep connected ubiquitously (i.e., anywhere and anytime).

On the basis of supporting the scenarios covered throughout this chapter harnessing real-world technology perspectives, we will use the wireless network emulation feature. With this in mind, we will be able to experiment with different wireless technologies by using a single laptop. Hence, given its importance, we will introduce the Mininet-WiFi wireless network emulator along with a comprehensive description of the features supported by this emulator in Section 1.2.

The remainder of this chapter is organized as follows. In Section 1.3, we will learn around mesh-oriented wireless technologies for the use of experimentation atop Linux-based systems; Section 1.4 introduces a set of dynamic routing protocols; finally, but not least, we will explore in Section 1.5 the two use cases: one of which is devoted to vehicular ad hoc networks (VANETs), while the other to unmanned aerial vehicles (UAVs).

1.2   Mininet-WiFi: a primer

Simulators, emulators, and testbeds emerge as the most common experimental platform tools for experimentally driven research, intending to evaluate network systems' functionalities and performance skills. All these evaluation tools differ in their different degrees of abstraction. However, aspects including scalability, reproducibility, and cost-benefit, among others, distinguish both simulation and emulation as the most preferred methods [1]. The choice among the most appropriate approach for experimentally driven research to evaluate the functionality and performance of a network is always a trade-off. However, as shown in Figure 1.1, the following list of benefits are considered if we compare the network emulation approach

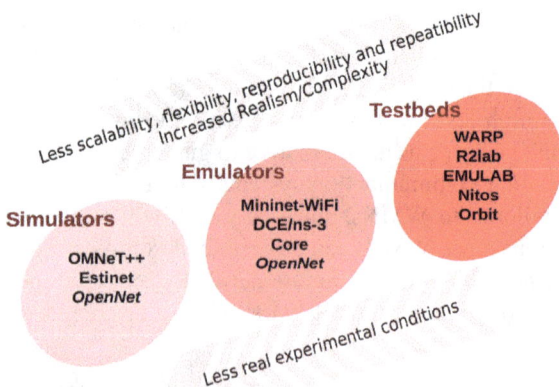

Figure 1.1   *Experimental platforms for wireless networks. Source: Adapted from [2]*

over simulators and physical testbeds: (i) it allows to run real code under realistic networking and computing conditions; (ii) it is easy to download and provides fast/interactive usage; and (iii) it provides more balanced evaluations.

The task of emulating networks offers more accurate and realistic outcomes. Thus, it has been widely used in performance evaluation experiments, protocol testing/debugging, as well as various activities concerning computer network architectures research. A researcher typically has several possible methods capable of evaluating and validating research data and network protocols, performing outcome analysis and other operations.

Figure 1.1 provides a nonexhaustive list of experimental platforms, such as OMNeT++ [3], Estinet [4], OpenNet [5], Core [6], WARP [7], R2lab [8], EMULAB [9], Nitos [10], and Orbit [11]. However, according to the literature, ns-3 [12] and Mininet-WiFi [2] are among the most popular approaches that both the academy and the industry use nowadays for wireless networking evaluation. However, we opt to use the Mininet-WiFi emulation tool all this chapter round for the reason of incorporating real-world network stacks, as well as allowing the use of third-party tools without modifications to the source code of these tools. If the reader is already a Linux user, he will certainly execute all of the proposed scenarios without further difficulties. If not, he will have the opportunity not only to know the Mininet-WiFi emulator but mainly to use tools that wireless network professionals adopt on Linux-based systems.

Written in Python, Mininet-WiFi is a Mininet-extended [13] wireless networks emulation tool. Mininet stands as a well-known emulator to researchers working in the field of software-defined networks [14, 15]. Mininet-WiFi allows users to set up and emulate WMNs on a single machine and has WiFi-native support. Through Mininet-WiFi, the user can virtualize stations, Access Points, and Mininet-supported node types (e.g., wired hosts and switches). In what concerns WMNs emulation support, Mininet-WiFi becomes essential since it supports multiple wireless mesh technologies and standards, especially those aimed at WiFi. The reader may write Python codes to emulate wireless networks with Mininet-WiFi or extend it with additional capabilities in a straightforward manner. Mininet-WiFi has the ability to emulate an entire network one-to-one in a laptop, whereas a process can represent each node in the network.

The emulation of the wireless network would not be possible without the mac80211 framework, which driver developers currently use to project drivers tailored to SoftMAC* wireless devices. More specifically speaking, Mininet-WiFi relies on the mac80211_hwsim module to simulate an arbitrary number of IEEE 802.11 radios for mac80211. The mac80211_hwsim module can be used to test most of the mac80211 functionalities and user space tools, such as `hostapd`, `wpa_supplicant`, and `iw` in a way that closely matches with the typical case of using real WLAN

---

*SoftMAC devices allow toward finer hardware control, allowing for the IEEE 802.11 frame management to be done in software for them, both parsing and generation of IEEE 802.11 wireless frames.

hardware. From the mac80211 viewpoint, mac80211_hwsim is yet another hardware driver, i.e., there is no need for changes to mac80211 to use this testing tool.

One of the advantages the mac80211_hwsim module brings is that it enables all the interface modes supported by mac80211—even if some mode is not supported by the physical network interface. We can confirm this assignment by running the iw list command, which will return the following output:

```
mininet − wifi > sta1 iw list
```

Supported interface modes :

- IBSS
- managed
- AP
- AP/VLAN
- monitor
- mesh point
- P2P − client
- P2P − GO
- outside context of a BSS

We can find among the supported interface modes listed above some that aimed at WMNs: (i) independent basic service set (IBSS)—corresponds to the traditional ad hoc mode interface; (ii) mesh point—stands to the mesh point interface that works with the open-source implementation of the ratified IEEE 802.11 seconds wireless mesh standard; (iii) P2P-client and P2P-GO that are used for WiFi-Direct link; and (iv) the outside context of a BSS (OCB) that IEEE 802.11p uses for wireless access in vehicular environments (WAVE) [16]—this mode has been enabled [17] by one of the authors of this chapter.

Figure 1.2 illustrates the components that participate in the Mininet-WiFi wireless network emulation, focusing on the wireless network stack of the Linux kernel driver. All the nodes use cfg80211 to communicate with the wireless device driver, a Linux IEEE

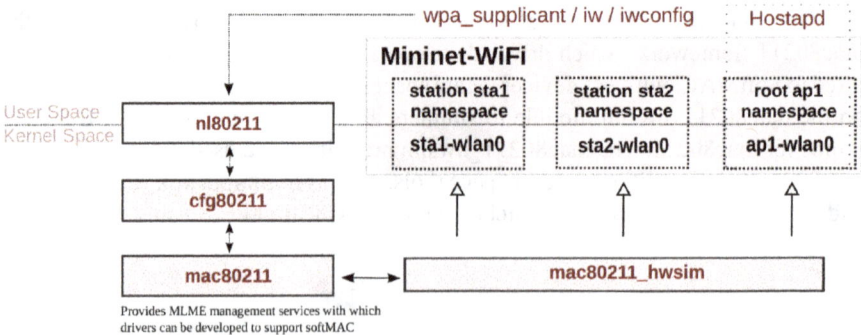

Figure 1.2   Components involved in the wireless network emulation

802.11 configuration API that provides communication between STAs and mac80211. This framework, in turn, communicates directly with the WiFi device driver through a netlink socket (or more specifically nl80211), which is used to set up the cfg80211 device and for kernel user space communication as well. This structure is essential, as the mac80211_hwsim module can automatically support any feature that is implemented for the kernel of the Linux system and Mininet-WiFi therefore.

Various WMN standards have been actively constituted for the last several years. Nevertheless, many more will undoubtedly come up—mainly due to the advent of the 5G and the Internet of Things (IoT) technologies. In the next sections, we will introduce five wireless network modes that Linux-based systems support aiming at wireless mesh emulation, namely IBSS, wireless distribution system (WDS), WiFi-Direct, IPv6 over low-power wireless personal area networks (6LoWPAN), IEEE 802.11p, as well as some of the most typical routing protocols used in WMNs. In addition to the detailed description of these standards and protocols operation, the reader will find pointers to guided instructions that will help in executing and visualizing different scenarios in practice. By concept, WMNs are dynamically self-organizing and self-configuring networks, whose nodes arranged in a meshed topology make up an ad hoc network. For this reason, we will often find the ad hoc nomenclature throughout this chapter.

1.3   Overview of wireless mesh technologies

1.3.1   IBSS (ad hoc)

An IBSS network, often called an ad hoc network, is a way to have a group of devices talking with each other wirelessly. It stands to a peer-to-peer (P2P) network example, in which all devices can talk directly to each other, with no inherent relaying. Since the IBSS network is a P2P network, the steps necessary to set up the WiFi link-layer should be the same on all devices. We do not have to worry about the necessary steps, as Mininet-WiFi assumes all the steps internally. However, at the level of curiosity, all steps are carried out with the support of iw [18], which is also used to afford creating the wireless interface in the IBSS operating mode.

In the computer networking field, we find two types of routing approaches: static and dynamic routing. On the one hand, routes are defined statically in static routing, as the name itself indicates. On the other hand, dynamic routing protocols are in charge of feeding the routing tables. The list of the main protocols projected for operating with wireless ad hoc networks routing includes ad hoc on-demand distance vector (AODV), optimized link state routing (OLSR), and better approach to mobile ad hoc networking (B.A.T.M.A.N.).

On the basis that routing protocols will be seen later, the instructions available at the link below can be used to understand how static routes can be created for the sake of allowing nodes to communicate with each other in a minimal wireless ad hoc network topology, as illustrated in Figure 1.3. Nevertheless, the iw tool will be used for the first time along this book to get some information about the ad hoc link.

https://github.com/mesh-book/instructions/blob/master/adhoc.md

Figure 1.3   Minimal wireless ad hoc topology

1.3.2   Wireless distribution system

The IEEE 802.11 standard defined two operation modes: infrastructure and ad hoc. In what concerns the infrastructure mode, one AP along with associated stations form a basic service set (BSS). Several BSSs can be connected with each other by a distribution system (DS), which forms an extended service set (ESS). Within an ESS, mobile stations roam and stay connected to the available network resource. DSs are usually wired links, and Ethernet is a typical example of DS. However, DSs can also be built with IEEE 802.11 wireless links, which is called WDS. In WDS-based WLANs, one AP (called base AP) is connected to the wired link, and one or more APs (called linked AP) can be linked with either the base AP or other linked APs. That is, WDS is a system enabling the wireless interconnection of APs in an IEEE 802.11 network using radio links. This is possible because WDS preserves the MAC addresses of client frames across links between APs.

The WDS introduces a set of requirements that the reader needs to be aware that although the service set identifiers (SSIDs) may be different, all base stations in a WDS cell must be (i) set to use the same radio channel and (ii) set with the same encryption method and the same encryption keys. WDS enables to provide two modes of connectivity:

- Wireless bridging, in which WDS APs communicate only with each other without allowing wireless STAs to access them, and
- Wireless repeating, where the APs communicate with each other and with the wireless clients.

Unfortunately, WDS may be incompatible with different products (even occasionally from the same vendor) since the IEEE 802.11-1999 standard does not define how to construct any such implementations or how stations interact to arrange frames of this format to be exchanged. For this reason, the Linux-supported WDS mode is not encouraged to be used—even though it is still supported and can still be experienced. In order to fix the incompatibility problem, even with the use of products developed by the same vendor, the Linux system implements the 4-address mode, as discussed in Section 1.3.2.1.

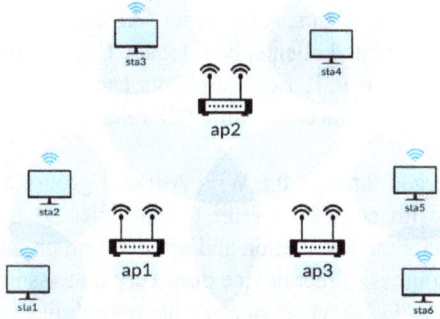

Figure 1.4   4-address network topology

1.3.2.1   The 4-address

The 4-address mode denotes how Linux-based systems (including embedded systems like OpenWRT) support WDS mode for mac80211 drivers. However, it is incompatible with other WDS implementations. That is, we will need all endpoints using this mode in order for WDS to work appropriately. Fortunately, all mac80211 drivers support 4-address mode if either master (e.g., access point) or managed (e.g., station) interfaces operation modes are supported.

In the case of a wired Ethernet, there are only two MAC addresses: the source and the destination (which is essentially the next hop address). Nevertheless, in the wireless LAN, while an IEEE 802.11 device is transmitting to a receiving device, either one (or both) of these devices may not be the actual source or destination of the Open System Interconnection (OSI) L2 traffic. Hence, this can create situations where we need four different distinct addresses: (i) transmitter address is used to send acknowledgments; (ii) receiver address indicates the receiver of the frame; (iii) source address is used only in a wireless DS as one AP forwards a frame to another AP and the source address of the original AP is contained here; and (iv) destination address is used for filtering by APs and the DS. The reader may want to refer to the IEEE 802.11-05/0710r0 document [19] for a complete view about 4-address.

A 4-address scenario can be emulated with the instructions available at the link below. This guided emulated scenario consists of testing the connectivity among stations connected to different APs that, in turn, are wirelessly connected to other APs with 4-address, as illustrated in Figure 1.4.

https://github.com/mesh-book/instructions/blob/master/4addr.md

1.3.3   WiFi direct

Initially called WiFi P2P, the WiFi Direct [20] has been proposed and standardized by the WiFi Alliance to facilitate the interconnection of nearby devices by supporting device-to-device (D2D) communications on WiFi channels. It has been designed following a hierarchical client-server architecture, where a single device manages all the communications within a group of devices. This means that devices willing to establish D2D communications must organize in groups, assuming the roles of group owners (GOs) and/or clients. A GO is an "AP-like" entity that provides BSS functionalities to the associated clients, both legacy (i.e., supporting standard IEEE 802.11 wireless access) and P2P. Legacy clients can communicate only with the GO by exploiting it as WLAN access, while P2P clients can also establish client-to-client communications.

WiFi Direct emerged through the WiFi Alliance group, a worldwide industry association charged with certifying WiFi technologies. It operates in frequency bands that do not require the installation and/or operation on the device, by working with two types of resources: direct device discovery that assumes the task of tracking devices that have WiFi Direct service, so that user choice with which to connect and Discovery, which list electronic devices.

WiFi Direct can be emulated, along with all the features supported by the Linux OS for this wireless technology, by following the instructions below. In particular, commands from `wpa_cli` [21] such as p2p_find, p2p_peers, and p2p_connect will be used.

https://github.com/mesh-book/instructions/blob/master/wifi-direct.md

1.3.4   IEEE 802.15.4 (6LoWPAN)

The 6LoWPAN protocol [22] is devoted to enabling IPv6 packets to be carried on top of low-power wireless networks, specifically IEEE 802.15.4. Several leading radio manufacturers have implemented IEEE 802.15.4, which specifies a wireless link for LoWPANs, characterized by small frame sizes, low bandwidth, and low transmit power. These make it widely used in embedded applications that generally require numerous low-cost nodes communicating over multiple hops to cover a large geographical area, as well as operating unattended for years on modest batteries.

As its primary task, 6LoWPAN adjusts IPv6 packets to wireless multihop communication's unique characteristics and requirements between low-power devices. The 6LoWPAN format defines how IPv6 communication is carried in 802.15.4 frames and specifies the adaptation layer's key elements. There are currently four basic header types defined in the standard: Dispatch Header, Mesh Header, Fragmentation Header, and the HC1 Header (IPv6 Header Compression Header) [23]. The only necessary headers are a Dispatch Header, an HC1 header, and the compressed IPv6 header for the simplest case. The three main mechanisms supported by 6LoWPAN are:

- **Header compression:** allows the transmission of IPv6 packets in as few as four bytes to ensure that the large IPv6 and transport-layer headers (Transmission Control Protocol (TCP)/User Datagram Protocol (UDP)) are reduced.
- **Fragmentation:** enables the fragmentation and reassembly of payloads larger than the size of the 802.15.4 frame (102 bytes of payload) to support the IPv6 minimum Maximum Transmission Unit (MTU) requirement.
- **Layer-two forwarding:** to deliver IPv6 datagram over multiple radio hops.

Experimenting 6LoWPAN with Mininet-WiFi is possible, thanks to the mac80215 4_hwsim module that the latest kernel versions of the Linux OS already embeds, as well as the Linux WPAN network development [24]. This initiative brought with it a tool called iwpan that is based on iw. Through the instructions below, we will have the opportunity to know iwpan and how 6LoWPAN currently works on Linux operating systems through the mac802154_hwsim module.

https://github.com/mesh-book/instructions/blob/master/6lowpan.md

1.3.5   IEEE 802.11p

Currently, industrial actors and investigated by regulatory organizations consider two competing technological approaches toward vehicle-to-everything (V2X) communications: Cellular-V2X (C-V2X—3GPP Release 14 [25]) and dedicated short-range communications (DSRC in the US [26] and European Telecommunications Standards Institute (ETSI) intelligent transport systems (ITS-G5) [27] in Europe). The C-V2X claims to have a wider range of applications in areas such as entertainment, traffic data, navigation, and, most notably, autonomous driving (especially New Radio (NR) C-V2X). The DSRC, on the other hand, has been explicitly designed for vehicular applications and, in particular, for collision prevention applications. Aside from that, these two technologies address identical use-cases while matching the same network, security, and application layers.

While the US, Europe, and Japan deploy DSRC-based V2X, C-V2X is gaining momentum in other regions, such as China. In the US, thousands of DSRC roadside units are equipped with DSRC V2X, whereas original equipment manufacturers began planning their deployment. The DSRC-based V2X standard is based on the IEEE 802.11p access layer developed for vehicular networks, while the C-V2X is based on the not yet ready 3GPP's LTE-V2X technology. The IEEE 802.11p (WAVE) represents the most mature set of standards tailored to DSRC/WAVE networks. The IEEE 802.11p amendment defines MAC and PHY layer protocols that DSRC nodes use. Although initially conceived by a task group formed in November 2004 and developed in 2010, the IEEE 802.11p is still an emerging family of standards intended to support wireless access in VANETs.

Experimenting with IEEE 802.11p on Linux-based systems is possible, thanks to the support for the outside the context of a basic service set (OCB). Ramon Fontes added support for the OCB and the 5.9 GHz band [17]. The OCB allows unicast, multicast, and broadcast data communication without any MAC sublayer setup and guarantees, at least for safety-related applications, in which noncoexisting BSS will operate on the 5.9 GHz DSRC band. Due to this contribution, experimenting with IEEE 802.11p requires the user to use the kernel version (at least) 5.8 of the Linux OS. Moreover, the 5.9 GHz band may need to be added to the WLAN regulatory domain of the system. A definition of the WLAN regulatory domain can be a bounded area under the control of a set of laws or policies. Currently, there are governing bodies in many countries around the world that follow a standard set by the Federal Communications Commission [28], ETSI [29], Japan, Israel, etc. By default, the computer may not enable the 5.9 GHz band. For this reason, it is necessary to enable it through a custom regulatory domain that will enable the 5.9 GHz band. This chapter does not provide instructions on how to enable 5.9 GHz, but the reader can follow the steps available at the Mininet-WiFi website [30]. In general, we will have to install some packages and then install and configure both wireless – regdb and Central Regulatory Domain Agent [31].

We can have our first scenario with IEEE 802.11p through the instructions available at the link below. There will be possible to identify if the system already supports 5.9 GHz or not.

https://github.com/mesh-book/instructions/blob/master/80211p.md

1.4   Routing protocols for WMN

The increase in availability and popularity of mobile wireless devices has led researchers to develop a wide variety of mobile ad hoc networking (MANET) protocols to exploit the unique communication opportunities presented by these devices. There are many routing technologies and routing protocols applied to MANETs, such as IEEE 802.11 seconds, OLSR, Babel, and B.A.T.M.A.N.

In this section, we will introduce these routing protocols. More specifically, we will introduce the main solutions of these protocols that Linux operating systems enable. Complementary, interesting discussions can be found in the wireless battle mesh [32] initiative. At the end of this section, we will find a comparative table that will allow us to have a comprehensive overview of each routing protocol.

1.4.1   IEEE 802.11s

The IEEE 802.11 seconds [33] stands for the first wireless LAN routing protocol we cover in this chapter. It is an IEEE 802.11 amendment for mesh networking whose development started in 2004. The final proposal was approved in July 2011 and published in late November 2011, being part of the IEEE standard 802.11-2012. It describes a WMN concept that introduces routing capabilities at the MAC layer. Path selection is used to refer to MAC-address-based routing and to differentiate it from conventional IP routing. IEEE 802.11 seconds requires hybrid wireless mesh protocol (HWMP) [34] to be supported as a default. However, other routing protocols (e.g., OLSR and B.A.T.M.A.N.) may be supported or even static routing.

According to [33], the IEEE 802.11 seconds not only helps interconnect BSSs wirelessly, and thereby fills the WDS gap, but also enables a new type of BSS, the so-called mesh BSS (MBSS). IEEE 802.11 seconds supports transparent delivery of unicast, multicast, and broadcast frames to destinations in- and outside of the MBSS (referred to as mesh in the following). Devices that form the mesh are called mesh stations (mesh STAs) that forward frames wirelessly but do not communicate with nonmesh stations. However, a mesh station may be collocated with other IEEE 802.11 entities.

The IEEE 802.11 seconds frame structure looks like IEEE 802.11, with the difference that IEEE 802.11 seconds extends data and management frames to provide for multihop, by an additional mesh control field that consists of a mesh time to

2 octets	2 octets	6 octets	6 octets	6 octets	2 octets	6 octets	2 octets	4 octets	0–7955 octets	4 octets
Frame control	Duration/ID	Address 1	Address 2	Address 3	Sequence control	Address 4	QoS control	HT control	Body	FCS
		Receiver address	Transmitter address	Mesh destination address						

Mesh control — 6, 12, 18, or 24 octets

1 octet	1 octet	4 octets	0, 6, 12, or 18 octets
Mesh flags	Mesh time to live (TTL)	Mesh sequence number	Mesh address extension

2 bits	6 bits
Address extension mode	Reserved

Mesh source address	Destination address	Source address

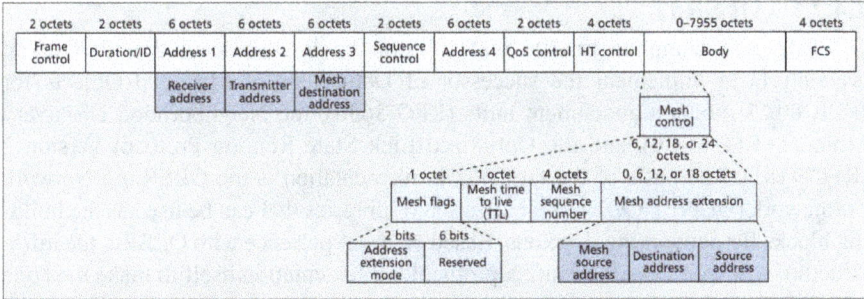

Figure 1.5   The IEEE 802.11 seconds mesh control field is part of the frame body and provides up to two more address fields [33]

live field, a mesh sequence number, a mesh flags field, and possibly a mesh address extension field, as illustrated in Figure 1.5.

In the Linux kernel, the IEEE 802.11 seconds amendment is supported by the open80211s [35] project, which is a reference implementation of the IEEE 802.11 seconds standard on Linux. The Open80211s is based on the mac80211 wireless stack and should run on any of the wireless cards that mac80211 supports. That is, it is a vendor-neutral implementation of IEEE 802.11 seconds for the Linux OS. This is possible mainly because IEEE 802.11 seconds introduces only minimal changes to the MAC layer, and it can be implemented in software and made to run even on legacy IEEE 802.11 cards. One of the major disadvantages of this protocol is that only smaller meshes under 32 nodes are supported.

The IEEE 802.11 seconds can be emulated through the instructions available at the link below. There, we will know the iw-supporting features, which can be used to obtain the mesh path and perform dump.

https://github.com/mesh-book/instructions/blob/master/mesh.md

1.4.2   OLSRd

The OLSR daemon (OLSRd) [36] (version 1) is an implementation of the OLSR protocol [37]. OLSR operates as a table-driven, proactive protocol, i.e., it exchanges topology information regularly with other network nodes. Each node selects a set of its neighbor nodes as "multipoint relays" (MPR). In OLSR, only nodes, selected as MPRs, are responsible for forwarding control traffic intended for diffusion into the entire network. MPRs provide an efficient mechanism for forwarding control traffic by reducing the number of transmissions required.

A wireless mesh scenario can be emulated through the instructions available at the link below. In it, we will use the route command that will allow us to view the routes created by OLSRd.

https://github.com/mesh-book/instructions/blob/master/olsrd.md

1.4.2.1   OLSRd2

OLSRd2 is a complete rewrite that follows the lessons learned with OLSRd (version 1) to implement the successor of Definitions of Managed Objects for IEEE 802.3 medium attachment units (RFC 3636), the Neighborhood Discovery Protocol (RFC 6130), and the Optimized Link State Routing Protocol Version 2 (RFC 7181). The basis of the OLSRd2 implementation is the OLSR.org Network Framework (OONF). OONF is a collection of libraries that can be used as the building blocks for networking daemons. Based on the experience with OLSRd, the infrastructure was split from the routing protocol implementation itself to make the code easier to reuse for other people.

The instructions available through the link below will show how to emulate a wireless mesh scenario using OLSRd2.

https://github.com/mesh-book/instructions/blob/master/olsrd.md

1.4.3   Babeld

According to [38], babeld is a loop-avoiding distance-vector routing protocol for IPv6 and IPv4 with fast convergence properties that is robust and efficient both in ordinary wired networks and in WMNs. Babeld implements a sophisticated scheduling scheme to handle outgoing messages. This means that outgoing messages are not sent immediately, since they can be delayed for up to a few seconds. This is done to avoid synchronization issues between nodes. It also allows the aggregation of multiple messages into a single UDP packet, which helps lower overhead.

Babeld is not the only implementation currently available for the babel routing protocol. Other implementations include Pybabel [39] and Sbabeld [40]. Although the instructions below enable us to test babeld, we will certainly be able to work with the other babel implementations if the reader knows the Linux OS and the programming language used by these other implementations.

https://github.com/mesh-book/instructions/blob/master/babel.md

1.4.4   B.A.T.M.A.N.

B.A.T.M.A.N. [41] is a routing protocol tailored to multihop mobile ad hoc networks, which is intended to replace the OLSR protocol. The B.A.T.M.A.N. algorithm's approach is to divide the knowledge about the best end-to-end paths between nodes in the mesh to all participating nodes. Each node perceives and maintains only the information about the best next hop toward all other nodes. Such routing schemes are best suited for low CPU consumption, therefore impacting less battery consumption for each node, making it an ideal candidate for routing in IoT.

At the time of reading this chapter, the user will be able to find two variant protocols of the B.A.T.M.A.N. working on Linux-based systems: B.A.T.M.A.N. daemon [42] (or batmand), which operates on Layer 3 (L3) of the OSI model; and B.A.T.M.A.N. advanced [43] (or simply batman − adv) that is an implementation of the B.A.T.M.A.N. routing protocol in the form of a Linux kernel module operating on Layer 2 (L2).

1.4.4.1   Batmand

As mentioned in Section 1.4.4, the B.A.T.M.A.N. daemon is built on OSI L3 using a user space daemon called the B.A.T.M.A.N. daemon, or simply batmand. Hence, it works by changing the routing table using a proactive routing method. This protocol's significant characteristic is that there is a decentralization of the knowledge about the entire topology. Unlike most other routing protocols, this protocol does not try to find the best path but, instead, uses a greedy approach to send the packet along the best neighbor to reach the destination. Through the link below, we can emulate a wireless network leveraging the batmand and learn how to use the batmand command.

https://github.com/mesh-book/instructions/blob/master/batmand.md

1.4.4.2   Batman-adv

The B.A.T.M.A.N. advanced (batman − adv), on the other hand, qualifies as L2 routing protocol that requires MAC addresses for routing. In batman − adv, protocols above L2 are unaware of the network's multihop nature, while in batmand, protocols are well aware of the network hops. Due to its implementation on the Ethernet layer, the network acts as if every node has a direct single-hop connection to the other node. Network topology with batman − adv can be found from the instructions available at the link below. There, we will also learn how to use batctl, a configuration, and debugging tool for batman − adv.

https://github.com/mesh-book/instructions/blob/master/batman-adv.md

1.4.5   Summary

Table 1.1 summarizes the routing protocols mentioned in Sections 1.4.1–1.4.4. In addition to the OSI Layer and the routing taxonomy, we can observe that the codes of all the routing protocols commented in this chapter are free for use. Regarding the routing management concerns, all the routing protocols, but the Open80211s, act proactively. In the Open80211s, the reactive component is the foundation of the HWMP, and it is based on the AODV protocol. However, its design was improved to include a radio-aware link metric to determine the best route instead of a simple hop-count known as Radio-Metric AODV (RM-AODV). Regarding the routing taxonomy, while open80211s introduces the radio-metric AODV approach, OLSR considers the link-state approach, and the Babel considers the distance vector approach. On the other hand, the B.A.T.M.A.N. incorporates a bio-inspired nature, which means it does not try to discover or calculate routing paths. Instead, B.A.T.M.A.N. tries to detect which neighbor offers the best path to each originator.

The last activity column, in turn, brings important insights regarding the code activity. It is worth mentioning that it has been a long time since open80211s has had updates. Moreover, OLSR is in hard freeze and no longer maintained. This sounds strange because it has a more recent update than OLSRv2; however, the updates made did not bring important changes to the OLSR code. We can also notice that batmand is falling behind while both Babel and batman − adv have an active community. To sum up, they appear to be more promising in terms of future updates and

Table 1.1   Comparison among the routing protocols until December 2020

	Open IEEE 802.11s	OLSR	OLSRv2	Babel	Batmand	Batman-adv
Availability	Open-source code					
Routing management	Reactive/ Proactive	Proactive				
Routing taxonomy	Radio-metric AODV	Link state	Link state	Distance vector	Bioinspired nature*	Bioinspired nature*
Last activity	Sep 2013	Jun 2020	Sep 2018	Dec 2020	Mar 2016	Dec 2020
OSI layer	L2	L3	L3	L3	L3	L2

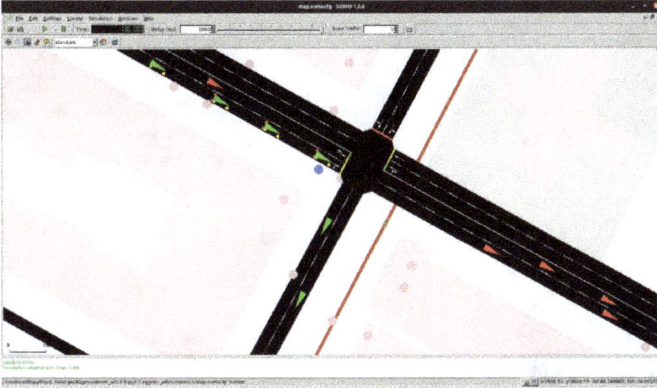

Figure 1.6   Vehicular emulation with SUMO

implementation of new features than other routing protocols. Lastly, open80211s and `batman − adv` are the only protocols to operate on layer two of the OSI model.

1.5 Experimental use cases

In this section, we present initial guidelines and pointers for experimenting with two use cases: the first one is about VANETs, whereas the second one refers to UAVs networks. The available scripts are not frozen, which means that the reader can also contribute to the scripts initially prepared for this section.

1.5.1 A realistic vehicular experimentation

VANETs are defined as a particular class of mobile ad hoc networks (MANETs) composed of vehicles equipped with wireless gadgets. VANETs feature unique functionalities that distinguish them from other mobile networks, such as the fast speed at which their nodes move and their highly modifiable topology, making the links created between networked vehicles happen for only a few seconds almost immediately. In addition to the support of IEEE 802.11p, Mininet-WiFi interworks with the simulation of urban mobility (SUMO), an open-source, highly portable, microscopic, and continuous multimodal traffic simulation package designed for the simulation of vehicular networks. SUMO provides a graphical interface called sumo-gui (see Figure 1.6) and can be used for many research purposes, such as forecasting, evaluation of traffic lights, route selection, or in the field of vehicular communication systems. Mininet-WiFi comes with a preconfigured map that was extracted from `OpenStreetMap` [44], a tool able to incorporate multiway maps from around the world, resembling Google Maps in this aspect. However, the user can select their own map from `openstreetmap` or even using maps generated by researchers around the world.

In order to emulate a realistic vehicular network with IEEE 802.11p, we can follow the instructions that the link below carries. The network topology contains

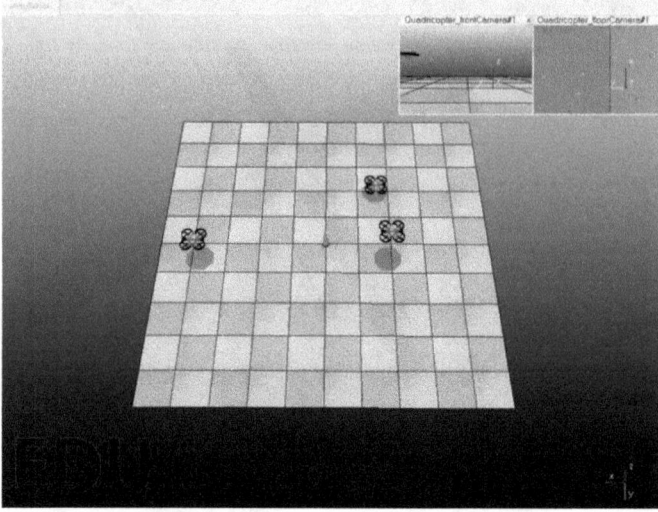

Figure 1.7   Drone emulation with CoppeliaSim

10 cars, and we can try to communicate them with the ping command and use iw to check the association among the nodes.

https://github.com/mesh-book/instructions/blob/master/sumo.md

1.5.2   Unmanned aerial vehicles

Let's talk about drone emulation right now. First of all, it is important to mention that no new wireless interface mode supported by mac80211 will be introduced in this section. Unlike the scenario aimed at vehicle networks where there is a specific wireless technology for cars, we can consider all the previously supported modes for the UAV communication, including the infrastructure mode, where UAVs can also work as APs. Recently, UAVs, also known as drones, have attracted significant attention attributed to their high mobility, low cost, and flexible deployment. Thus, UAVs may potentially overcome the challenges of the Internet of Everything, which is presented as an omnipotent extension of the IoT. Moreover, it is an important component of the upcoming wireless networks that can potentially facilitate wireless broadcast and support high-rate transmissions, enabling ubiquitous and enhanced broadband services as well as smart/autonomous navigation [45]. Indeed, there are many cases where UAVs can be applied, and many of them have been explored by researchers around the world.

In the UAV emulation, we will use Mininet-WiFi with the CoppeliaSim robot simulator [46]. CoppeliaSim is a simulation tool that allows experimentation with virtual robots and provides a graphical interface (see Figure 1.7) that allows creating and editing different simulated models, as well as designing the environment with the necessary elements. Similar to what happens with the integration between Mininet-WiFi and SUMO, the integration with CoppeliaSim passes only by capturing the UAVs' position. That is, once Mininet-WiFi is aware of the UAV position that Mininet-WiFi performs in the entire network part. Therefore, everything that concerns UAVs regarding their mobility and characteristics is part of the resources supported by CoppeliaSim.

The instructions that the link below contains will allow us to emulate a UAV network using the IBSS interface mode introduced previously. The nodes will have mobility defined by CoppeliaSim, and we will use for the first time the batmand − adv routing protocol, which Section 1.4.4.2 introduces in more detail.

https://github.com/mesh-book/instructions/blob/master/uav.md

1.6   Conclusion

In this chapter, we intend to cover valuable experimental alternatives to wireless mesh technologies supported by the Linux operating system at the emulation. More specifically, we showcase the features supported by Mininet-WiFi wireless network emulator to understand how mesh wireless networks can be experienced with this emulator. In addition, we present relevant dynamic routing protocols that can be applied in the context of WMNs. Finally, we introduce two network scenarios focused on vehicular networks and drones. To follow up on the topics covered in this chapter, we invite the readers to collaborate with the open-source tutorials presented throughout this chapter through suggestions and sharing of experiences through the source code page. Regarding future work, we seek to expand the possibilities so that new features can be supported for the emulation of WMNs through collaborative efforts around the Linux operating system kernel.

References

[1]   Imran M., Said A.M., Hasbullah H. 'A survey of simulators, emulators and testbeds for wireless sensor networks'. *International Symposium on Information Technology*. 2010;2:897–902.

[2]   Fontes R.R., Afzal S., Brito S.H.B., Santos M.A.S., Rothenberg C.E. 'Mininet-WiFi: emulating software-defined wireless networks'. 2015 11th International Conference on Network and Service Management (CNSM); 2015. pp. 384–9.

[3]   Varga A., Hornig R. 'An overview of the OMNeT++ simulation environment'. Proceedings of the 1st International Conference on Simulation Tools and Techniques for Communications, Networks and Systems & Workshops. Simutools'08. ICST, Brussels, Belgium, Belgium: ICST (Institute for Computer Sciences, Social-Informatics and Telecommunications Engineering); 2008. pp. 1–60.

[4]   Wang S.Y., Chou C.L., Yang C.M. 'EstiNet openflow network simulator and emulator'. *IEEE Communications Magazine*. 2013;51(9).

[5]   Chan M., Chen C., Huang J., Chan M., Chen C., Huang J. 'OpenNet: A simulator for software-defined wireless local area network'. IEEE Wireless

Communications and Networking Conference; WCNC, Istanbul, Turkey; 2014. pp. 3332–6.

[6] Ahrenholz J., Danilov C., Henderson T.R., Ahrenholz J. 'CORE: a real-time network emulator'. IEEE Military Communications Conference. Institute of Electrical and Electronics Engineers (IEEE); 2008.

[7] WARP. WARP Project [online]. 2021. Available from http://warpproject.org [Accessed 4 Jan 2021].

[8] Testbeds OFI. R2lab Testbed [online]. 2021. Available from http://r2lab.inria. fr/ [Accessed 4 Jan 2021].

[9] Hibler M., Ricci R., Stoller L., *et al.* Large-scale virtualization in the emulab network testbed. USENIX Annual Technical Conference. ATC'08; Berkeley, CA, USA; 2008. pp. 113–28.

[10] Pechlivanidou K., Katsalis K., Igoumenos I., Pechlivanidou K., Katsalis K., Igoumenos I. NITOS testbed: a cloud based wireless experimentation facility. 26th International Teletraffic Congress (ITC); 2014.

[11] Raychaudhuri D. 'Orbit: open-access research testbed for next-generation wireless networks'. *Proposal submitted to NSF Network Research Testbeds Program*. 2003.

[12] Carneiro G. NS-3: network simulator 3. UTM Lab Meeting April; 2010. pp. 4–5.

[13] Lantz B., Heller B., McKeown N. 'A network in a laptop: rapid prototyping for software-defined networks'. Proceedings of the 9th ACM SIGCOMM Workshop on Hot Topics in Networks; 2010.

[14] Feamster N., Rexford J., Zegura E. 'The road to SDN: an intellectual history of programmable networks'. *ACM SIGCOMM Computer Communication Review*. 2014;44(2):87–98.

[15] Haleplidis E., Pentikousis K., Denazis S., Hadi Salim J., Meyer D., Koufopavlou O. Software-Defined Networking (SDN): Layers and Architecture Terminology, RFC 7426. 2015.

[16] Jiang D., Delgrossi L. Towards an international standard for wireless access in vehicular environments. IEEE Vehicular Technology Conference; 2008. pp. 2036–40.

[17] Fontes R. *mac80211_hwsim: add support for OCB* [online]. 2020. Available from https://github.com/torvalds/linux/commit/7dfd8ac327301f302b030720 66c66eb32578e940 [Accessed 4 Jan 2021].

[18] Wireless Linux. *The iw tool* [online]. 2021. Available from https://wireless. wiki.kernel.org/en/users/documentation/iw [Accessed 4 Jan 2021].

[19] *IEEE. 4-address format* [online]. 2021. Available from http://www.ieee802. org/1/files/public/802_architecture_group/802-11/4-address-format.doc [Accessed 4 Jan 2021].

[20] Alliance W.-F. Discover Wi-Fi Direct. 2021. Available from http://www.wi-fi. org/discover-and-learn/wi-fi-direct [Accessed 4 Jan 2021].

[21] Archlinux. *WPA Supplicant* [online]. 2021. Available from https://wiki.arch-linux.org/index.php/wpa_supplicant [Accessed 1 Jan 2021].

[22] Hui E.J., Corporation A.R., Thubert P. *Compression format for IPv6 datagrams over IEEE 802.15.4-based networks* [online]. 2011. Available from https://tools.ietf.org/html/rfc6282 [Accessed 1 Jan 2021].

[23] Mulligan G. 'The 6LoWPAN architecture'. Proceedings of the 4th Workshop on Embedded Networked Sensors; 2007. pp. 78–82.

[24] Linux. *wpan* [online]. 2021. Available from https://linux-wpan.org [Accessed 4 Jan 2021].

[25] 3GPP. *3GPP Release 14* [online]. 2021. Available from https://www.3gpp.org/release-14 [Accessed 4 Apr 2021].

[26] Shulman M., Deering R. 'Vehicle safety communications in the United States'. Conference on Experimental Safety vehicles; Citeseer; 2007.

[27] ETSI T., Systems I.T. 'Intelligent transport systems (ITS): access layer specification for intelligent transport systems operating in the 5 GHz frequency band'. *EN*. 2013;302(663):V1–21.

[28] Coase R.H. 'The Federal Communications Commission'. *The Journal of Law and Economics*. 1959;2:1–40.

[29] ETSI. *European Telecommunications Standards Institute* [online]. 2020. Available from https://www.etsi.org/ [Accessed 4 Jan 2021].

[30] Fontes R. *Mininet-WiFi* [online]. 2021. Available from https://mininet-wifi.github.io/ [Accessed 4 Jan 2021].

[31] kernel W. *wireless-regdb* [online]. 2021. Available from https://wireless.wiki.kernel.org/en/developers/regulatory/wireless-regdb [Accessed 4 Jan 2021].

[32] Battlemesh. *Wireless Battle Mesh* [online]. 2021. Available from https://www.battlemesh.org/ [Accessed 4 Jan 2021].

[33] Hiertz G.R., Denteneer D., Max S., *et al.* 'IEEE 802.11s: the WLAN mesh standard'. *IEEE Wireless Communications*. 2010;17(1):104–11.

[34] Joshi A. HWMP Protocol specification. 2006. Available from https://mentor.ieee.org/802.11/public/06/11-06-1778-01-000s-hwmp-specification.doc [Accessed 02 Jan 2021].

[35] Open80211s. *Open80211s Source Code* [online]. 2021. Available from https://github.com/o11s/open80211s/ [Accessed 4 Jan 2021].

[36] OLSR. *OLSR* [online]. 2021. Available from olsr.org [Accessed 4 Jan 2021].

[37] Clausen PJ T., Hipercom P. *RFC3626* [online]. 2003. Available from https://tools.ietf.org/html/rfc3626 [Accessed 2 Jan 2021].

[38] Chroboczek J. *RFC6126* [online]. 2011. Available from https://tools.ietf.org/html/rfc6126 [Accessed 2 Jan 2021].

[39] Clausen PJ T., Pybabel H.P. Pybabel source code. 2021. Available from https://github.com/fingon/pybabel/ [Accessed 2 Jan 2021].

[40] sbabeld. 2003. Available from https://github.com/jech/sbabeld/ [Accessed 3 Jan 2021].

[41] Seither D., König A., Hollick M. 'Routing performance of wireless mesh networks: a practical evaluation of BATMAN advanced'. IEEE 36th Conference on Local Computer Networks; 2011. pp. 897–904.

[42] Open-Mesh. *B.A.T.M.A.N. daemon documentation overview* [online]. Available from https://www.open-mesh.org/projects/batmand/wiki [Accessed 4 Jan 2021].

[43]   Open-Mesh. *B.A.T.M.A.N. advanced documentation overview* [online]. Available from https://www.open-mesh.org/projects/batman-adv/wiki [Accessed 4 Jan 2021].

[44]   OpenStreetMap. *OpenStreetMap* [online]. Available from https://www.open-streetmap.org [Accessed 4 Jan 2021].

[45]   ZhaoC., CaiY., LiuA., ZhaoM., HanzoL. 'Mobile edge computing meets mmwave communications: joint beamforming and resource allocation for system delay minimization'. *IEEE Transactions on Wireless Communications*. 2020;19:2382–96.

[46]   Robotics C. CoppeliaSim. Available from https://www.coppeliarobotics.com/ [Accessed 4 Jan 2021].

Chapter 2

A sink-oriented routing protocol for blue light link-based mesh network

Luca Davoli[1,2], Massimo Moreni[3], and Gianluigi Ferrari[1,2]

The need to leverage "smart" mechanisms to route data among heterogeneous devices is a key aspect in modern scenarios and applications, especially those targeting the integration of existing systems in the Internet of Things (IoT)-oriented environments. To this end, the exploitation of the Bluetooth Low Energy (BLE) protocol, especially its advertisement channels, allows a large amount of devices to interact, collect, and exploit data for future-proof applications. Therefore, the definition of routing protocols exploiting BLE advertisement channels and being able to target different classes of BLE nodes is useful for heterogeneous IoT scenarios.

2.1 Introduction

Among the widely adopted technologies equipping devices used every day by almost everyone, one of the most interesting and pervasive (due to its widespread diffusion) is BLE, which has emerged as a major low-power wireless technology. In detail, BLE allows low-energy communication in heterogeneous environments and scenarios, allowing sensors, actuators, smartphones, wearables, etc., to exchange data and perform further actions. Moreover, thanks to its wide availability, even standardization entities – such as the Bluetooth Special Interest Group (SIG) [1] and the Internet Engineering Task Force (IETF) [2] – are defining adaptation mechanisms to facilitate the interaction of BLE devices within IoT-oriented scenarios – e.g., the definition of the support of IPv6 over BLE [3] to be applied in constrained devices.

Nevertheless, BLE networks are traditionally organized with a star topology, so they suffer from a major drawback related to limited coverage range [4]: this could sometimes represent a limitation in various scenarios (e.g., urban, agricultural, industrial), where alternative topologies may improve and simplify the communication

[1]Internet of Things (IoT) Laboratory, Department of Engineering and Architecture, University of Parma, Parco Area delle Scienze, Parma, Italy
[2]things2i s.r.l., Parco Area delle Scienze, Parma, Italy
[3]TCI Telecomunicazioni Italia s.r.l., Via Parma, Saronno, Italy

and interaction of devices [5]. As an example, an alternative approach can rely on BLE-based wireless mesh networks (WMNs) [6, 7], which require the adoption of mesh-oriented mechanisms for end-to-end communications, allowing to overcome coverage limitations of a star topology [8, 9].

On the basis of these advantages, in this chapter, we discuss a BLE-oriented routing protocol for Point-to-Point, MultiPoint-to-Point, and broadcast communications, to be applied to heterogeneous contexts where over-the-air operations and update mechanisms are needed. In particular, BLE devices, sensing the environment and injecting their collected data inside the BLE-based WMN, will be the main actors of the network itself. The proposed routing protocol is deployed on BLE devices composed of a host micro-controller – performing all the operating tasks – and a BLE System-on-Chip (SoC) managing all the networking functionalities. The host can request the SoC to forward to the BLE WMN specific packets through which each receiving node can perform particular operations. Finally, the WMN (based on the BLE protocol) exploits the BLE advertisement channels transmitting the traffic with a flooding approach – encrypting the traffic through a symmetric key-based algorithm – and is composed of *on-field* devices, in charge of performing sensing and actuating tasks, management devices, denoted as sink nodes or gateways (GWs),* and external nodes interested in obtaining data from the sink nodes for further actions.

The remainder of this chapter is organized as follows. In Section 2.2, an overview of existing mesh-oriented BLE-based solutions is presented. In Section 2.3, the proposed BLE-based routing protocol is detailed and analyzed, while its main downlink and uplink data collection operations are discussed in Sections 2.4, 2.5, and 2.6, respectively. In Section 2.7, two representative use cases supported by the proposed routing protocol are shown. Finally, in Section 2.8 we draw our conclusions.

2.2   Related works

The possibility to rely on BLE-based communications is attractive in different contexts, ranging from domestic scenarios to urban and industrial ones. This has triggered different (also standardization) initiatives aimed at enabling the interaction among devices in BLE-based WMNs. In particular, SIG and IETF have guided this process, proposing the Bluetooth Smart Mesh Working Group [10] and the adaptation of the IPv6 Low-Power Wireless Personal Area Networks protocol [11] to support IPv6 over BLE networks [3, 12, 13]. To this end, two main BLE-oriented categories exist, namely (i) flooding-based solutions and (ii) routing-based solutions.

---

*In the following, the terms "sink" and "GW" will be used alternatively, but with the same meaning.

- Flooding-based solutions do not perform any kind of routing among the nodes composing the network, while instead broadcasting packets over BLE advertising channels. Examples of these solutions are proposed in Kim *et al.* [14], where authors defined BLEmesh, a bounded flooding mechanism limiting re-broadcasting in intermediate nodes by only admitting, on the basis of the expected transmission count (ETX) [15], a subset of these nodes to perform broadcasting operations. Similarly, in Gogic *et al.* [16] authors analyze how to keep energy consumption low and, at the same time, to bound latency and packet delivery ratio.

- Routing-based solutions adopt a routing protocol for packet forwarding and transmitting data over BLE data channels. These solutions can be further separated into *static* and *dynamic* routing mechanisms. Examples of *static* routing schemes are discussed in Maharjan *et al.* [17], where a wireless sensor network-oriented static tree topology involving nodes with different roles and with a 2-byte addressing space is proposed. Unfortunately, this solution lacks a mechanism for re-building the network after a component (such as a node or a link) fails and suffers from the single-node failure problem, being tree-oriented. The static routing solution proposed in Patti *et al.* [18], denoted as real-time BLE, suffers from scalability for both master and slave nodes, as it keeps only a default route and an alternative route as a backup, and with a master able to establish a connection with at most another master. On the other hand, examples of *dynamic* routing schemes are proposed in Sirur *et al.* and Reddy *et al.* [9, 19], where a routing-based solution, denoted as BLE mesh network and similar to the adaptation layer between BLE and RPL protocol [20], exploiting a directed acyclic graph structure – inspired by the IPv6 Routing protocol for low-power and lossy networks (RPL) [21] – for transmitting routing messages via advertising channels, is proposed. Similarly, in Mikhaylov and Tervonen [22] a BLE-based mesh solution, denoted as MultiHop Transfer Service and based on on-demand routing over the Generic ATTribute GATT layer, is discussed. Moreover, in Guo *et al.* [23] an on-demand routing protocol targeting the formation of scatternets – network topologies composed of interconnected piconets – is defined. Finally, in Balogh *et al.* [24] authors leverage attributes, characteristics, and GATT services to apply Named Data Networking [25, 26] to support BLE-based WMNs.

In addition to the above solutions, the interest in applying the mesh paradigm to BLE networks has led also to the definition of proprietary network solutions, with commercial examples given by CSRmesh [27], OpenMesh [28], Wirepas Mesh [29], MeshTek [30], EtherMind [31], and solutions from Estimote [32] and NXP [33].

Given the above characterization, the BLE-based mesh-oriented routing protocol proposed in this chapter follows the possibility of addressing BLE nodes in a routing-based way, even exploiting the use of advertisement channels (as in flooding-based solutions).

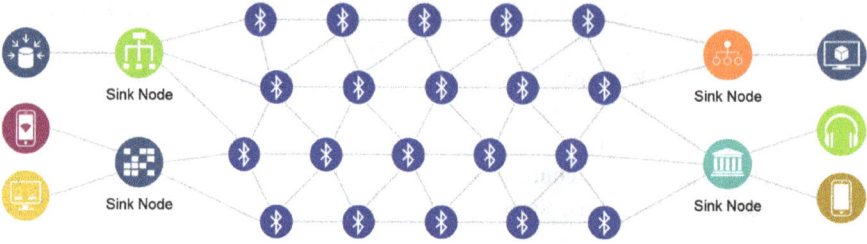

Figure 2.1   Example of a BLE-based WMN

2.3   Sink-oriented routing protocol

The proposed BLE-based routing protocol is designed to operate in WMNs composed of different BLE-enabled devices (as shown in Figure 2.1) such as

- *on-field* BLE devices, based on commercial SoCs and in charge of performing sensing (e.g., through external sensors connected directly to the sensing board) and actuating tasks (e.g., through the activation of specific registries on the host micro-controller handling the outputs)
- management devices, denoted as sink nodes or GWs, equipped with BLE functionalities and in charge of requesting *on-field* devices to perform actions and return knowledge and data
- external BLE-enabled entities, denoted as "smart" devices, interested in obtaining data from the nodes composing the BLE WMN – in particular, interacting with the sink nodes – to perform further actions (e.g., for visualization or data analysis purposes)

2.3.1   Receive message (RMS) packet

In order to simplify the management of a BLE-based WMN, the proposed routing protocol allows only one type of network packet to flow inside the BLE-based WMN: the RMS packet. In detail, an RMS packet is used to carry out requests and responses, thus containing all the information required to identify the type of action to be performed by *on-field* nodes and the body of the packet. As shown in Figure 2.2, the structure of an RMS packet is composed of the following fields:

- $\mathbb{U}$ represents the identifier of the packet (as an incremental integer value).
- $\mathbb{M}$ contains the current hop counter ($\mathbb{M}$ is reduced by one at each receiving node), can assume a value in the range $1 \div 10$, and is used to determine if the

$\mathbb{U}$	$\mathbb{M}$	RMS	$\mathscr{S}_{ADDR}$	$DEVICE_{TYPE}$	$ADDR_{TYPE}$	$\mathscr{T}_{ADDR}$	$\mathbb{P}$

Figure 2.2   Representation of the structure of an RMS packet carrying information inside the BLE-based WMN

packet should be forwarded to the next-hop node or if it needs to be discarded and not re-transmitted inside the WMN.

- RMS identifies the packet type (RMS) traveling along the BLE-based WMN.
- $\mathscr{S}_{\text{ADDR}}$ contains the ID of the BLE node that built the RMS packet (it is not changed by intermediate nodes).
- $DEVICE_{\text{TYPE}}$ represents the BLE device's type (e.g., light, actuator, sensor).
- $ADDR_{\text{TYPE}}$ specifies the address type (e.g., unicast, group).
- $\mathscr{T}_{\text{ADDR}}$ corresponds to the address of the target node, on the basis of the address type $ADDR_{\text{TYPE}}$.
- $\mathbb{P}$ represents the payload of the RMS packet to be sent on the WMN.

From an operational point of view, the RMS packet is checked by the BLE node's operating system *daemon*, in turn verifying the destination address to decide if the RMS message is addressed to the BLE node itself or if it has a broadcast address and needs to be re-transmitted in the WMN. Moreover, the daemon filters and eliminates message duplicates. If the RMS message has a broadcast address and the value of $\mathbb{M}$ to be inserted in the outgoing packet is greater than zero, then the RMS message is sent out on the BLE WMN.

2.4   Topology construction (downlink)

The routing protocol proposed in this chapter is based on the construction of routing tables inside each BLE node composing the network itself. The downlink-oriented topology construction task is performed through the definition of a particular RMS packet, denoted as BLE Originator Message (BOM) that is created by the sink and propagates inside the WMN exploiting the flooding feature of the mesh protocol itself. For the sake of simplicity, a single sink node is considered in this chapter, but the proposed routing protocol can be extended to multiple sinks/GWs without any further modification.

In detail, as shown in Figure 2.3, at periodic time intervals the sink instantiates a BOM packet containing the command identifier b and the ID of the sink node that originated the topology construction, denoted as ID_{GW}. Then, the BOM message is sent out on the WMN and, when it is received by the BLE devices, each receiving

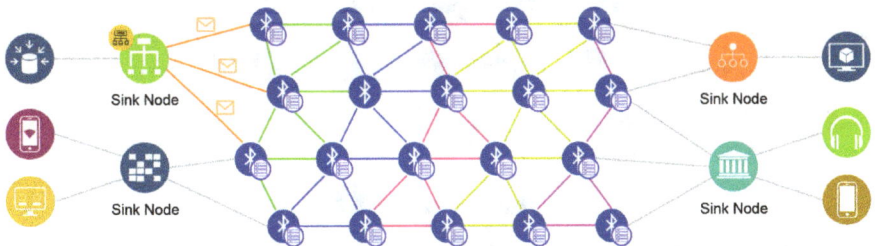

Figure 2.3   BOM packets flooding into the BLE-based WMN for topology construction

(intermediate and final) BLE node verifies if the distance among the originator sink node and itself is better with respect to the already existing routing rules and, should this be the case, an internal routing table's update operation is performed by the BLE node itself. In particular, the routing table contains the ID of the sink node, denoted as ID_{GW}, and a ranking identifier highlighting, for the same sink node, which is the best routing rule to be applied to the traffic.

2.5   Data collection – request (downlink)

When a sink node is interested in collecting data from the nodes composing the BLE-based WMN, thus starting an asynchronous information harvesting operation, it sends a request to them. In detail, as shown in Figure 2.4, the GW – that, as detailed in Section 2.1, is used in this chapter as a synonym of the term "sink" – will send a particular message, denoted as GET packet, containing (i) the command type g, (ii) the data class (denoted as k) that should be collected by the BLE nodes, and (iii) the ID(s) of the node(s) having to reply with their k class-related data.

To this end, when a GET request is received by a BLE node participating in the WMN activities, the proposed BLE-oriented mesh protocol can cope with the following cases:

- In the case the sink node is interested in targeting all the nodes composing the WMN, it needs to issue a *broadcast* GET request.
- In the case the sink node is interested in targeting a specific node, it needs to send a *unicast* GET request.
- In the case the sink node is interested in targeting a subset of the BLE nodes composing the WMN, it needs to issue a *group* GET request.

Then, each BLE node undertakes multiple controls on the GET packet and, if all controls are positive, each BLE node prepares a response message, denoted as SEN packet and described in Section 2.6, for each data to be sent to the requesting GW.

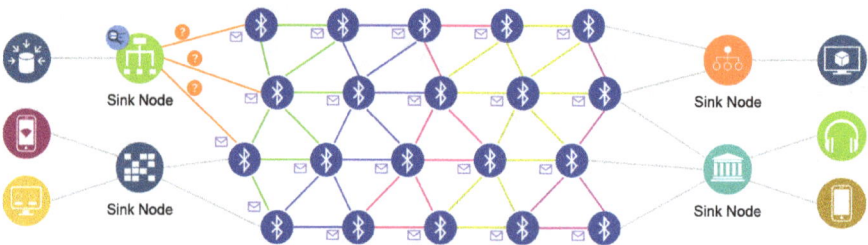

Figure 2.4   *GET packets sent by the sink node into the BLE-based WMN, and response packets, denoted as SEN messages, sent back from the BLE nodes to the requestor GW*

2.6   Data collection – response (uplink)

When a network node needs to send a value to a sink node, either because this action has been requested by the sink (as a "stimulated" action) or because of a "proactive" behavior of the specific *on-field* node, an SEN packet (as shown in Figure 2.4) needs to be prepared. This SEN packet is addressed to the GW and contains (i) the command type s, (ii) the data class (denoted as k) that is being sent from the BLE node to the sink ID_{GW}, and (iii) the payload of interest.

   Then, when the SEN packet is received by the neighboring nodes (thanks to the flooding behavior of the WMN itself), each neighbor node checks if it is the target of the data and, should this be the case, it processes the received packet. Otherwise, the neighbor node retrieves the next-hop node for the referred sink node ID_{GW} from its internal routing table and forwards the incoming data to this next-hop node.

2.7   Use cases

In order to further highlight the scalability of the proposed BLE-based routing protocol, the following two use cases are discussed based on the structure of the RMS packets discussed in Section 2.3.1.

2.7.1   Network topology reconstruction

In the proposed BLE-based WMN, the SEN packet has been defined to be useful also to reconstruct (at sink level) the topology of the BLE WMN. As shown in Figure 2.5, this task is performed by reserving a particular data class k, which each BLE node composing the network is aware of. Upon receiving a GET message with this particular data class k, each BLE node will reply preparing an SEN packet targeting the corresponding sink. To this end, the payload contained in the SEN packet will contain the ID of its next-hop BLE device (denoted as $ID_{NEXT-HOP}$) toward the sink node ID_{GW}. In this way, the sink node ID_{GW} will be aware of the overall network topology and will improve its knowledge in terms of "backup" routing information, represented by the hop counters contained in the received SEN packets, for further processing tasks.

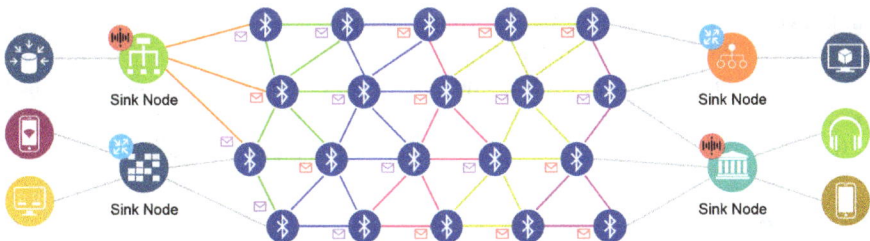

Figure 2.5   *Network topology reconstruction and sensing of BLE devices performed in the BLE-based WMN*

2.7.2   Sensing of BLE devices in the neighborhood

In addition to routing functionalities, the proposed mesh-oriented routing protocol supports sensing of BLE devices through BLE signal parameters (e.g., received signal strength indicator, RSSI), as shown in Figure 2.5. More precisely, since at fixed time instants, BLE devices emit advertisement packets (providing, to the receiver node sensing these packets, their Medium Access Control, MAC, address, RSSI value, and additional information), there is the possibility of considering these devices and their information for further uses. In detail, each time a BLE advertisement packet from a new node is detected by a BLE node of the WMN, an admission decision is performed taking into account both the RSSI of the packet and its information content. If the admission check is positive, then the new node is admitted to the WMN for additional activities and tasks.

2.8   Conclusions

In this chapter, we have proposed a routing protocol for BLE-based WMNs able to address different classes of devices and to build a true network topology, allowing to perform different types of operations and tasks. In detail, heterogeneous entities composing the BLE WMN have been discussed, together with an overview of the classes of packets that can be used in the network, in order to implement a "controlled" flooding mechanism. The proposed protocol embodies the advantages of the flooding mechanism – i.e., simplicity and absence of connection establishment requirements – with those of routing-based schemes and advertisement channel-based ones. On the basis of these characteristics, further research activities can be carried out, such as IPv6 support integration, comparisons with approaches based on the use of data (rather than advertisement) channels, experimental performance evaluation, and security improvements.

Acknowledgements

This work was supported by TCI Telecomunicazioni Italia s.r.l. Luca Davoli and Gianluigi Ferrari received funding from the [1] research and innovation program ECSEL Joint Undertaking (JU) under grant agreement No. 876038, InSecTT project – "Intelligent Secure Trustable Things." The work of Luca Davoli was also partially funded by the University of Parma, under "Iniziative di Sostegno alla Ricerca di Ateneo" program, "Multi-interface IoT sYstems for Multi-layer Information Processing" (MIoTYMIP) project. The JU received support from the [1] research and innovation programme and the nations involved in the mentioned projects. The work reflects only the authors' views; the European Commission is not responsible for any use that may be made of the information it contains.

References

[1]   Bluetooth Special Interest Group (SIG) [online]. 2021. Available from https://www.bluetooth.com/ [Accessed 5 Apr 2021].

[2]   Internet Engineering Task Force (IETF) [online]. 2021. Available from https://www.ietf.org/ [Accessed 5 Apr 2021].

[3]   Nieminen J., Savolainen T., Isomaki M., *et al.. IPv6 over BLUETOOTH(R) low energy. Internet Engineering Task Force (IETF) [online].* 2015. Available from https://tools.ietf.org/rfc/rfc7668 [Accessed 25 Feb 2022].

[4]   Di Marco P., Chirikov R., Amin P., Militano F. 'Coverage analysis of Bluetooth low energy and IEEE 802.11ah for office scenario'. IEEE 26th Annual International Symposium on Personal, Indoor, and Mobile Radio Communications (PIMRC); Hong Kong, China; 2015. pp. 2283–7.

[5]   Pei Z., Deng Z., Yang B., Cheng X. 'Application-oriented wireless sensor network communication protocols and hardware platforms: a survey'. 2008 IEEE International Conference on Industrial Technology; 2008. pp. 1–6.

[6]   Darroudi S.M., Gomez C. 'Bluetooth low energy mesh networks: a survey'. *Sensors.* 2017;17(7):1467.

[7]   Hernandez-Solana Angela., Pérez-Díaz-De-Cerio D., García-Lozano M., Bardají A.V., Valenzuela J.-L. 'Bluetooth mesh analysis, issues, and challenges'. *IEEE Access.* 2020;8:53784–800.

[8]   Ghori M.R., Wan T.-C., Sodhy G.C. 'Bluetooth low energy mesh networks: survey of communication and security protocols'. *Sensors.* 2020;20(12):3590.

[9]   Sirur S., Juturu P., Gupta H.P., *et al.* 'A mesh network for mobile devices using Bluetooth low energy'. *2015 IEEE SENSORS.* 2015:1–4.

[10]   *Bluetooth Radio Versions* [online]. 2021. Available from https://www.bluetooth.com/learn-about-bluetooth/radio-versions/ [Accessed 25 Feb 2022].

[11]   Kushalnagar N., Montenegro G., Schumacher C., Internet Engineering Task Force (IETF). *IPv6 over low-power wireless personal area networks (6LoWPANs): overview, assumptions, problem statement, and goals [online].* 2007. RFC 4919. Available from https://tools.ietf.org/rfc/rfc4919 [Accessed 25 Feb 2022].

[12]   Basu S.S., Baert M., Hoebeke J. 'QoS enabled heterogeneous BLE mesh networks'. *Journal of Sensor and Actuator Networks.* 2021;10(2):24.

[13]   Luo B., Sun Z., Pang Y., *et al.* 'Neighbor discovery for IPv6 over ble mesh networks'. *Applied Sciences.* 2020;10(5):1844.

[14]   Kim H., Lee J., Jang J.W. 'BLEmesh: a wireless mesh network protocol for Bluetooth low energy devices'. 2015 3rd International Conference on Future Internet of Things and Cloud; 2015. pp. 558–63.

[15]   Couto D.S.J.D., Aguayo D., Bicket J., Morris R. 'A high-throughput path metric for multi-hop wireless routing'. *Wireless Networks.* 2005;11(4):419–34.

[16]   Gogic A., Mujcic A., Ibric S., Suljanovic N. 'Performance analysis of Bluetooth low energy mesh routing algorithm in case of disaster prediction'. *International Journal of Computer and Information Engineering.* 2016;10(6):1075–81.

[17]   Maharjan B.K., Witkowski U., Zandian R. 'Tree network based on Bluetooth 4.0 for wireless sensor network applications'. 2014 6th European Embedded Design in Education and Research Conference (EDERC); Milan, Italy; 2014. pp. 172–6.

[18]   Patti G., Leonardi L., Lo Bello L. 'A Bluetooth low energy real-time protocol for industrial wireless mesh networks'. IECON 2016 – 42nd Annual Conference of the IEEE Industrial Electronics Society; Florence, Italy; 2016. pp. 4627–32.

[19]   Reddy Y.K., Juturu P., Gupta H.P., *et al.* 'Demo: a connection oriented mesh network for mobile devices using Bluetooth low energy'. Proceedings of the 13th ACM Conference on Embedded Networked Sensor Systems. SenSys'15; 2015. pp. 453–4.

[20]   Lee T., Lee M., Kim H., Bahk S. 'A synergistic architecture for RPL over BLE'. 2016 13th Annual IEEE International Conference on Sensing, Communication, and Networking; London, UK; 2016. pp. 1–9.

[21]   Winter T., Thubert P., Brandt A., *et al. RPL: IPv6 routing protocol for low-power and lossy networks. Internet Engineering Task Force (IETF) [online].* 2012. Available from https://tools.ietf.org/rfc/rfc6550 [Accessed 25 Feb 2022].

[22]   Mikhaylov K., Tervonen J. 'Multihop data transfer service for bluetooth low energy'. 2013 13th International Conference on ITS Telecommunications (ITST); Tampere, Finland; 2013. pp. 319–24.

[23]   Guo Z., Harris I.G., Tsaur L., Chen X. 'An on-demand Scatternet formation and Multi-Hop routing protocol for BLE-based wireless sensor networks'. IEEE Wireless Communications and Networking Conference (WCNC); New Orleans, LA, USA; 2015. pp. 1590–5.

[24]   Balogh A., Imre S., Lendvai K., Szabó S. 'Service mediation in Multihop Bluetooth low energy networks based on NDN approach'. 2015 23rd International Conference on Software, Telecommunications and Computer Networks (SoftCOM); Split, Croatia; 2015. pp. 285–9.

[25]   Zhang L., Afanasyev A., Burke J., *et al.* 'Named data networking'. *ACM SIGCOMM Computer Communication Review.* 2014;44(3):66–73.

[26]   Jacobson V., Smetters D.K., Thornton J.D., *et al.* 'Networking named content'. Proceedings of the 5th International Conference on Emerging Networking Experiments and Technologies. CoNEXT '09; 2009. pp. 1–12.

[27]   CSRmesh Development Kit [online]. 2021. Available from https://www.qualcomm.com/products/csrmesh-development-kit [Accessed 5 Apr 2021].

[28]   nRF OpenMesh (formerly nRF51-ble-broadcast-mesh) [online]. 2021. Available from https://github.com/NordicPlayground/nRF51-ble-bcast-mesh [Accessed 5 Apr 2021].

[29]   Wirepas Mesh [online]. 2021. Available from https://wirepas.com/what-is-wirepas-mesh/ [Accessed 5 Apr 2021].

[30]   MeshTek [online]. 2021. Available from https://meshtek.com/meshtek-platform/ [Accessed 5 Apr 2021].

[31]   EtherMind Bluetooth 5.2 [online]. 2021. Available from https://www.mindtree.com/ethermind [Accessed 5 Apr 2021].

[32]   Estimote [online]. 2021. Available from https://estimote.com/ [Accessed 5 Apr 2021].

[33]   Bluetooth Smart/Bluetooth Low Energy [online]. 2021. Available from htt-ps://www.nxp.com/products/wireless/bluetooth-low-energy:BLUETOOTH-LOW-ENERGY-BLE [Accessed 5 Apr 2021].

Chapter 3

Body sensor networks—recent advances and challenges

Shama Siddiqui[1], Anwar Ahmed Khan[2], and Indrakshi Dey[3]

3.1 Introduction

Recent advances in the domains of microelectronics, biomaterials, wireless communications, and pervasive systems have revolutionized modern healthcare systems. Today, we enjoy the perks of remote health monitoring and management, facilitated by wireless sensor networks (WSNs), and more precisely, by body sensor networks (BSNs). WSN is defined as a network of sensors that often operate autonomously to monitor the environment or places for various parameters of interest, such as temperature, pressure, humidity, air or water pollution levels, movement, etc. WSN may comprise hundreds of nodes that sense data from the environment on a periodic or continuous basis. These data are then reported to a central sink or cluster head, which subsequently processes and forwards it to the base station for storage and distribution to the desired stakeholders. Due to the facility of remote monitoring and control, WSN has given rise to various applications from the domains of target tracking, intrusion detection, industrial automation and control, habitat monitoring, environmental monitoring, and health monitoring.

The WSN customized for monitoring body parameters is known as BSNs, wireless BSNs, or wireless body area networks (WBANs). BSN is defined as a network that collects body parameters using physiological sensors; these sensors may be implanted within the human body, embedded with wearable devices or placed around the body [1]. Different types of sensors are used to facilitate BSN, based on the requirement of physical signals to be collected. BSNs exist at the core of broader healthcare architecture, often referred to as the Internet of Medical Things (IoMT). IoMT holds the potential to bring together various stakeholders and systems of the healthcare sector such as clinical facilities, research laboratories, emergency services, and specialized care units. The major aim of BSN and IoMT is to offer healthcare-monitoring and management facilities to individuals in real time to

[1]DHA Suffa University, Karachi, Pakistan
[2]Millennium Institute of Technology and Entreprenuership, Karachi, Pakistan
[3]Maynooth University, Ireland

ensure high quality of service (QoS) delivery. As a result, these technologies play a crucial role in improving the life quality and expectancy of populations belonging to various demographic regions alike, thus reducing the healthcare disparities.

Over the past decade, there have been various advances in BSNs. Heterogeneous sensors placed on, within, or near the human body mainly for the purpose of vital sign monitoring or fitness tracking [2] have been proposed, and wearable devices have particularly become common. The wearable devices are integrated with simple physiological sensors and interact with smartphone and web applications. Hence, using wearable devices, the users can easily track their state of health. Furthermore, the health statistics are also communicated to the physicians over wireless links for real-time monitoring. Initially, it was thought that wearable devices will mainly be useful for elderly and disabled patients, but today, these devices are heavily used by healthy people of every age, mostly for the purpose of fitness tracking. In addition to the wearable BSNs, there have been numerous architectures developed for implanted BSNs, which may detect the abnormal health states even earlier. In the implanted BSNs, the sensor and actuator nodes are implanted within the human body and they communicate wirelessly through the body tissue. Since implanted nodes are near to the human organs, the delay for data collection from these nodes is much lower as compared to the on-body or wearable devices. Moreover, cross-disciplinary techniques such as deep learning, advanced data analytics, and artificial intelligence (AI)-aided action realization also support BSN applications to significantly enhance the healthcare service quality. The major target of integrating machine learning and AI techniques with BSN is to optimize the performance such that the quality of healthcare delivery may improve in terms of timeliness and accuracy.

With the emerging applications of BSNs, on the one hand, the quality of healthcare is expected to improve but, on the other hand, numerous challenges have also risen. Most BSN applications impose stringent requirements for QoS parameters such as reliability, latency, energy efficiency, and low processing overhead. The nodes must be available at the time when critical information needs to be communicated and energy consumption must be dealt with wisely; in case the nodes are available during the entire day but their battery source expires just when an emergency occurs, the major purpose of having BSN will not be served. Therefore, the nodes deployed in or around the human body must be robust to ensure timely and continuous delivery of crucial health data. Also, there are strict size requirements for the BSN nodes and these should also be biocompatible. Due to the tiny size and biocompatible material of BSN nodes, it becomes a challenge to provide them with the functionality of processing bulky data generated by emerging BSN applications. Similarly, the power consumption and recharging is also more critical for BSN as compared to WSN, because kinetic or heat energy makes the major source of energy for BSN. Due to the regular use, it is not possible for the designers or users to recharge or replace the batteries very often, which has raised the challenges for energy harvesting to power the BSN nodes. Moreover, the data collected by BSN is of personal nature and user safety and confidentiality/privacy must be maintained. Hence, the specialized design and application requirements of BSN motivate the need for custom-designed frameworks and protocols, as those earlier developed for wireless mesh networks do not fit most application scenarios of BSN.

This chapter presents a detailed review of BSN applications, components, architectures, communications models, network topologies, network layers, security threats, solutions, and, finally, the open research directions and opportunities. Based on the concepts covered in this chapter, the reader may identify the present trends in BSN and directions for future development.

3.2   Applications of BSN

BSN provides crucial assistance to the users for both medical and nonmedical applications through offering remote-monitoring and -reporting facilities. Modern BSNs could trigger injection/pumping of medicine in the human body as required (either automatically based on threshold parameter monitoring or upon receiving instructions from the remote service providers). There are various chronic diseases, which can benefit from continuous remote monitoring as they could offer an insight into the lifestyle patterns of users. Based on the data collected from BSN, it becomes more realistic for medical experts to devise healthcare strategies and lifestyle/fitness recommendations for the users. Furthermore, there are also nonmedical applications of BSN where wearable or implantable nodes can help to track physiological parameters, movement, body postures, or any other parameters of interest. In this section, we briefly discuss the recent medical and nonmedical applications of BSN.

3.2.1   Medical applications

Today, the populations requiring remote monitoring have been increasing all over the world. According to WHO, the global elderly population (60 years or above) would reach approximately 2.1 billion by 2050. With the increase in the elderly population, the global healthcare burden would increase at an alarming rate as this sector of the population often suffers from chronic diseases such as cardiovascular, diabetes, cancer, asthma, Parkinson's, and others; some of these diseases can be even fatal for elderly if timely care is not provided. Also, there have been certain diseases whose risk could be reduced through appropriate lifestyle management. For example, in 2019, 32% of all deaths worldwide were associated with cardiovascular diseases (CVD) and 85% of these deaths were reported to occur due to heart attack and stroke. Similarly, 1.5 million deaths were directly caused by diabetes in 2019, and WHO reported an increase of 5% in the rate of premature mortality between 2000 and 2016. Both for CVD and diabetes, maintaining a healthy diet and physical activity levels could reduce the risk. Considering these facts and the need for continuous patient monitoring, the major focus of BSN has been on healthcare applications as it promises to reduce the delay in collection of critical data, which, in turn, enhances the quality of healthcare service delivery. Instead of the patient needing to physically visit the clinic, the BSN reports his crucial health data remotely. Since BSN transmits quantitative health data (blood pressure, blood glucose levels, pulse rate, oxygen saturation levels, etc.) on a continuous basis, telehealth applications have become a reality by reducing the service time as well as cost.

BSN nodes facilitate remote health monitoring and can be regarded as minia-ture base stations that collect and transmit vital parameters [3]. With the usage of BSN, highly personalized and customized care delivery becomes possible because the network is configured as per individual needs; for example, a person needing help with pulse rate monitoring will be provided with pulse oximetry (SPO$_2$) sensor which can be simply worn as a ring, such as O$_2$ ring [4]. In this ring, SPO$_2$ sensor is used for noninvasively measuring oxygen saturation level in the human blood. If the oxygen saturation becomes too low, the patient may face a life-threatening situation; therefore, the ring can be configured to sense and send pulse rate and SPO$_2$ data over regular periodic time intervals. In case abnormal values are detected, an alert will be generated on the smartphone of the patient himself as well as on those of physicians and family members. If an emergency situation is identified, immediate assistance can be provided to the patient from any of the remote contacts. In a similar fashion, the patients with diabetes, blood pressure, or heart diseases can be provided with specific sensors, which can measure the physiological parameters that could monitor their health state.

Various innovative uses of BSN have also become common for clinical appli-cations. In this regard, sleep monitoring is an interesting application. Sleep is an essential requirement for the human body for maintaining appropriate mental and physical health. Several disorders and health issues such as anxiety and depression are often linked with a lack of quality sleep (referred to as insomnia). If not treated, there could be serious consequences occurring due to lack of sleep, including but not limited to sleeping at the workplace or during driving, narcolepsy (neurological disorder affecting the control of wakefulness and sleep), and increased risk for CVD. Physicians often suggest conventional medicine and alternative therapeutic tech-niques to help the patients sleep well. Polysomnography is an important diagnos-tic technique often used by medical experts for monitoring sleep disorders. In this technique, sleep data is stored and analyzed by experts. The data is conventionally acquired through using wired sensors, while BSN offers flexibility for monitoring sleep data wirelessly. Wireless wearable biopotential sensors are used to detect the sleep state of the user and coordinate with the sink node and remote server for the storage of data. Therefore, physicians can collect data easily from remote locations without disturbing the routine activities of users.

BSN has often been recommended for asthma patients. Asthma is a very com-mon health problem and could occur in people of all ages. The disease is categorized by narrowed and swollen airways of the patients, which may also produce mucus. During a typical asthma attack, breathing becomes difficult and the patient may suf-fer from cough, wheezing, tightness of chest, and shortness of breath. It could be fatal in certain serious situations if not treated timely. Along with the internal issues with the patient's lungs, asthma is often triggered due to external factors such as air pollution, presence of dust and fumes in the environment, weather conditions, etc. As soon as asthma attacks, a patient is required to inhale a bronchodilator or take terbutaline orally or intravenously to return to the normal breathing state. BSN helps the patients to detect the presence of allergens in the environment and also by reporting their health state to the service providers in real time. Therefore, BSN could be configured in different ways for asthma patients, such as for injecting medicine within the body or to arranging ambulance in case of emergency.

Novel BSN solutions have been proposed increasingly for elderly, independent living, and physically impaired populations. Real-time fall detection, prevention, and alarm systems are easily being managed with BSN. Ultrasonic or similar sensors can be worn by users, which could generate an alarm to alert them about obstacles on their way so the fall may be prevented. Furthermore, in case a fall still occurs, remote contacts can be informed along with the emergency service providers. Such BSN systems are very useful for blind populations, those with injuries, or elderly who may fall due to weakness or unconsciousness. Also, wearable sensors can also be used for identifying and recording the posture of patients; such information can be used by experts for developing effective exercise and posture management strategies.

Implanted BSN has also introduced innovative health-monitoring applications for users. To facilitate implanted or intra-body BSN, novel communication technologies and signal propagation models have been developed. We discuss some of these later in the chapter. The implantable devices being near to the internal organs and tissues possess the capability to generate early warning signals about diseases, which could become fatal at a later stage, for example, cancer. Some of the common implanted devices which could be integrated with BSN include pacemakers, neuro-stimulators, drug pumps, baclofen pumps, and cardiac defibrillators. Sensors having the capability to monitor cancer cells can also be integrated with implanted BSN; these sensors can also detect tumors, which could eliminate the need for biopsy and could significantly increase the speed of detection. Implanted BSNs coordinate with on-body devices which, subsequently, transmit the data of sensors to remote stations for real-time actions and decisions.

BSN has also been used for behavioral-monitoring and fitness-tracking [5] applications. For fitness tracking and lifestyle monitoring, physiological sensors are used to detect exercise patterns. Fitness-tracking application is particularly useful for patients who are at risk of developing obesity and associated diseases such as diabetes and cholesterol. In this application, wearable devices such as Fitbit and/ or sensors embedded with smartphones sense activity data by measuring parameters like the number of steps walked, variations in pulse rate, calories burnt, etc. The smart food-intake-monitoring applications are also being integrated with BSN where the users receive recommendations on the type and duration of activity based on their health history and calories consumed each day. The data generated by sensors and smart applications does not only help users themselves toward developing healthy habits but also lead the physicians and nutritionists to make informed decisions as they no longer need to rely exclusively on information provided by the patient; rather, they can monitor the entire trend of activity level and its impact on the patient's health by analyzing the data generated by BSN. Moreover, the specialized BSN with a combined focus on biokinetic and physiological sensing may help the athletes and sports persons to improve their performance by optimizing their movements suggested by AI algorithms [6]. On the other hand, dense sensing concepts can be used for monitoring human–object interactions [7], which could shed light on the behavior of the user. Dense sensing refers to collecting data from multiple miniaturized sensors that are connected to objects surrounding the user. Instead of physiological sensors, which could provide an insight into the health parameters of an individual or the impact of exercise on the health state, dense sensing focuses more on the interaction of the user with the environment. For example, if there is a need to monitor the movement of a person in their home, ultrasonic sensors could be installed in different rooms, which could detect the presence. Dense sensing can particularly be used for the elderly and children who need to be monitored for their movement and location.

In addition to the primary usage of BSN for remote monitoring of critical physiological parameters, there have been several emerging applications for healthcare. Today, BSN is being proposed for deep brain stimulation, prosthetic actuation, drug delivery, and heart regulation. For these applications, the sensors are required to be inserted surgically in the affected areas and send the sensed data to the external devices. Once the devices begin sending signals, they can be controlled externally for regulating the functions of the brain or heart by actuation of prosthetic. Similarly, remote drug delivery can also be facilitated with the continuous monitoring of physiological parameters. In this application, the implanted actuators can either pump a prefilled drug based on the detection of abnormal values and transmit this info to the remote stations, upon receiving signals from external devices, or release the medicine automatically into the human body. These emerging areas will not only benefit patients with chronic diseases but extend the service for a diverse set of stakeholders such as space and deep-sea explorers, fire responders, and soldiers whose life may be at risk due to the nonavailability of physicians in the physical proximities.

3.2.2   Nonmedical applications

Various nonmedical applications of BSN have also emerged, followed by the establishment of technology in the healthcare sector. For example, BSN can be used to monitor the fatigue of users. Physical fatigue refers to the failure of muscles to maintain the optimal or best physical performance. The user under fatigue feels exhaustion if he engages in physical activity. BSNs can play a crucial role in identifying fatigue in people such as soldiers and athletes during training activities. BSNs can even be used to track the fatigue of soldiers on the battlefield, where the need for remote monitoring becomes even more crucial; if any soldier is found to have high levels of fatigue, he may be replaced by another. Timely replacement of soldiers can improve the performance of the army. Similarly, BSN can be used during training activities not only for soldiers but also for firefighters, police forces, disaster management teams, and athletes. In these applications, sensors such as an accelerometer (motion sensor) and lactic acid sensor may be used to detect the level of energy and fatigue. Based on the collected data, the trainers may subsequently develop more efficient training strategies. Also, BSN could be used during the selection of athletes and sports persons; during trial physical activities, the candidates may be advised

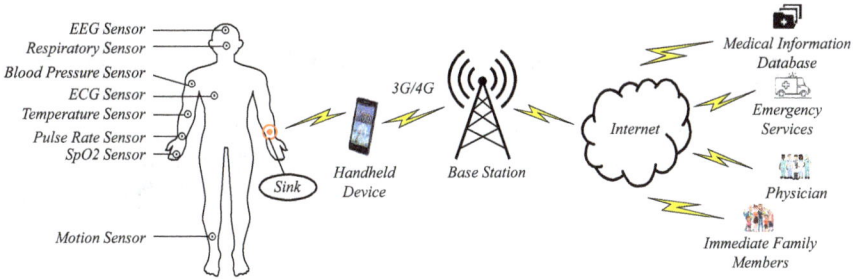

Figure 3.1   Architecture

to wear BSN which sends signals about their fatigue, indicating their fitness level and physical performance. The best performing candidates can then be identified using information about their stamina rather than just based on the visual and timing observations.

3.3   Body sensor networks—overview and components

3.3.1   Overview

Figure 3.1 shows the fundamental BSN architecture, where data are collected from a patient's body using different types of sensors and are reported to remote locations. The usual BSN components that directly interact with the user (to collect data) include sensor nodes and sink. The sensors can be configured to sense data continuously or at periodic intervals. Mostly, the sensors generate alert signals upon detecting abnormal values of physiological parameters [8]; the abnormal values are identified based on predefined criteria of thresholds. Data from all the sensor nodes in a BSN are collected at the sink node and transmitted to the handheld device using technologies such as Bluetooth (if the sensor nodes are wearable). Also, advanced communication technologies such as intra-body molecular communication may be used for the transmission of data from implanted intra-body sensors to the sink [9]. The location/placement of sensor and sink nodes within the body is decided based on the monitoring requirement of a specific patient and also on several performance aspects of networks such as tolerable path loss, delay, etc. These data are subsequently taken to the handheld device and after performing some local processing, it is transmitted to the base station over 3G/4G/5G connections. Finally, the data are made available via the internet to various stakeholders including but not limited to medical information databases, physicians, emergency service providers, and immediate family. In this way, not only the cost and time required for emergency care reduces but also the ubiquitous medical records are maintained over network servers. Hence, a rightly configured BSN could facilitate various stakeholders of the healthcare sector.

3.3.2   Components

BSN continuously monitors the body parameters of the users and allows them to perform their daily routine activities without any hindrance. For efficient working, all the mandatory components of BSN should be configured according to the customized needs of individual users. This section highlights the major BSN components required for achieving the desired objectives of remote health monitoring or specific parameter monitoring for nonmedical applications.

3.3.2.1   Sensors

Sensors are the major components of BSN, which exist at the physical layer. BSN sensors are miniature in size and low on power requirements to fit the needs of monitoring human body parameters. The quality and precision of sensing have significantly advanced today due to the advancements in the areas of nanotechnology, signal processing, and microelectromechanical systems. The BSN sensor nodes are designed to collect, process, store, and forward data in a single- or multi-hop fashion using either wired or wireless communication modes. Based on the application requirements, each sensor node may be configured to do some local processing or may just need to transmit all the collected data to the cluster head/coordinator node. Three common communication technologies are used by the sensors: wireless such as Bluetooth or Zigbee, radio-frequency identification (RFID), and ultrawideband (UWB). Moreover, the implanted sensors using the human body as the transmission medium have also been proposed for applications requiring early detection and higher precision.

Sensors used in BSN are broadly categorized into three types based on the sensed parameters and location, physiological (implantable or wearable), biokinetic, and environmental/ambient. Physiological sensors are used to monitor various health conditions by measuring ambulatory blood pressure, blood oxygen, blood glucose, body temperature, and signals related to respiratory inductive plethysmography, electrocardiography (ECG), electroencephalography (EEG), and electromyography (EMG). For the physiological category, the wearable sensors must be compact so as not to cause hindrance for the usual human activities; these sensors are now integrated into wearable devices and clothes. On the other hand, the implantable sensors also need to be biocompatible and noncorrosive in addition to being tiny. Implantable sensors are generally inserted surgically under medical supervision or can be taken in the form of a pill. Once inserted, the implanted sensors can be used for monitoring body parameters from outside the body using mobile applications and specialized interfaces. Furthermore, the implanted sensors can also send an alert in case of any abnormal or emergency situation, upon which the external agent may guide the implanted actuators to take appropriate action such as pumping medicine. Biokinetic sensors are used to monitor human movement by measuring acceleration and angular rate of rotation. Finally, the ambient sensors are also used in integration with wearable and/or implanted BSN sensors. These sensors often measure parameters such as environmental temperature, light intensity, humidity, sound pressure, etc.

Sensors can also be classified based on the amount of data they collect or the frequency of data collection. The sensors such as EEG, ECG, EMG, accelerometers, gyroscopes, auditory, and visual sensors are classified into the first category, and they collect time-varying signals continuously. These sensors provide the advantage of continuous real-time monitoring, but at the same time, their power consumption and bandwidth requirements are very high. Therefore, in most scenarios, the use of these sensors has only been limited to the healthcare facilities rather than opting for regular usage in BSN. The second category includes temperature, humidity, blood pressure, blood glucose, and blood oxygen saturation level sensors, which monitor the physiological parameters at discrete time intervals. It is realistic to collect data less frequently from the second category of sensors because the parameters they monitor change slowly; thus, it becomes possible to send these sensor nodes to sleep after each data collection and transmission cycle. Therefore, the second category of sensors is mostly integrated into a BSN for regular use.

Mostly, the data are collected from multiple sensors in BSN in order to reduce the chances of error. A single BSN node may comprise different sensors, or sensors at different locations (such as intra-body, on-body, and ambient) may be used to send their data at the sink node/coordinator and subsequently to the handheld device. The coordinator then aggregates data collected from different sources and take the appropriate decisions. These decisions could be of various types; for example, the user may be guided to take some action such as drinking water or taking a pill, or the decision could trigger some predefined automated actions. For the automated actions, we need actuators, which are defined next.

3.3.2.2 Actuators

Actuators are also proposed to be used in integration with sensor nodes of BSN to trigger automated actions remotely. An actuator refers to a component that performs a mechanical action upon receiving a control signal; it is often used for performing actions such as opening a valve. In BSN, the actuator can be used to automate tasks in the intra-body, on-body, and ambient environment. For example, an actuator can be placed near a glucose sensor for automated pumping of insulin in the body of a diabetic patient. Similarly, in case the temperature of a user is found to be too high, the actuator installed with fans or air conditioning system of the room may be triggered to switch the appliances on. In order to ensure the reliable and delay-intolerant action of actuators, the BSN shall be well-designed and customized to the needs of individual users. Furthermore, the log of actions taken by actuators must also be maintained by the server to be used in future decision-making.

3.3.2.3 Types of nodes

The nodes in BSN can be categorized as sensing, relay, sink, coordinating, and gateway. The sensing nodes are placed at appropriate locations of the human body based on specific monitoring requirements; these nodes could be on-body or implanted within the body. Some sensing nodes can also act as a relay; the relay nodes can generate their own data or may only be used for forwarding data received by neighbors to the cluster head/sink node. There is a single sink node facilitating data collection and storage from all the sensor nodes. Based on the location of each sensor node and the nature of the monitoring application, the sensing nodes may transmit data to the sink using either single- or multi-hop topologies. The sensing nodes may be given a fixed route to forward their data through relay nodes; however, mostly, the selection of relay is made as a dynamic decision. The sensor nodes may choose the relay nodes based on a number of criteria; for example, the level of energy/power present in the relay nodes can be set as selection criteria for each relay node. We discuss some examples of relay selection/routing algorithms used in BSN in Section 3.5.3. The relay nodes often broadcast information about their energy level and queue length; the sensor nodes may calculate the expected waiting time for transmitting their packets and hence, the optimal selection of relay nodes becomes possible.

The sink node/cluster head may serve various purposes based on the protocol definition; it may assign time division multiple access (TDMA) slots to the sensing nodes so they may take turns to send their data, it may collect data from all the sensors and aggregate it before sending to the coordinator (smartphone, PDA, tablet, or any other similar device), or it may trigger some automated action based on the values being sensed. The coordinator node then transmits the data to the gateway node, which helps to connect the BSN with the internet and hence with the outer world/other networks. At times, the coordinator node can also serve as a gateway node.

It is crucial to note that sensing, sink, and coordinator nodes all work in collaboration with each other. Any protocol governing the operation of BSN defines the functionalities of each of these node types. The problems specific to BSN such as data redundancy and delayed transmission are dealt with by designing collaborative protocols. For example, the sink may run data fusion protocols so the cost of transmitting data over the internet may reduce. Although high energy efficiency may be achieved by suppressing the redundant transmissions, the delay may increase. Therefore, data collection and aggregation schemes are strategically designed to optimize both the crucial performance parameters of delay and energy consumption.

3.3.2.4 Antennas

The increased deployment of BSN has raised the need for implantable and wearable antennas that could help the sensor nodes to communicate. For implantable nodes, the challenge is to transmit through human tissues, and microstrip antennas have been recommended for this purpose. These antennas have been found to overcome the constraints incurred by the intra-body environment made up of human tissues and operate at 402 MHz. Microstrip antennas have been designed in different shapes and forms based on the application requirements, such as stretchable/flexible antennas for transmitting unchanged resonant frequencies so energy harvesting could be possible.

Moreover, a large number of proposals have been made for wearable antennas, which refer to the antennas using textile as their patch or substrate. Such antennas need to be compact, lightweight, and flexible and could be either linearly, dually, or circularly polarized. The wearable antennas should not be influenced largely by the factors associated with the human body such as mobility. The efficiency of wearable antennas largely governs the electrical performance of BSN. Here, it is important to note that the properties of the textile material may affect antenna performance. For example, a lower loss tangent of the substrate material has been found to be associated with a narrower bandwidth. Therefore, there is a need to carefully design and fabricate textile-based antennas. The common antenna types being used as wearable include microstrip antennas, slot antennas printed dipoles antennas, and loop antennas.

Several attempts have been made to customize the wearable antennas for integration with different clothing items and materials. For example, an antenna with harmonic suppression placed on the belt has been shown to facilitate medical communication with reliable accuracy [10]. Microstrip antennas integrated with Denim and stacked photo paper has also been proposed, which use radiating patch-conducting layer and silver fabric as a ground [11]. Slit-loaded textile antennas (with denim and wash cotton substrates) have also been proposed to introduce circular polarization which assists transmission, catering to the mobility needs of the human body [12]. The use of slit changes the resonant frequencies and circular polarization is achieved which improves the performance of the textile antenna in terms of gain, voltage standing wave ratio, and return loss. The antenna circuits have also been proposed for efficient RF energy harvesting from the sources available in the environment such as Wi-Fi, mobile base stations, Bluetooth, television transmitters, mobile phones, and laptops.

3.4 BSN architecture

The entire BSN architecture can be divided into three communication sections or modules. The first section involves interaction between communicating nodes located within the human body or within 2 meters of the human body and is coined as intra-BSN communication. The second section involves communication between several BSNs and nodes that route information from the BSNs to a central data processing unit or a gateway and is termed as inter-BSN communication. The final section involves communication between the gateway and the application end points and this part of the network architecture is referred to as beyond-BSN communication.

3.4.1 Intra-BSN communication

Communication within an intra-BSN can be achieved over different modes like nanomechanical, acoustic, chemical, electromagnetic (EM) waves, and molecules. Of these, the two most promising paradigms are the molecular and EM approaches, the mechanism for both will be detailed in this subsection. However, traditional techniques are not suitable as we move down to the micro- or nanoscale within the human body owing to the difference in size, complexity, and energy consumption. Therefore, the basic blocks of a traditional communication system like transmitter, channel, receiver, etc. are redefined using new terminologies suitable for the nanoscale environment.

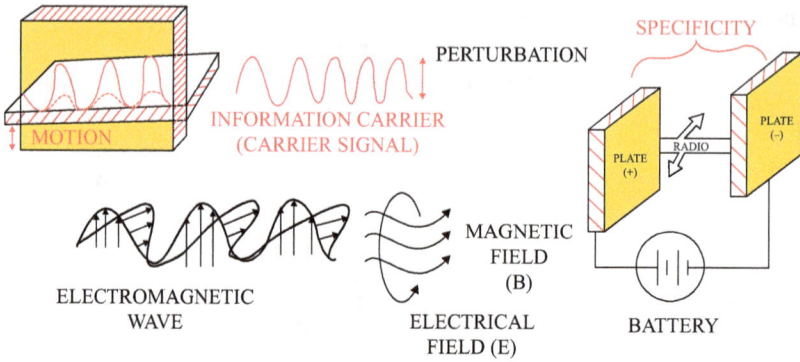

Figure 3.2   *Nanoscale system model representation for the EM wave-based communications in intra-BSN architecture*

3.4.1.1   System model

The system model for the intra-BSN architecture consists of the five fundamental blocks. A diagrammatic representation of the nanoscale system model along with the abovementioned blocks is presented in Figure 3.2 for the EM wave-based communication and in Figure 3.3 for the molecular communication environment. The constituent blocks are,

- Message Carrier—Wave or particle that is used for transporting the information.
- Motion Component—Force that enables movement of the message carrier through the medium.
- Field Component—Swarm motion, nonturbulent fluid flow, chemical gradient, or microtubules that guide the flow of message carrier.

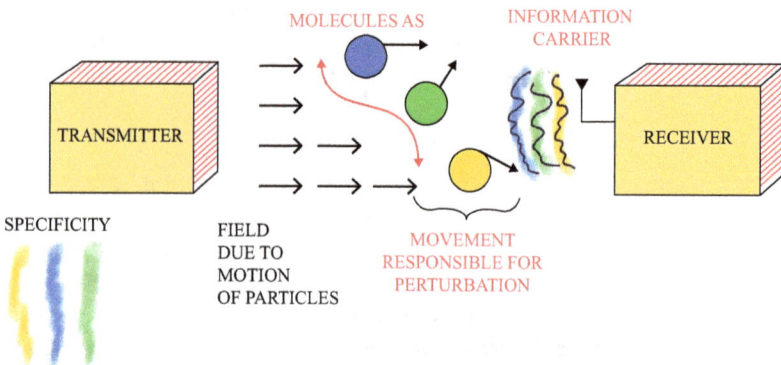

Figure 3.3   *Nanoscale system model representation for the molecular communications in intra-BSN architecture*

- Perturbation—Variations created in the message carriers in order to represent a signal; such variations can be created by controlling the concentration of the molecules, for example.
- Specificity—Reception of the message carriers by a target characterized by the shape of the molecules or its affinity to the target.

Based on the mode of transmission, it is important to describe the medium of environment in order to design the full-fledged communication techniques and systems. For EM-based systems, the terahertz (THz) frequency range has been recommended owing to their high-speed transmission capability over short distances. Frequencies ranging from THz band (0.1–10 THz) to the optical frequency bands (infrared, 30–400 THz, and visible, 400–750 THz) are preferred for communication between plasmonic nanoantennas.

3.4.1.2   EM-based intra-BSN

Propagation model
Path loss experienced by THz wave inside human tissue can be divided into three parts: the spread path loss, PL_S, the absorption path loss, PL_A and the scattering path loss, PL_C, which can be mathematically expressed as,

$$PL_T(dB) = PL_S(f, d)(dB) + PL_A(f, d)(dB) + PL_C(f, d)(dB) \tag{3.1}$$

where f is the frequency while d stands for path length. Expansion of wave within a medium results in the spread path loss,

$$PL_S(f, d) = (4\pi d/\lambda_g)^2 = (4\pi \eta_d f d/c)^2 \tag{3.2}$$

where $\lambda_g = \lambda_0/\eta_r$ is the wavelength in the present medium, λ_0 is the free-space wavelength, and d is the distance over which the wave travels. Owing to the spherical nature of the traveling EM power, the isotropic expansion is denoted by $4\pi d^2$ and $4\pi(\eta_d f/c)^2$ is the frequency-dependent receiver aperture term.

The attenuation resulting from the energy spent on internal kinetic energy to excite the molecules in the medium results in the absorption path loss and can be given by,

$$PL_A(f, d) = 1/\tau(f, d) = e^{\alpha(f)d} \tag{3.3}$$

where τ is the transmittance of the medium, α is the absorption coefficient, and d is the distance. The nonuniformity in the environment causes deflection of the traveling beam and this results in the scattering path loss, which can be mathematically expressed as,

$$PL_C(f, d) = e^{\mu_C d} \tag{3.4}$$

where μ_C is the scattering coefficient and d is the traveling distance.

Noise model
The total noise power spectral density S_N can be constructed by summing up the power spectral densities of the atmospheric noise S_{N_0}, self-induced noise S_{N_1},

and noise due to other sources/devices S_{N_d} and can be mathematically expressed as, $S_N(d, f) = S_{N_0} + S_{N_1} + S_{N_d}$

where

$$S_{N_0} = \lim_{d \to \infty} K_B T_0 (1 - e^{-\alpha(f)d})(c/\sqrt{4\pi f_0})^2 \tag{3.5}$$

and

$$S_{N_1} = S(f)(1 - e^{-\alpha(f)d})(c/\sqrt{4\pi f d})^2 \tag{3.6}$$

where d is the distance between source and target, f is the frequency of operation, K_B is the Boltzmann's constant, T_0 is the medium temperature, $\alpha(f)$ is the absorption coefficient, c is the speed of light in vacuum, f_0 is the center frequency of the frequency band of operation, and $S(f)$ is the power spectral density of the transmitted signal. The third noise component is contributed by other sources and can be attributed to many factors like background noise due to different tissue types of varying refractive indices, fierce fluctuation of THz communication-induced noise, etc.

Achievable transmission rate

Assuming additive colored Gaussian noise at the receiver and a binary asymmetric channel, the achievable information rate can be given by,

$$IR_{\max(sec)} = \frac{B}{\beta} \{ \max_x \{ H(X) - H(X|Y) \} \} \tag{3.7}$$

where X is the message sent by the transmitter, Y is the noisy version of the message received, $H(X)$ is the entropy of the message X, $H(X|Y)$ is the conditional entropy of X given Y, B represents the bandwidth of operation, and β is the ratio of the symbol interval T_s to the pulse length T_p.

3.4.1.3 Molecular communication-based intra-BSN

Propagation model

Two types of phenomena can be experienced over this kind of propagation environment: free diffusion and assisted diffusion. In the case of the free-diffusion phenomenon, Brownian motion is responsible for the movement of the information molecules through body fluid. The propagation process in this case can be modeled using the Weiner process with Fick's second law governing the flow through the following equation [13],

$$\frac{\partial C}{\partial t} = D\nabla^2 C \tag{3.8}$$

where C denotes the concentration of molecules, D is the diffusion coefficient of medium given by $D = K_B T / 6\pi \eta r_m$, ∇^2 denotes the squared-differential operator given in Cartesian coordinates $\{x, y, z\}$, $\nabla^2 = i \frac{\partial^2}{\partial x^2} + j \frac{\partial^2}{\partial y^2} + k \frac{\partial^2}{\partial z^2}$, T is the temperature of operation in Kelvin, η is the viscosity of the medium, r_m is the radius of the information molecule, and K_B is the Boltzmann constant.

In the case of assisted diffusion, the propagation phenomenon can be described using the expression,

$$\frac{\partial C}{\partial t} = D\nabla^2 C - V_x \frac{\partial C}{\partial x} - V_y \frac{\partial C}{\partial y} - V_z \frac{\partial C}{\partial z} \tag{3.9}$$

with drift velocities V_x, V_y, and V_z in $+x$, $+y$, and $+z$ directions, respectively.

Noise model

Noise in the molecular communication environment can be attributed to two different sources: inherent and external. The inherent noise results from the random arrival of emitted molecules at older bit intervals. Such inherent noise can be described either by inverse Gaussian (IG) distribution as,

$$N_T \sim \text{IG}(l/v, 2l^2/D) \tag{3.10}$$

where N_T is the noise at the first arrival time, l is the communication distance, and D is the diffusion coefficient for a positive drift velocity $v > 0$ or by the binomial distribution as,

$$N_B \sim \sum_{i=1}^{B-1} \text{Binomial}(N, F(d, (B-i)T_B, (B-i+1)T_B)) \tag{3.11}$$

where N_B is the noise at Bth bit interval due to previous $(B-1)$ bit intervals, N is the number of transmit molecules at the start of the first-bit interval, B is the number of bit intervals, T_b is the length of one-bit interval, d is the distance between the transmitter and the receiver, and $F(., ., .)$ is the fraction number of molecules received at the target.

The external noise results from a combination of different sources and factors encompassing biochemical, thermal, physical, sampling, and counting noise. Interaction between information molecules and the environment results in biochemical noise. Changing surrounding environment temperature results in stochastic thermal agitation, causing thermal noise. The viscous medium exerts a physical force on the movement of the molecules, resulting in physical noise. Randomness and discreteness of the molecular movement result in varying counts of the molecules at the receiver contributing to counting noise. Discreteness of the molecules and their unwanted perturbation suffered at the emission while modulating the concentration cause sampling noise.

Achievable transmission rate

If we consider a point transmitter that encodes a message in the release time of a molecule, the molecule will be fully absorbed by the receiver. The initial arrival time of the molecule denotes the actual arrival time of the molecule. If the release time from the transmitter is X and the arrival time at the receiver is Y, then

$$Y = X + N_T \tag{3.12}$$

where N_T is the noise experienced at the initial arrival time at the receiver boundary. Using the traditional definition of channel capacity,

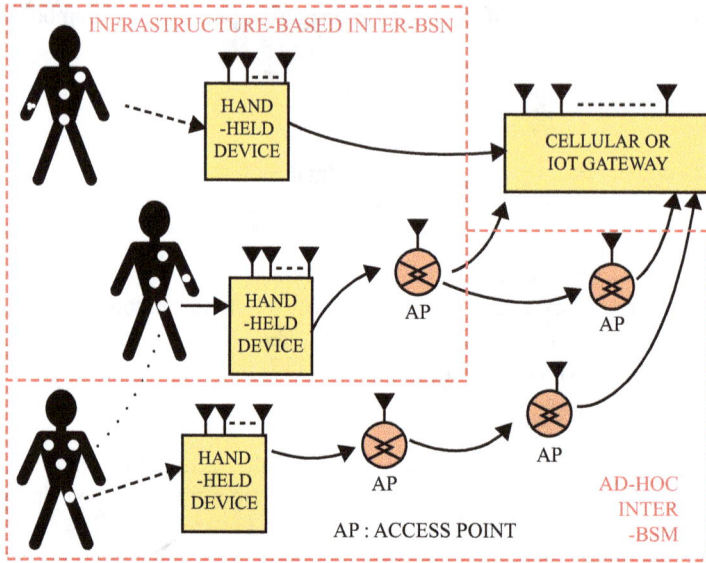

Figure 3.4   *An example sets of inter-BSN communication over infrastructure-based and ad hoc-based connection frameworks*

$$C = \max_{f_X(x):E[X]} I(X, Y) \tag{3.13}$$

we can derive the expression for the achievable information rate over the molecular environment.

3.4.2 Inter-BSN communication

The portion of the wireless network infrastructure used to connect more than one BSNs represents the inter-BSN communication network. Inter-BSN can be deployed based on the available infrastructure (fixed) or on an ad hoc basis (variable). In the infrastructure-based scenario, the BSNs communicate with each other over a centralized unit (or gateway), providing management and security control; the diagrammatic representation of which is presented in Figure 3.4. In an ad hoc-based scenario, the BSNs do not communicate with each other. Each BSN sends its individual information using multiple access points (APs) between the BSN and the data center. The APs used for routing the information to the data center are chosen on an ad hoc basis depending on the location of BSN and the corresponding environmental conditions, like instantaneous signal-to-noise ratio. The info-graphic representation of this kind of inter-BSN architecture is portrayed in Figure 3.4.

3.4.2.1 System model

The overall system model of the inter-BSN architecture involves three fundamental kinds of nodes.

- Sensor/Actuator Nodes—Nodes around the human body consisting of environmental sensors, wearable sensors, and handheld devices extract information from the BSN in a format that can be transmitted over the air.
- Router Nodes—Nodes consisting of APs, routers, and relays forward their information received from the handheld devices to the gateway.
- Gateway—Central data center capable of managing the received information and then forwarding it to the relevant application end point.

3.4.2.2 Propagation model

The propagation path loss over an inter-BSN communication scenario can be mathematically expressed as,

$$P(d, \alpha)|_{dB} = G_o(d, \alpha)|_{dB} + F|_{dB} \tag{3.14}$$

where $G_o(d, \alpha)$ is the mean channel gain over a distance d between the transmitter and the receiver and an angle α between the line-of-sight (LOS) and non-LOS (NLOS) components received at the receiver, and F accounts for the multipath fading experienced. Several experimental campaigns reveal that F follows closely the Nakagami distribution;

$$f(F; \mu, \omega) = \frac{2\mu^\mu}{\Gamma(\mu)\omega^\mu} F^{2\mu-1} e^{-\frac{\mu}{\omega}F^2} \tag{3.15}$$

where μ is the shape factor, ω is the shape parameter, and $\Gamma(\cdot)$ is the incomplete Gamma function. An important phenomenon observed in inter-BSN architectures is the body shadowing effect. This effect results from the increase in path loss owing to the presence of additional obstacles like the human body in the path of EM waves traveling through the environment. Body shadowing effect is encompassed within the term $G_o(d, \alpha)|_{dB} = G_s(d)|_{dB} + S(\alpha)|_{dB}$, where $G_s(d)|_{dB}$ is the mean channel gain over the angle α, $G_s(d)|_{dB} = E_\alpha\{G_o(d, \alpha)|_{dB}\}$, and $S(\alpha)|_{dB}$ is the shadowing component that varies around the mean channel gain $G_s(d)|_{dB}$. Inter-BSN frameworks can be a mesh structure where multiple transmission of data results in extended radio coverage supporting patient's mobility and emergency response systems can be quickly deployed through flexible network structures. APs can be added as needed to the path of information transmission.

3.4.3 Beyond-BSN communication

The portion of the network that routes information from the gateway or the central processing platform to the relevant application point constitutes the beyond-BSN communication framework.

3.4.3.1 System model

The system model for this part of the network comprises basically two different types of communication entities: the gateway or central platform and the end nodes, and the communication takes place over the air using radio-frequency waves and

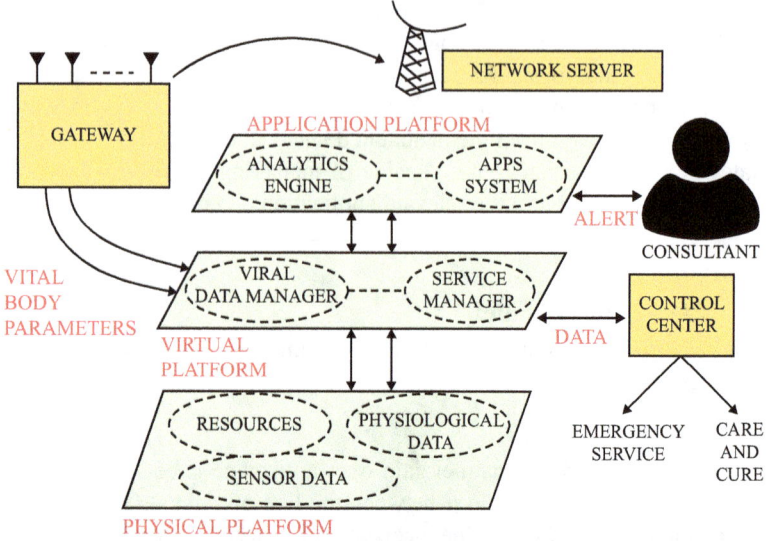

Figure 3.5   *A detailed representation of the framework commonly used in beyond-BSN communications*

existing Internet of Things (IoT) or cellular backhaul network. A graphical representation of this part of the framework is presented in Figure 3.5. It is worth mentioning here that the beyond-BSN network part is designed depending on the application at hand. For example, depending on the body parameters, the relevant person may be advised to take appropriate self-care measures to make the situation normal. While, in some other scenarios, the person under observation may need medical intervention or social support. The application end point can also serve as a database where continuous monitoring data are stored.

3.4.3.2   Propagation model

If the beyond-BSN activities are communicated over the macro- or micro-cellular network, the propagation path loss can be determined by,

$$PL = P_T - \text{RSSI} + G_T + G_R \tag{3.16}$$

with a transmit power P_T emanated from the gateway equipped with transmit antennas of gain G_T and G_R is the receive antenna gain at the application end node. RSSI stands for the received signal strength indicator, measured in dB at the receiver node. If an IoT network setup is chosen for the BSN backhaul, the propagation path loss can be given by,

$$PL_{\text{overall}} = P_{T_{\text{LOS}}}(L_{\text{LOS}}) + (1 - P_{T_{\text{LOS}}})(L_{\text{NLOS}}) + \xi \tag{3.17}$$

where ξ is the small-scale fading loss, L_{LOS} is the large-scale loss over the LOS path, L_{NLOS} is the large-scale losses experienced over the NLOS path, and $P_{T_{\text{LOS}}}(L_{\text{LOS}})$

is the probability of experiencing LOS communication link between the gateway and the intended end node.

3.4.3.3 Noise model

Both in the case of the inter-BSN and beyond-BSN sections of the framework, the noise over the mobile radio reception link can be characterized by independent Gaussian random variables following the distribution,

$$f_N(n) = \frac{1}{\sigma\sqrt{2\pi}} e^{-\frac{n^2}{2\sigma^2}} \qquad (3.18)$$

where n is the in-phase component of the received signal. The changing velocities of the human bodies when they are mobile can also result in Doppler shifts to the transmitted frequencies originating from wearable sensors or on-body sinks. Then the signal wave will be observed to have frequency f_d, for a signal arriving at an angle ϕ, mathematically expressed as, $f_d = f_m \cos\phi$, where $f_m = \frac{<v(t)>}{\lambda_g}$ is the maximum Doppler shift at the signal traveling velocity and carrier wavelength λ_g. The experienced Doppler shift can be related to the noise model in (3.18) through the following relation,

$$\sigma = \sqrt{\int_{-f_m}^{f_m} S(f)df} \qquad (3.19)$$

where σ is the variance of the noise experienced and $S(f)$ is the power spectral density of the received signal.

3.4.4 BSN network topologies

From a network topological point of view, communication nodes belonging to a BSN can be classified into two different types: peripheral nodes and central nodes. The nodes that are in charge of collecting information from other nodes deployed within and outside the human body and processing it to activate appropriate actions constitute the group of *central nodes*. Central nodes can be the gateway, cloud processing platform, data fusion center, etc. The nodes that are used to sense activities within the human body monitor the environment and transfer information from the body and environment to the processing center constitute the group of *peripheral nodes*. Peripheral nodes can be implantable devices, wearable sensors, handheld devices, etc.

Based on the concept of peripheral and central nodes, IEEE TG6 [14] has recommended three types of network topologies for implementing BSNs. They are,

- Star—Each peripheral node has a direct connection to the central node;
- Mesh—Not all peripheral nodes are directly connected to the central node; they can be connected to the other peripheral nodes;
- Hybrid—Combination of mesh and star topologies with clusters of peripheral nodes connected to a central node.

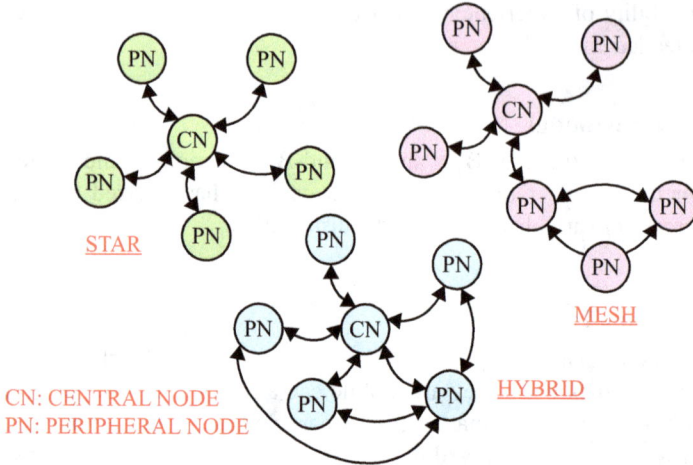

Figure 3.6   Representation of different network topologies used for BSNs

A diagrammatic representation of these different network topologies is presented in Figure 3.6.

3.5   BSN network layers

To design an efficient BSN, it is crucial to consider the functions and requirements of each layer in the communication stack. This section describes the fundamental functions, requirements, and challenges for each layer of BSN.

3.5.1   Physical layer

The physical layer exists at the bottom of the communication stack in BSN nodes. The major functions supported by this layer include sensing information/physiological parameters from the nodes encoding/decoding, generation, and removal of preambles for specifying the transmission medium, reception and transmission of bits, and synchronization. In addition, appropriate channel selection and characterization are also major responsibilities of the BSN physical layer. Bluetooth, Zigbee, and UWB are the most common physical layer schemes used for connecting the BSN nodes. Other emerging physical layer schemes include RuBee, Sensium, and Zarlink. We define each of these schemes briefly in this section.

Bluetooth is a commonly used technology for establishing wireless connections over short ranges. The operating frequency of the Bluetooth standard is 2.4 GHz, which is categorized in industrial, scientific, and medical bands. Since the BSN nodes are located in a closed range, they are often connected via Bluetooth. This technology can be used for single-hop communication between the sensor and relay nodes and also for multi-hop communications. Very often, Bluetooth has been seen to report health data from smart devices such as Fitbit to the associated smartphone application.

Zigbee has been another common protocol for BSN. The protocol is developed by IEEE 802.15 Task Group 4 and has been adopted as a standard for wireless personal area networks. Zigbee operates at two distinct physical bands 2.4 GHz and 868/915 MHz. The data rate for the 2.4 GHz band is up to 250 kbps, whereas it is 20 kbps for 868 MHz and 40 kbps for 915 MHz. In the past, Zigbee technology had widely been used for home automation applications. Later, the standard was modified to operate at low power and low duty-cycle due to which it became ideal for BSN applications. The protocol is also low in cost and, therefore, it has been one of the top choices for establishing connectivity between BSN nodes.

UWB was standardized by IEEE 802.15.6 Task Group 6 and it has become a famous technology for BSN due to its unique features. This scheme works using the band 3.1–10.6 GHz and operates numerous benefits to BSN such as low power consumption, anti-multipath capability, and large bandwidth. UWB was specifically designed for short-range communications. It has two operating bands: a low band at the frequency range of 3.2448–4.7424 GHz and a high band at the range of 6.24–10.2336 GHz. Thus, UWB offers a bandwidth of 499.2 MHz.

RuBee is based on the IEEE standard 1902.1. The features due to which RuBee has been regarded as a suitable technology for BSN are its high security level, efficient transmission distance, long battery life, and stable operation. The protocol operates at a low frequency of below 450 kHz. It offers a dual-way wireless protocol where long-wave magnetic signals are used for the transmission and reception of data. Another unique characteristic of RuBee is that the allowed data size is 128 bytes only, which, on the one hand, may limit the applications to send some important data; however, on the other hand, this feature enables the protocol to prolong the battery lifetime. Finally, another important characteristic of RuBee is that its operating frequency cannot be attenuated by liquid or metal; hence, this protocol can easily be deployed in environments where RFID may not function due to attenuation and other losses.

Sensium is another physical layer technology that offers medical experts an opportunity to monitor their patients continuously (and remotely) at a low cost. Sensium facilitates the on-body applications where sensor nodes may send low data rate information. As with the standard BSN architecture, the nodes using Sensium can send their data to multiple locations including sink node, smartphone, laptop, PDA, tablet, or any other dedicated device for health status monitoring. Subsequently, the data are transmitted to remote locations over an internet connection. Sensium nodes work on the TDMA mechanism and they are kept in either sleep or standby until their assigned time slot occurs.

Zarlink is a physical layer technology developed for offering connectivity to medical implants. The technology combines Reed–Solomon coding and cyclic redundancy check to access a link with very high reliability. The implants working on the Zarlink frequency band can be inhaled in the form of capsule which disintegrates once it reaches the stomach. The world's first camera capsule was developed using Zarlink for the purpose of gastrointestinal tract examination; the system was designed to avoid the need for complex procedures of colonoscopy and endoscopy. Using the Zarlink communication radio, the implanted node was configured to send two images per second. Like Sensium, the current consumption of Zarlink nodes is also very low by keeping the nodes in a sleep state for most of the time.

3.5.2   Medium access control layer

Maintaining energy efficiency has been regarded as the most critical challenge for WSN and BSN. The energy required for different devices needs to be minimized to avoid frequent replacement of batteries, particularly for the implanted devices. It has been reported by various experimental case studies that approximately 80% of the BSN energy is consumed in communication processes. The medium access control (MAC) layer being the direct contact point with the radio of the nodes has been considered ideal for developing techniques that could reduce the overall energy consumption. Therefore, we see that most of the MAC protocols developed for BSN focus on optimizing energy usage. MAC layer protocols mainly govern the channel access for each node in the BSN. Broadly, MAC protocols can be categorized as reservation-based or contention-based; for the reservation-based protocols, the cluster head generally sends a beacon to all the nodes to mark the beginning of each cycle and the nodes with data can reserve slots for sending their data. On the other hand, for contention-based protocols, the nodes generally run a random back-off counter, and the node whose counter expires first gains the channel access. Like the traditional wireless networks, the WSN and BSN also utilize the mechanisms of Request to Send, Clear to Send, and Network Allocation Vector.

The MAC protocols for BSN can use either a reservation- or contention-based scheme, or even a hybrid of the two. Often, the hybrid MAC schemes are preferred because reservation-based schemes cannot be directly deployed due to the requirements of clock synchronization, high bandwidth utilization, and lack of scalability; since the nodes in BSN are mostly heterogeneous, it is not possible to satisfy all of these requirements. On the other hand, the protocols based only on contention may also not serve the purpose of timely and reliable transmission of data from different sensors as the sensors with critical data may not get a transmission opportunity due to the channel being occupied by the sensors transmitting data of lower priority. Therefore, the protocol designers often develop customized MAC protocol for each BSN scenario rather than using the standard protocols in their original form. However, the standard protocols are used to date for providing the fundamental guidance and blueprints.

The standard protocol developed for BSN is IEEE 802.15.6, which supports short-range and low-power wireless communication. This protocol supports several Phy layers including narrowband, UWB, and human body communications layers. The data rate offered by IEEE 802.15.6 may be up to 10 Mbps. An interesting feature of this standard is its capability to reduce the signal radiation absorption rate resulting from the movement of the human body. Another common MAC and Phy layer protocol for BSN is Zigbee (IEEE 802.15.4); this protocol also offers low power and reliable communication between low data rate wireless devices. Zigbee also requires low battery power which is an essential requirement for BSN nodes operating on short-range frequencies. Moreover, Bluetooth low energy standard has also been used for BSN.

In addition to the above-described standard protocols, there are tens of MAC protocols being proposed for BSNs. A large base of such protocols has focused on energy efficiency by proposing diverse duty-cycling, synchronization, sleep scheduling, contention window prioritization, channel polling, data aggregation, multichannel access, wake-up radio, and other schemes. As previously discussed, the fundamental goal of BSN MAC protocols is to minimize the delay and energy consumption, so a mix of various techniques has often been suggested. For example, a recent protocol FROG-MAC has been developed to address the requirement of differentiated QoS, particularly delay of biomedical traffic [15]. The protocol proposed to fragment the low-priority data and send the higher-priority data without fragmentation. The higher-priority data are allowed to sense the channel in between the transmission of fragments of lower-priority data; this approach resulted in lowering the delay for higher-priority traffic as these packets do not have to wait for the complete transmission of lower-priority data. Another protocol called energy efficient distributed queuing random access (EE-DQRA) has been proposed specifically for BSN [16]. The concept of distributed queuing has been deployed for designing EE-DQRA so that the utilization of radio channels can be optimized. The protocol segregates channel access for the nodes with lower- and higher-traffic loads; instant channel access is used for lower-traffic loads and reservation process is used for higher-traffic loads. This way, both throughput and delay are improved by efficiently scheduling transmission. Similarly, various other approaches have been used for designing MAC protocols so the emerging needs of scalability and reliability in BSN can also be addressed along with energy efficiency.

Link quality, particularly for the implanted BSN nodes, may impose major challenges for the timely delivery of data. Path loss faced by the signal as it travels through the complex body tissues may worsen the entire performance of BSN. In this regard, the emerging technologies for intra-body communication such as molecular communication and THz communications (using the band 0.1–10 THz) have been proposed. Molecular communication deals with the transmission of information from the nanoscale intra-body sensors to the external sinks. In this mode of communication, transmitters, receivers, and/or actuators are deployed in the human body using a predefined application-specific topology, and molecules are used as information carriers. The transmitters may encode the message as a number, type, or release time/pattern of the molecules; this message is subsequently sensed and decoded by the receiver nodes [17]. Molecular communication is expected to be used for intra-body communication in the near future because it reduces the hazards associated with the transmission of EM signals across the human tissue; however, the communication mechanism is very slow and challenges exist to develop efficient protocols and strategies which could utilize its benefits. Similarly, the THz communication within an intra-body environment promises negligible latency and higher throughput. Due to the unique challenges of scattering, reflection, and high path loss associated with THz, there is a requirement of developing novel MAC protocols.

3.5.3   Network layer

The major function of the BSN network layer is to enable BSN nodes to connect with others using appropriate routing algorithms. Also, this layer enables the BSN to be connected with other networks including other BSNs and the backhaul IoT network. It is often assumed that the routing protocols designed for BSN are simple because there are several defined paths between the nodes, and the routing algorithms are mostly based on reducing the energy consumption using data aggregation. Thus, the packets generated by each sensing node are forwarded to the sink using a specific routing algorithm. Although the sensing nodes may forward their data directly to the sink or coordinator, due to the short range and limited processing capacity of these nodes, multi-hop communication from source to sink node is often practiced. Thus, the major task for the network layer in BSN is to use an appropriate routing protocol to select the best possible next hop node based on various predefined criteria.

Considering the resource-constrained BSN nodes, the routing algorithms are also designed to optimize energy usage in the network. There have been various protocols that identify the nodes with maximum residual energy and suggest the routing path using such nodes. In addition to the residual energy of nodes, the routing protocols also consider equalization among the neighbor nodes and their distance from one another. Furthermore, the route which could let the packet reach the sink node via a lesser number of nodes is also selected by some protocols. The context-aware routing or routing based on previous knowledge has also been commonly used for BSN [18]; in this approach, the weights are assigned to each node based on the parameters such as residual energy, link stability, delay, and distance. These strategies help to enhance the lifetime of each node such that all can achieve a similar average lifetime.

Energy Aware Link Efficient Routing approach deals with developing green methods for selecting the next hop. Unlike the conventional wireless networks or WSNs, it is crucial for BSN to focus more on the quality of the link rather than the shorter path; the shortest-path algorithms may end up consuming more energy in the BSNs. Therefore, in most routing approaches that have been customized for BSN, network and path cost models are used to compute the energy efficiency of each link, and the most energy-efficient next hop link is chosen by the nodes. The cost functions are mostly comprised of the energy level of the node, its distance from the sink, link efficiency, and transmission power. In general, it is considered that a path with a lesser number of hops to the sink should always be selected for energy-efficient routing; however, for BSN, there may be situations where a path with a higher number of hops may be used in order to ensure energy-balancing for the entire network.

In some schemes, BSN nodes are also segregated based on their energy levels for energy-aware routing. The nodes with lower levels of energy only send critical data and are also not chosen as relay nodes. This results in improved network performance as the probability of packet loss decreases. Energy-aware routing has also been integrated with security as it is one of the most critical aspects of BSNs. Rather than forwarding data to every node, several trusted nodes are selected as relays and the reputation of each node may be monitored based on different criteria such as the number of node failures in the past and the possibility of receiving malicious data from each node. For this purpose, advanced security algorithms have been linked with the routing protocols of BSN.

An important challenge for the network layer is to deal with the mobility of BSN users. As a person moves, the connectivity of BSN would alter and the network performance could degrade. To deal with this, adaptive energy-efficient multi-hop routing protocols exist, which also cater to mobility needs. Since the routing challenges further increase for the mobile users due to the change in posture and the subsequent impact on the position of sensors, the routing protocols dealing with mobility often segregate the data based on their urgency. For example, the data with low priority may be sent using multi-hop routing, whereas urgent or on-demand data may be forwarded directly to the sink. Opportunistic routing algorithms have also been proposed for the mobile users of BSN. In this method, the sink is often kept at a location such that it could make a LOS connection with the BSN worn over or implanted inside the body. If the sink node is kept on the hand, as the patient walks, the hand would be moving back and front. When the hand comes in front of the body, the BSN can directly send the data to the sink taking advantage of the LOS communication opportunity. On the other hand, when the hand moves at the back of the body, the data moves via relay node instead of directly going to the sink to avoid the communication cost. In the future, if it becomes possible to predict user's mobility (possibly through machine learning and pattern recognition techniques), the link failures could be timely managed. Furthermore, knowledge about mobility would also lead to reliable resource allocation and routing for all intra-, inter-, and beyond-BSN environments.

3.5.4 *Application layer*

The application layer of BSN is responsible for presenting the data to stakeholders in their desired format and can be uniquely designed for each BSN application. The application layer hides the complexity of the sensing modules and processes and provides an easy-to-use interface to the users. The protocols deployed for application management play a key role to make the use of software and hardware at lower layers of the BSN protocol stack. Commonly, the protocols used at this layer of BSN serve the purposes of data aggregation, data dissemination, and data collection. The three most famous application protocols for BSN are the sensor management protocol (SMP), sensor query and data dissemination protocol (SQDDP), and task assignment and data advertisement protocol [19].

SMP is used by the sensor network administrators to perform software operations such as defining rules for data aggregation, managing synchronization and mobility of sensor nodes, turning nodes on and off, and querying the sensor networks for their configuration and status. Moreover, SMP is also used for managing the security of BSN by performing tasks of key distribution and authentication. SQDDP provides application interfaces to the users for issuing queries, responding to them, and collecting incoming responses. Instead of issuing queries to particular nodes, these are often generated for the entire network or a part of it. Since BSNs are limited in resources, generating queries simultaneously for multiple nodes offers the advantage of energy conservation.

The common protocols used for BSN deploy customized application layer according to their needs and specific features. For example, Zigbee, which has been designed on IEEE 802.15.4 standard and has commonly been used for BSN application, defines three application sublayers, the application objects (AO), the ZigBee device object (ZDO), and the application support sublayer (APS). Zigbee has been regarded as an ideal protocol for BSN because the complete protocol stack is already implemented and can be used for any BSN application without needing modification. The application sublayers have been designed in a way that users can deploy them for their use cases without needing to know the concepts of network-ing and thus, offers almost all the functions any user may look for. The AO layer is composed of manufacturer-defined applications which are provided according to the user-specific requirements. Using this sub-layer, a single node can handle up to 240 applications. ZDO comprises the functions which are offered by the device for a smooth BSN operation; this layer defines the role of various devices such as network coordinator, gateway, router, etc. Finally, the APS communicates with the relevant application, sends messages between nodes, and also communicates with the trust center (location where security and privacy of BSN node can be controlled from). APS manages an APS information base which holds crucial information about sys-tem security. Moreover, APS also facilitates the process of node discovery so that the neighboring devices may know about its presence.

WirelessHART project has also defined specific features for application layers. The protocol establishes communication between gateway and devices following the method of commands and responses. The application layer has been designed to perform the tasks of parsing the messages, extracting the number of commands, exe-cution of commands, and generation of corresponding responses for each command. Another protocol Incremental Join Algorithm (IJA) uses a customized application layer to collect data from sensors using the commands SELECT and AGGREGATE. In IJA, a distributed database is maintained at the nodes to collect and aggregate data from the BSN. As previously mentioned, data aggregation is among one of the key tasks for the application layer as it saves energy due to lesser transmission of over-head bytes. BSNs are considered ideal for the process of data aggregation, because very often, multiple nodes may have similar data to send, and also, the actual data content may be small as compared to the associated header bytes. When some pack-ets from single or multiple nodes are aggregated, the cost of sending multiple header bytes is saved. Application layers of various protocols have been configured to per-form data aggregation based on predefined criteria. Therefore, in addition to energy conservation, well-designed application layer protocols for BSN can also play a cru-cial role in bandwidth optimization by controlling the amount of data transmitted.

Today, along with data aggregation and dissemination, several novel responsi-bilities have emerged for the application layer of BSN. For example, data analytics has emerged as one of the major application layer challenges for BSN data. Learning from the data generated by BSN nodes and identifying patterns in this data with-out conflicting with the patients' rights of privacy and confidentiality has become a serious concern. On the one hand, there is a clear opportunity for medical researchers and physicians to identify useful patterns from the data collected from patients using advanced machine learning algorithms. On the other hand, the patients' privacy and security could be at serious risk. Thus, innovative techniques for integrating data analytics technologies have been developed for the BSN application layer.

3.6 Security threats and solutions for BSN

As the applications of BSN have increased, the security threats for these networks have also evolved a lot in the recent past. Since the information propagating in BSN is about human health, the effects of any tampering could be life-threatening. Similarly, delay in transmission of crucial data occurring due to damage of resources could affect the overall system reliability and result in severe negative consequences for human health. Various types of security threats and attacks have been reported for stealing or compromising the data from BSNs. These threats could be active or passive; here, active attack refers to the attack where data and system resources are damaged or modified; on the other hand, no modification of data or damage to system resources occurs in passive attacks and the target does not even know about such attack unless there is a system to monitor the node identities. For BSN, the list of active attacks includes message corruption or data modification, impersonation attack, replaying, and forged base station attack, and passive category often refers to eavesdropping. We briefly define each of these attacks in this section.

3.6.1 Active security threats for BSN

From the active category, message corruption or data modification is the major security threat for BSN, which not only compromises the reliability of the network but could also cause life-threatening situation to the user. In this attack, the attacker deletes, adds, or modifies data content that are generated from the BSN nodes. When modified data reaches the destination (medical experts or central servers), the processing results in false diagnosis and harmful recommendations. More advanced the BSN would be, the higher the risk associated with the message corruption. For example, if there is a BSN, which only transmits alerts to the remote stations, in case of message corruption, the system would not send an alert at the time of emergency; in such a situation, the patient would not receive immediate help from the remote service providers; however, there would not be any direct risk to the patient. On the other hand, if an advanced version of BSN is considered where the system does not only send an alert to the emergency service provider upon detection of abnormal values of physiological parameters but, at the same time, also pumps medicine into the patient's body. For the case of message corruption in such a system, the risk will be much higher as the intruder will be able to inject high doses of medication through implanted pumping devices. Moreover, the attacker may also make data unusable for processing at the coordinator device and server. The partial or full deletion of data may also occur resulting in the loss of valuable information about a patient's health from a remote medical information database.

Impersonation attack can also cause serious security threats for BSN. In this attack, the attacker appears with the identity of some legitimate user in the communication system. If an attacker appears as the patient or any other stakeholder of the BSN, such as a medical expert, he may easily convince the other party to believe in their fake identity and can get access to the crucial health information. Also, the attacker can use impersonation to provide inaccurate data about the physiological parameters of patients to the remote server, resulting in misleading diagnosis and harmful recommendations. Similarly, the patient may find treatment advice from an impersonated attacker if it appears from the legitimate and usual remote physician.

Replay attack can also be used in BSN. In this attack, the attacker captures data from a BSN node and sends the same data to the receiver. Since the receiver (medical expert) believes the message to be coming from a reliable source (BSN sensor), it replies to the message which could contain health advice or some other useful information for the patient. The attacker can subsequently modify this message received from the medical expert and can forward it to the sender. Also, the attacker could resend the message from the medical expert to the BSN node several times so its energy source could deplete and it may stop working.

Forged BS attack refers to the impersonation of the base station or sink node by the attacker. When the BSN nodes find request packets or beacons from the attacker in this attack, they consider the message from a legitimate source and start to transmit their data to the attacker instead of their legitimate sink. This way, the attacker may become capable to take over the entire network and collect data from all the sensor nodes that are a part of BSN.

3.6.2 Passive security threats for BSN

The common passive attack for BSN is eavesdropping. An attacker may listen to the information being transmitted from BSN nodes to the remote stations. This information may then be used or sold, which may cause loss of user's confidentiality. The crucial health data can be shared with third-party organizations such as insurance agencies, resulting in negative consequences for users with high health risks.

3.6.3 Security solutions

To deal with the above-mentioned security threats, a number of security solutions specific to the requirements of BSN have been proposed. It is important to note that security solutions for BSN need to be lightweight so they could be adopted by BSN nodes of limited computing capabilities. As BSN comprises very few nodes compared to WSN, which could have hundreds of nodes, some of the security threats such as sinkhole, wormhole, or hello flood are not as common, and the solutions must be designed to target the attacks discussed above. The security mechanisms must ensure privacy, authenticity, and integrity of patients' information. The BSN must ensure that only legitimate nodes send encrypted packets and the frequency of data collection should be set efficiently; for most BSN scenarios, the freshness of information collected should be considered as an important security parameter.

3.7 Opportunities and open research directions

Due to the increasing use and acceptance of BSN, various new inter-disciplinary directions have been evolved for research. Although various BSN challenges such as energy efficiency, delay optimization, and reliability have been addressed by recent protocols, the evolution in the domains of big data, communication and engineering technologies, biomaterials, microelectronics, IoT, pervasive computing, machine learning, edge and fog computing, blockchain, and software-defined networks has given birth to new opportunities.

New sensors are being developed, which will give further rise to the applications of BSN. Since BSN sensors have the core requirements of small size and long battery life, it is expected that the sensors in use today may be considered obsolete in the near future. For example, recently, a novel radio-frequency heartbeat sensor has been proposed to measure systolic and diastolic blood pressure based on the observation of multipoint near-field [20]. This sensor can easily be worn by the users over their clothing for 24/7-monitoring and hence eliminates the need for conventional arm-cuff monitors which had not been convenient for integration with BSN. These trends show that in the future, there will be numerous opportunities for engineers, scientists, and technicians for exploring and designing innovative BSN applications. There will be a market need for manufacturing innovative sensors and devices to detect various physiological parameters with high reliability and low latency.

An interesting area for making BSN more practical and cost-effective is to develop clothing which could offer sensing capability for body parameter monitoring. BSN has already been integrated with clothing that receives power from the smartphone (appropriately) placed in the pocket via near-field communication [21]. There could be various sensors distributed across the clothing, connected through conductive threads used for embroidery. In addition to the integration of sensors with fabric, it is also crucial to design efficient wearable antennas integrated with clothing. Such antennas will require optimization of size and cost along permittivity and loss tangent such that the transmission could be effective without interfering the health outcomes of the human body. Design of antenna array and efficient integration of multiple antennas on clothing without disturbing the routine activities of users is also an important open research topic. It is to be noted that the wearable antennas will be used not only for medical BSN but also for other emerging applications such as those for military, athletes, and space communication.

BSN sensors provide highly individualized data for the users and identification of patterns can be used for guiding them to a better quality of health and life. With the increasing number of users and fog devices, the generation of big data will continue to increase at a tremendous rate, which shall further give rise to even more ubiquitous BSN. At present, the major focus of BSN has been to develop individual insights about the fitness status and activity level. Also, the recommendations for exercise, sports, and other fitness activities have already been provided to the users based on their data collected over a certain period. In the future, the sense of social accountability may also be integrated with the data collected from BSN.

The use of BSN for pattern recognition will require efficient mapping to the health and disease history of the individual user. Standard values of physiological parameters may not be sufficient for tracking the health of users. For example, the patients with lung issues will be more prone to the air pollution level as compared to the general population; a similar concept is valid for other population groups such as the elderly. This implies that there is a need to precisely identify the threshold level of physiological parameters for different diseases, each for a different population group. In this context, there is an opportunity for cross-disciplinary researchers to form teams and conduct clinical trials where BSN sensors can be used to collect information, and the medical experts may guide about healthy ranges of each parameter for patients belonging to different groups.

As the applications of BSN continue to grow, power generation will increasingly be sought from harvesting sources. For energy harvesting, there will be a need for infrastructure as well as customized protocols to offer a solution to each BSN stakeholder according to their needs. The nodes, antennas, and other equipment shall be designed such that the components could participate efficiently in the energy harvesting process. Similarly, there is a growing opportunity for the protocol designers to innovate MAC and routing schemes which could optimize the energy performance of BSN. There is room for improvement in both the routing and MAC layers of BSN at present. As previously discussed, despite various proposals on optimizing the energy consumption of routing, the relay nodes are likely to deplete their energy levels quickly as they need to keep forwarding data from the sensor nodes. Hence, further modifications in these protocols are required.

Similarly, at the MAC layer, even in the presence of advanced schemes for catering to the requirements of urgent heterogeneous traffic generated by BSN, the QoS still needs improvement in terms of reliability. Therefore, the diverse cross-disciplinary techniques would provide researchers with an opportunity to continue developments in the domain of BSN. For example, the existing schemes could be combined with data analytics and big data techniques. The patterns of energy consumption as well as overall performance of the BSN in terms of data delivery efficiency (in terms of reliability and latency) may be analyzed to develop more efficient protocols. In the recent past, such opportunities were not available because the protocols' designers mostly worked with the limited amount of data generated during a clinical or simulation study. However, today, advanced data analytics techniques have provided the opportunity to conduct analysis on the actual user data and to optimize the performance of protocols for each user in a customized fashion.

Maintaining the security of crucial health information has been regarded as one of the major challenges for BSN. All across the world, a large base of users is reluctant to opt for BSN only because they do not want to risk their privacy and confidentiality. As per the general understanding of the public, a person becomes totally vulnerable to theft of health information as soon as he starts to use BSN. Although various encryption algorithms have been proposed, the existing systems do not guarantee end-to-end security for the BSN data. Therefore, there is a golden opportunity for network security experts to develop security frameworks and algorithms to protect the privacy of BSN users by proposing lightweight yet effective security mechanisms.

The efficiency of BSN is expected to tremendously increase by combining the emerging technologies. For example, the combination of IoT with edge and fog computing technologies promises to reduce the end-to-end transmission delay by moving the processing power closer to data. This is because the delay caused due to the transportation of data from a BSN node to the central cloud location is significantly reduced by processing the data at a local node (edge node). At the same time, user privacy and confidentiality also improve by using these distributed computing technologies. Similarly, the integration of advanced machine learning and AI techniques with healthcare technologies has been an interesting area of research lately. Since biomedical sensors generate data at a fast pace, the present big data analytics and pattern recognition techniques would not be able to accommodate all the data collected over fog devices during a short span of time. There is still a research gap to develop techniques, which could help in understanding and interpreting the heterogeneous, diverse, comprehensive yet isolated BSN data.

Fusion of data from multiple sensors and identification of patterns is also a challenge as it could help the medical experts to assess the health risks and decide on the treatment strategies. In this context, advanced machine learning algorithms are needed which could resolve the bottlenecks such as identifying the most important pieces of data for each individual BSN user and setting the data collection frequency, reducing the volume of data by applying real-time preprocessing, synchronizing different data streams such that meaningful information along with its context may be extracted, developing new real-time algorithms, and deploying solutions that could implement the knowledge gained from collected data. There are opportunities for developing hundreds of machine learning algorithms as the possibilities of deploying and customizing BSN for healthcare applications seem to be endless today.

Data analytics associated with BSN does hold opportunities for bringing positive health outcomes not only for individuals but also for societies. BSN promises to revolutionize the caregiving system and remote health monitoring for wider societal levels. Modern BSN will collect data from sensors deployed at different locations, including ambient sensors as well. Although the data collected from multiple sensors may be imprecise and it is hard to interpret, once successful interpretation and pattern recognition is done, the analytics would be helpful for wider populations. For example, the data collected from elderly or cancer patients may help the medical experts to develop effective treatment strategies for similar groups of patients. Hence, the issues of dealing with the aging population all across the world and the patients with cancer, heart diseases, chronic diseases, and mental health issues can be offered a better quality of life.

3.8 Conclusion

Recent advances in BSN promise to revolutionize the healthcare sector. This chapter presented a detailed review of fundamental concepts of BSN, the understanding of which is critical for the design and development of BSN solutions for medical and nonmedical applications. Basics of BSN architecture and communication modules have been discussed along with the corresponding system, propagation, and noise models. The operation and mathematical formulation of performance parameters of intra-BSN, inter-BSN, and beyond-BSN modules have been presented. The requirements, for example, recent protocols and challenges associated with each layer of the BSN communication stack, have been discussed. Emerging security threats for BSN have been highlighted along with the famous security solutions. Finally, the opportunities and open research directions for BSN have been detailed. In conclusion, there is a need to manage challenges such as energy efficiency, size, cost, reliability, and biocompatibility of BSN nodes. Furthermore, state-of-the-art customizable protocols are also needed to support emerging BSN applications.

References

[1]   Lai X., Liu Q., Wei X., Wang W., Zhou G., Han G. 'A survey of body sensor networks'. *Sensors*. 2013;13(5):5406–47.

[2]   ul Islam S., Ahmed G., Shahid M. 'Implanted Wireless body area networks: energy management, specific absorption rate and safety aspects'. *Ambient Assisted Living and Enhanced Living Environments*. Elsevier; 2017. pp. 17–36.

[3]   Pramanik P.K.D., Nayyar A., Pareek G. 'WBAN: driving e-healthcare beyond telemedicine to remote health monitoring: architecture and protocols'. *Telemedicine Technologies*. Elsevier; 2019. pp. 89–119.

[4]   'O2Ring Oximeter Continuous Ring Oxygen Monitor.' *Wellue [online]*. Available from https://getwellue.com/pages/o2ring-oxygen-monitor [Accessed June 2021].

[5]   Li D., Wang X. 'On monitoring and detecting abnormal physiological state of athletes from Internet of bodies'. *Internet Technology Letters*. 2021;4(3).

[6]   Farrokhi A., Farahbakhsh R., Rezazadeh J., Minerva R. 'Application of Internet of things and artificial intelligence for smart fitness: a survey'. *Computer Networks*. 2021;189(4):107859.

[7]   Yu Z., Wang Z. 'Sensor-based behavior recognition'. *Human Behavior Analysis: Sensing and Understanding*. Springer; 2020. pp. 17–25.

[8]   Liu X., Zhao M., Liu A., Wong K.K.L. 'Adjusting forwarder nodes and duty cycle using packet aggregation routing for body sensor networks'. *Information Fusion*. 2020;53(1):183–95.

[9]   Yang K., Bi D., Deng Y., *et al.* A comprehensive survey on hybrid communication in context of molecular communication and terahertz communication for body-centric nanonetworks'. *IEEE Transactions on Molecular, Biological and Multi-Scale Communications*. 2020;6(2):107–33.

[10] Subramani P., Al-Turjman F., Kumar R., Kannan A., Loganthan A. 'Improving medical communication process using recurrent networks and wearable antenna S11 variation with harmonic suppressions'. *Personal and Ubiquitous Computing*. 2021;89(12):1–13.

[11] Jattalwar N., Balpande S.S., Shrawankar J.A. 'Assessment of Denim and photo paper substrate-based microstrip antennas for wearable biomedical sensing'. *Wireless Personal Communications*. 2020;115(3):1993–2003.

[12] Sreemathy R., Hake S., Sulakhe S., Behera S. 'Slit loaded textile microstrip antennas'. *IETE Journal of Research*. 2020;4(12):1–9.

[13] Berg H.C. *Random Walks in Biology*. Princeton: Princeton University Press; 1993.

[14] 'IEEE 802.15 WPAN Task Group 6 (TG6) Body Area Networks'. *IEEE 802.15. [online]*. Available from https://https://www.ieee802.org/15/pub/TG6.html [Accessed June 2021].

[15] Khan A.A., Siddiqui S., Ghani S. 'FROG-MAC: a fragmentation based MAC scheme for Prioritized heterogeneous traffic in wireless sensor networks'. *Wireless Personal Communications*. 2020;114(3):2327–61.

[16] Pandey A.K., Gupta N. 'An energy efficient distributed queuing random access (EE-DQRA) MAC protocol for wireless body sensor networks'. *Wireless Networks*. 2020;26(4):2875–89.

[17] Akyildiz I.F., Pierobon M., Balasubramaniam S. 'Moving forward with molecular communication: from theory to human health applications [point of view'. *Proceedings of the IEEE*. 2019;107(5):858–65.

[18] Abdu A.I., Bayat O., Ucan O.N. 'Designing insistence-aware medium access control protocol and energy conscious routing in quality-of-service-guaranteed wireless body area network'. *International Journal of Distributed Sensor Networks*. 2019;15(1):155014771881584.

[19] Filipe L., Fdez-Riverola F., Costa N., Pereira A. 'Wireless body area networks for healthcare applications: protocol stack review'. *International Journal of Distributed Sensor Networks*. 2015;2015(10):1–23.

[20] Hui X., Conroy T.B., Kan E.C. 'Multi-point near-field RF sensing of blood pressures and heartbeat dynamics'. *IEEE Access*. 2020;8:89935–45.

[21] Masuda Y., Noda A., Shinoda H. 'Body sensor networks powered by an NFC-coupled smartphone in the pocket'. 2018 40th Annual International Conference of the IEEE Engineering in Medicine and Biology Society (EMBC); 2018. pp. 5394–7.

Chapter 4

Seamless IoT mobile sensing through Wi-Fi mesh networking

Antonio Cilfone[1], Luca Davoli[1], Laura Belli[1], and Gianluigi Ferrari[1]

The research activity in the field of wireless mesh networks (WMNs) has been extremely active in the past years, leading to the design and implementation of different protocols and architectures. Moreover, due to their flexibility, WMNs have often been considered for Internet of things (IoT) applications, in order to provide seamless connectivity in scenarios where traditional infrastructure-based connectivity is not available (e.g., rural or industrial areas). In this chapter, an IoT-oriented mesh infrastructure for WMNs, based on the Better Approach To Mobile Ad-Hoc Networks (B.A.T.M.A.N.) protocol, is presented, with the aim to support mobility of nodes and also to allow the integration of non-mesh IoT nodes, enabling them to access the network and transmit data collected from the environment in a "transparent" way.

4.1 Introduction

In the context of networking, an important role is played by WMNs, in which the nodes can dynamically connect to each other through multi-hop communications. This enables the mobility of the nodes composing the backbone of the WMN itself [1]. Moreover, due to the absence of a static infrastructure, the network topology can evolve: for instance, nodes can be dynamically added, removed, or displaced, still guaranteeing connectivity. Therefore, the network deployment phase is faster and less expensive than that of centralized or infrastructure-based networks. This makes WMNs very attractive for IoT scenarios [2, 3].

The above characteristics highlight how WMNs are one of the more flexible and scalable networks approached for smart scenarios in large areas (e.g., smart agriculture

[1]Department of Engineering and Architecture, University of Parma, Parco Area delle Scienze, Parma, Italy

monitoring [4, 5]). In fact, over the past years, the potential of WMNs has continuously grown, becoming a reality in several scenarios. Thanks to the ease of deployment and scalability, several mesh networks, based on various radio technologies (besides Wi-Fi), have found applications in military, industrial, public safety, surveillance, and distributed sensing scenarios [6–10]. In general, WMNs are attractive when the geographic area that needs to be covered is not easily accessible and/or traditional connectivity strategies are not economically convenient or practically feasible [11].

In this chapter, we describe a WMN where the backbone is composed of different wireless mesh nodes based on IEEE 802.11 standard and on Raspberry Pi (RPi) 3 model B [12] nodes. The network allows external (non-mesh) nodes to join and also supports mobility for both mesh and non-mesh network nodes. The proposed infrastructure, due to its flexibility, can be used in several IoT scenarios to collect data through the use of mobile nodes—equipped with sensors and/or actuators—that move in the monitored area and access the network to transmit collected information or to execute received commands. Each mobile node, implemented using an RPi 3 model B, is not part of the mesh network but uses the mesh backbone as a client and is unaware of the internal organization of the WMN itself. Relying on the approach proposed in [13], the nodes composing the backbone of the WMN have two different IEEE 802.11 interfaces in order to separate the backbone tier from the network access tier. Moreover, the B.A.T.M.A.N. version IV routing algorithm [14], which is natively available in the Linux kernel, has been chosen to route the traffic inside the backbone network. This algorithm has been developed by the German Freifunk community to overcome the limitation of the optimized link state routing protocol [15] and is specifically designed to fit WMN scenarios.

The rest of this chapter is organized as follows. In Section 4.2, a background describing the most relevant protocols and standards adopted in building mesh networks is provided. Section 4.3 is devoted to the detailed description of the proposed WMN implementation and a preliminary experimental setup. Finally, in Section 4.4 we draw our conclusions, highlighting possible applications, in the field of IoT, of the proposed architecture.

4.2 Background

4.2.1 IEEE 802.11s basics

The demand for larger wireless infrastructures has led, in the past decade, to the development of an amendment of the IEEE 802.11 standard [16] specifically designed for Wi-Fi mesh networking, denoted as IEEE 802.11 seconds [17], which introduces new frame forwarding and routing capabilities at the MAC layer, together with new inter-networking and security techniques, in order to support mesh capabilities. The IEEE 802.11 seconds standard does not change L1 (PHY layer) of IEEE 802.11 but just modifies L2 (MAC layer). The most important novelty introduced is that the traffic routing is performed at L2 instead of L3 (network layer) so that nodes in the network can have direct knowledge of their "radio neighborhood."

In an IEEE 802.11 seconds mesh network, also named as mesh basic service set (MBSS), there are different logical components. Besides a sufficient number of "mesh stations" (mesh STAs), there are other mesh points (MPs) with augmented functionalities. While one type of enhanced MPs, denoted as mesh Access Points (MAPs), acts as APs for classical IEEE 802.11 stations, there exist other components, denoted as mesh portal points (MPPs), performing as gateways (GWs) toward an external (typically wired) network. Therefore, each entity composing the mesh network relies on a specific ISO/OSI stack implementation. Moreover, only mesh STAs have mesh functionalities (e.g., formation of the MBSS, path selection, and forwarding); thus, a mesh STA is not a member of an independent BSS (IBSS) or an infrastructure BSS, with mesh STAs not directly communicating with non-mesh STAs. In order to enable communication between mesh BSS and other BSSs, in fact, a mesh node can communicate with non-mesh nodes through the distribution system using the *mesh gate*, which is the logical component that enables the integration between mesh BSS and infrastructure BSS. In order to enable also the communication between the mesh BSS and non-IEEE 802.11 local area networks (LANs), such as wired LANs, another logical component is used, namely the *portal*. In the following, we assume that non-mesh nodes can communicate with mesh STAs through the MAPs.

4.2.2   IEEE 802.11s routing algorithm

One of the key aspects of a WMN is the traffic organization process handled by the specific routing protocol chosen for the WMN itself. The goal of that routing protocol is to discover and manage the best routes connecting pairs (or, more generally, groups) of nodes, according to one or more link- or route-based metrics (e.g., hop number, link quality, throughput). Moreover, inside a mesh network, all the devices should use the same path metric and routing protocol and, to this end, IEEE 802.11 seconds defines a default behavior for both, which, however, can be replaced by other custom solutions. The default metric, called "airtime metric," indicates the total cost of a link by taking into account some parameters (such as data rate, overhead, or frame error rate) measured by transmitting a 1 kbyte frame. The default routing algorithm is the hybrid wireless mesh protocol [18], based on the Ad-hoc On-demand Distance Vector protocol [19] combined with a proactive tree-based solution, in which a mesh station (typically acting as MPP) propagates routing messages to all mesh stations in order to establish and maintain the links.

4.2.3   B.A.T.M.A.N.

B.A.T.M.A.N. advanced (batman-adv in the following) is a proactive L2 routing protocol for WMNs and, namely, the "wireless" version of the B.A.T.M.A.N. protocol (originally designed for wired networks), which also supports roaming of mobile nodes [20]. More in detail, it keeps the information about the existence of nodes in the mesh network that are accessible via single-hop or multi-hop communication links. The batman-adv approach consists in allowing each node to determine, for each destination in the mesh network, its best next-hop, which can be identified as a GW to communicate with the destination node without requiring the knowledge of the complete route. In this way, there is no need for transmitting and keeping information about the whole topology at each node, as each node performs routing independently of the other ones. Therefore, each node needs to keep updated for each destination, the best next-hop; this significantly reduces the amount of control traffic and makes synchronization faster. Therefore, such behavior is similar to that specified by the software-defined networking (SDN) paradigm [21, 22], in detail looking at the data plane, where network nodes do not need to take care of the whole network topology—to be known only by the SDN controller(s) at the control plane and on which all the traffic-related decisions (e.g., based on Traffic Engineering strategies [23, 24]) will be taken.

In the version used in this chapter, B.A.T.M.A.N. IV [25], in order to perform the discovery of its neighbors, every B.A.T.M.A.N.-based node periodically broadcasts an OriGinator message (OGM), corresponding to a 12 byte UDP payload (for a total packet size equal to 52 bytes, including IP and UDP headers). The OGM has relevant information, such as a sequence number which is used to (i) distinguish new OGMs, (ii) guarantee that OGMs are not counted twice, and (iii) discover if a neighbor is a GW toward Internet or not. At the same time, by sending OGMs, each node informs its link-local neighbors about its existence [14].

Having to maintain B.A.T.M.A.N. as light as possible, each B.A.T.M.A.N.-based packet is encapsulated into a single UDP packet and consists of an OGM and zero or more attached Host Network Announcement (HNA) messages—HNA is a message type used to announce a GW to a network. The formats of the OGM and the HNA message are shown in Figure 4.1a and b, respectively.

As the default path metric, B.A.T.M.A.N. uses the transmission quality (TQ) metric, based on expected transmission count [26], to find a trade-off between a short (in terms of hops) route and a (potentially) long route with good links.

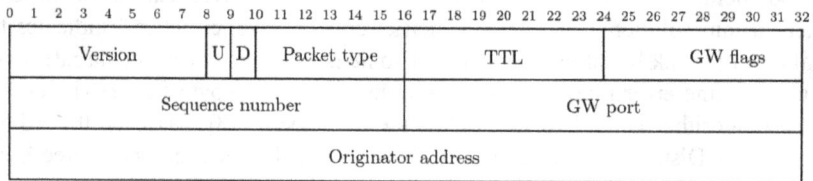

(a)

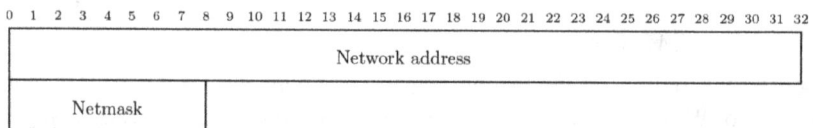

(b)

Figure 4.1   Packet formats: (a) OGM message and (b) HNA message

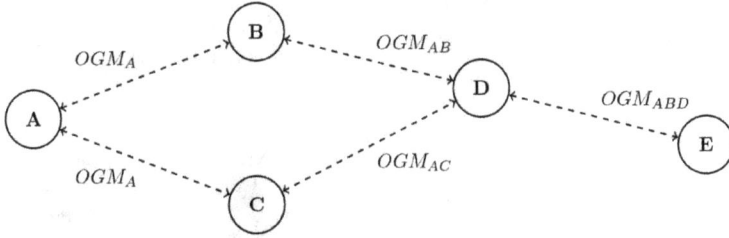

Figure 4.2   OGM rebroadcasting process

During OGMs' broadcasting, a node also counts the OGMs received from a given neighbor: this value is denoted as receive quality (RQ) and its calculation takes place considering a sliding window of 64 bits (which leads to 2^{64} possible entries). The sliding window keeps track of the last received sequence numbers of OGMs and the current received from each node in the network. The in-window sequence numbers are those that fit in the window below the current sequence number. If an out-of-range sequence number is received, it is set as the current sequence number and the sliding window is moved accordingly. Sequence numbers that are no longer in the sliding window are deleted. Neighbors rebroadcast received OGMs so that nodes more than one hop away get information about the existence of far nodes, as shown in Figure 4.2. In order to avoid overcrowding the network, each node resends only OGMs received from its neighbor with the best TQ metric.

In particular, a B.A.T.M.A.N.-enabled node evaluates the TQ metric of a generic neighbor *i* as the fraction of its OGMs that are correctly received by this neighbor as follows:

$$TQ = \frac{EQ}{RQ} \tag{4.1}$$

where echo quality (EQ) corresponds to the number of received broadcasts of its own messages within the sliding window. Finally, the best hop is determined by applying penalties for asymmetric links and taking into account the number of hops needed to reach the destination node.

4.3   Mesh network implementation

Our goal is to carry out an experimental evaluation of a WMN which (i) relies on a mesh backbone composed of B.A.T.M.A.N.-based nodes, (ii) allows the integration of non-B.A.T.M.A.N.-enabled devices as external clients that communicate with both mesh and non-mesh nodes, and (iii) allows the mobility of these external clients. In the proposed IoT-like architecture, the mobile nodes are non-B.A.T.M.A.N. devices—in detail, based on RPi boards—equipped with sensors or actuators and aiming at collecting data or executing commands using the WMN infrastructure to send and receive information. Moreover, the MPP and MAPs are implemented using an RPi 3 board, which embeds a Quad Core @ 1.2 GHz Linux-based Single-Board Computer with 1 GB RAM and on-board IEEE 802.11b/g/n interface.

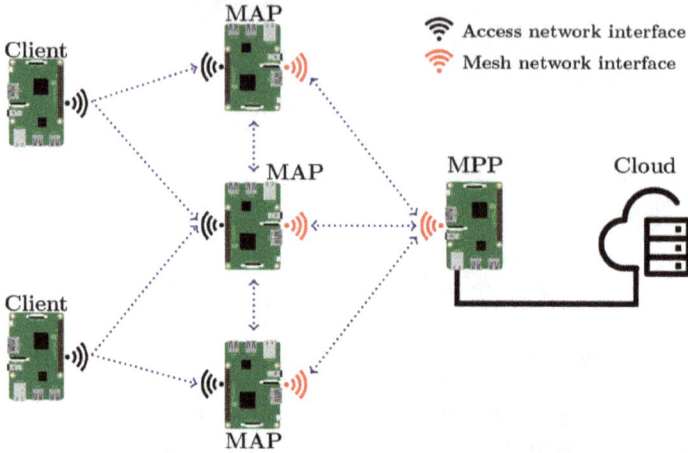

Figure 4.3   Multi-hop mesh network architecture

Focusing on the network topology, the overall IoT architecture is composed of two network tiers: the B.A.T.M.A.N.-based mesh backbone network, composed of four MAPs and one MPP; and a set of non-B.A.T.M.A.N.-based (namely, mesh-unaware) client nodes which can be used to collect data of interest from the deployment environment.

4.3.1   Proposed mesh backbone network

The backbone network, as shown in Figure 4.3, is composed of an MPP and one or more MAPs, all implemented on top of RPi boards. The MPP, acting as a GW, is the only node with a direct connection to the Internet (through an Ethernet cable) and has a single wireless interface, denoted as bat0, which is reserved to execute B.A.T.M.A.N. and to build the backbone network among all MAPs.

MAPs, instead, are "completely wireless" nodes equipped with the built-in IEEE 802.11 interface and an external IEEE 802.11 dongle—in other words, they have two IEEE 802.11 interfaces. Moreover, since a key goal of the proposed architecture is the support of the roaming functionality of a mobile node among the backbone nodes, the MPP is the only node running a DHCP server, whose aim is to distribute IP addresses to mesh-unaware nodes, *regardless* of their connection point. More in detail, the IP addresses' distribution is carried out through the presence of some specific daemons, namely *DHCP relays*, running on the access interface of the MAPs. In this way, when a new client joins the network and asks for an IP address, the request is forwarded to the MPP, which releases a new IP address and sends it back to the requester. In Table 4.1, the network configurations of a B.A.T.M.A.N. MPP and a generic MAP are shown, with reference to IP classes in which they are reachable and the services that will run on their network interfaces.

In order to enable connectivity to external mobile clients, each MAP node runs (i) a hostapd daemon on its wlan1 interface, turning this network interface

Table 4.1   *Backbone mesh network configuration*

	Interface	**Network**	**IP Class**	**Services**
MPP	eth0	LAN	192.168.1.0/24	
	bat0	Mesh	192.168.3.0/24	DHCP server, batctl
MAP	bat0	Mesh	192.168.4.0/24	DHCP relay, batctl
	wlan1	Access	192.168.2.1	hostapd

into an AP and an authentication server, and (ii) a DHCP relay, operating through the dhcp-helper daemon, which is used to dynamically assign IP addresses to external clients. Furthermore, the DHCP requests are forwarded from wlan1 to the bat0 interface and, once the request reaches bat0, it is finally sent to the DHCP server following the proper multi-hop route foreseen by the B.A.T.M.A.N. protocol. More in detail:

- the DHCP client broadcasts the packets in the subnet 192.168.2.0/24 generated through the *hostapd* daemon;
- the DHCP relay agent receives the broadcast and transmits it to the DHCP server(s) using a unicast transmission, thus being able to route the DHCP request to other DHCP servers not strictly in the same local network;
- the DHCP relay agent stores its own IP address in the giaddr field of the DHCP packet, as shown in Figure 4.4, and specifies, through the option82 field [27], that the request is coming from the subnet 192.168.2.0/24 in such a way that the DHCP server can lease the proper address.

As previously introduced, in the proposed B.A.T.M.A.N.-based mesh network, external nodes, potentially mobile, can use the mesh backbone as external clients,

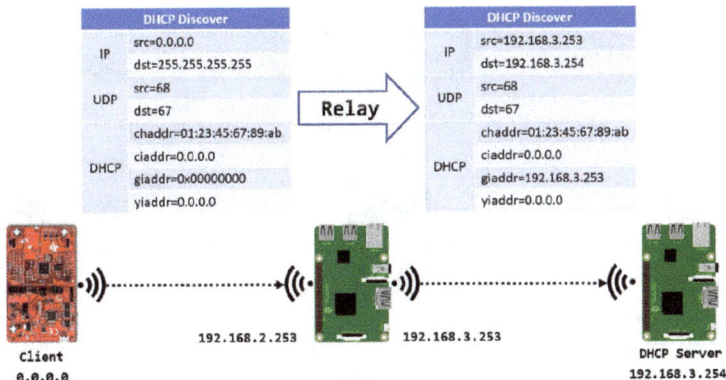

Figure 4.4   *DHCP relay message exchange*

in order to reach the Internet or communicate with other nodes. In the proposed architecture, these nodes have to simply connect to the Wi-Fi network generated through the wlan1 interface of the nearest MAP. Due to the presence of DHCP management functionalities, external B.A.T.M.A.N.-unaware clients are able to connect in a seamless way, without performing any additional configuration or installing specific software.

The main fields of the DHCP messages are the following:

- chaddr, containing the MAC address of the client requesting the IP address;
- ciaddr, containing the IP address of the client, which is 0.0.0.0 in the case that the client has no IP address;
- giaddr, containing the IP address of the GW, namely the DHCP agent relaying requests from the client;
- yiaddr, containing the IP address leased for the client.

The most relevant fields are chaddr and giaddr, which are used to forward the DHCP requests and the DHCP responses in a proper way. In particular, a MAP uses the chaddr to understand which client is the destination of the DHCP response containing the leased IP address. The DHCP server uses the giaddr to determine the subnet from which the DHCP relay agent received the broadcast, then allocates an IP address to this subnet. When the DHCP server replies to the client, it sends the reply to the giaddr address, again using a unicast transmission, as shown in Figure 4.5. The message is then routed to the correct node, eventually following a multi-hop route. Then, the DHCP relay agent retransmits the response to the local network.

In order to properly route the traffic arriving from external non-mesh clients, as well as to allow them to connect to the Internet, the following routing rules have been defined for MPP and MAPs.

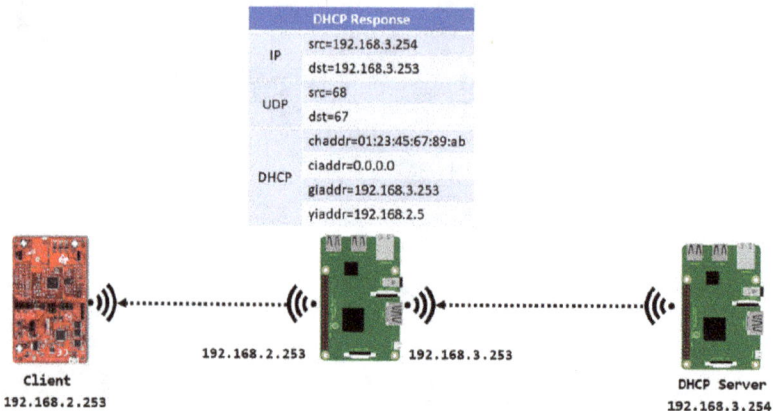

Figure 4.5   DHCP response

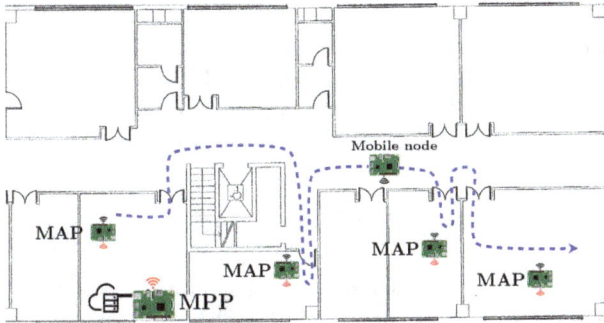

Figure 4.6   Experimental setup scenario. The B.A.T.M.A.N. interfaces are represented in red color, while regular wireless interfaces are represented in black.

With regard to the MPP:

- the traffic outgoing to the wired network (through cth0) is NATted, flowing outside the wireless network with only one public IP address, namely the IP address of the eth0 interface statically assigned by the network administrator;
- the traffic coming from bat0 is sent to eth0;
- the traffic coming from eth0 is sent to bat0 only for already established traffic flows.

With regard to the MAPs:

- the traffic outgoing to the backbone with source address 192.168.2.0/24 is sent through bat0 and is NATted in order to flow outside with only one IP address, namely the IP address of the bat0 interface of the considered MAP;
- all the traffic coming from wlan1 to bat0 is accepted;
- the traffic coming from bat0 and with destination address 192.168.2.0/24 is accepted only for already established traffic flows;

In order to test the performance of the proposed IoT mesh architecture, some connection tests have been performed in an indoor environment, as shown in Figure 4.6. More in detail, the experimental setup includes six double-interface Wi-Fi nodes, deployed in different rooms of the building and configured as follows:

- one MPP node connected to the Internet through the eth0 and a B.A.T.M.A.N. interface, to communicate with the mesh backbone;
- four MAP nodes, each one located in a different room;
- one mobile node moving on a path that enters and exits from all the 4 covered rooms.

The described deployment has been chosen to experimentally verify the roaming activity of the mobile node, which can thus connect to the network in a seamless and transparent way (as well as it happens, on high layers, to IoT nodes joining Web of Things contexts [28]) and be completely unaware of the mesh network infrastructure existing "behind the surface" of the publicly available Wi-Fi network.

4.4   Conclusions and application scenarios

In this chapter, we have proposed a WMN architecture based on the B.A.T.M.A.N. protocol, supporting the integration of non-B.A.T.M.A.N. external mobile clients for seamless IoT mobile sensing. In detail, the proposed mesh backbone allows external mesh-unaware clients to connect and send their collected data toward Wi-Fi clients in a transparent way, thus allowing to extend the normal Wi-Fi coverage with a multi-hop approach. To this end, the first experimental results obtained with a preliminary setup (composed of six IoT nodes) in an indoor environment seem to be promising, highlighting the flexibility of the proposed approach. Therefore, this mesh-oriented architecture seems to be suitable for several IoT applications, where traditional connectivity strategies are not employable or not economically suitable.

Looking at alternative scenarios in which such a mesh-oriented architecture may fit and be useful for extending the coverage from indoor to outdoor contexts, one example can involve the smart agriculture and rural areas-monitoring scenarios. To this end, IoT-like technologies and paradigms are nowadays rising a certain interest from different "players" in this field (e.g., technology developers, system integrators, and farmers) and is used for real-time data collection and actuation, as shown in Figure 4.7. Therefore, farmers can make conscious decisions on the basis of data sensed from their agricultural fields (e.g., soil temperature and humidity, wind speed, soil moisture, pH value [29]). Then, due to their geographical extension and the presence of natural obstacles, in rural areas, the deployment of a WMN can be the best solution to provide connectivity. Moreover, in order to reduce infrastructure costs, the use of mobile nodes (such use drones [30], as shown in Figure 4.8) to periodically perform environmental data collection campaigns and surveillance activities in the monitored area, joining the mesh network as a mobile external client, can represent another example fitting the characteristics of these networks. Finally, other possible mobile nodes can be represented by tractors, animals, or other entities that need to be monitored.

Another relevant scenario that can take advantage of the proposed architecture involves smart industries and smart infrastructures. In particular, our approach can be

Figure 4.7   Smart agriculture application scenario

Figure 4.8   Application scenario involving flying drones communicating and collecting data from the environment

beneficial for all those large industrial environments where, due to their geographical peripheral position, as well as the presence of obstacles, it is not possible to rely on cellular networks or standard Wi-Fi connectivity, as well as on the adoption of alternative long-range low data-rate wireless protocols (e.g., LoRa and LoRaWAN). Therefore, as shown in Figure 4.9, the possibility to deploy the proposed WMN network can allow a data collection in different manufacturing areas, through both fixed nodes (e.g., sensors linked to machines) and mobile nodes (e.g., industrial vehicles moving inside the manufacturing plants and environments). Finally, the dataset built from these monitoring campaigns can then be used to perform high-level activities based on these data, such as predictive maintenance, failure prevention, quality control, and so on.

Finally, on the basis of the heterogeneity of the illustrative reference scenarios and contexts, it would be interesting to analyze how such mesh-oriented environments would evolve in the presence of a large amount of involved (mesh-aware and non-mesh) devices (e.g., through network emulators [31–33]), as well as analyzing the data generated by them (to be processed locally or outsourced to external entities through Edge and Cloud Computing paradigms [34, 35]).

Figure 4.9   Smart industry application scenario

Acknowledgments

This work received funding from the European Union's Horizon 2020 research and innovation program ECSEL Joint Undertaking (JU) under grant agreements: No. 783221, AFarCloud project—"Aggregate Farming in the Cloud;" No. 876038, InSecTT project—"Intelligent Secure Trustable Things;" No. 876019, ADACORSA project—"Airborne Data Collection on Resilient System Architectures." The work of Luca Davoli was also partially funded by the University of Parma, under "Iniziative di Sostegno alla Ricerca di Ateneo" program, "Multi-interface IoT sYstems for Multi-layer Information Processing" (MIoTYMIP) project. The JU received support from the European Union's Horizon 2020 research and innovation program and the nations involved in the mentioned projects. The work reflects only the authors' views; the European Commission is not responsible for any use that may be made of the information it contains.

References

[1]   Zhang Y., Luo J., Hu H. *Wireless Mesh Networking: Architectures, Protocols and Standards*. Boca Raton: Auerbach Publications; 2007.

[2]   Liu Y., Tong K.F., Qiu X., Liu Y., Ding X. 'Wireless mesh networks in IoT networks'. 2017 International Workshop on Electromagnetics: Applications and Student Innovation Competition; 2017. pp. 183 5.

[3]   Belll L., Cirani S., Davoli L. 'Design and deployment of an IoT application-oriented testbed'. *Computer*. 2015;48(9):32–40.

[4]   Aliev K., Moazzam M., Narejo S., Pasero E., Pulatov A. 'Internet of plants application for smart agriculture'. *International Journal of Advanced Computer Science and Applications*. 2018;9(4).

[5]   Codeluppi G., Cilfone A., Davoli L., Ferrari G. 'VegIoT garden: a modular IoT management platform for urban vegetable gardens'. 2019 IEEE International Workshop on Metrology for Agriculture and Forestry (MetroAgriFor); 2019. pp. 121–6.

[6]   Moore J.P.T., Bagale J.N., Kheirkhahzadeh A.D., Komisarczuk P. 'Fingerprinting seismic activity across an Internet of things'. 2012 5th International Conference on New Technologies, Mobility and Security (NTMS); 2012. pp. 1–6.

[7]   Han K., Zhang D., Bo J., Zhang Z. 'Hydrological monitoring system design and implementation based on IOT'. *Physics Procedia*. 2012;33:449–54.

[8]   Belli L., Davoli L., Medioli A., Marchini P.L., Ferrari G. 'Toward industry 4.0 with IoT: optimizing business processes in an evolving manufacturing factory'. *Frontiers in ICT*. 2019;6:17.

[9]   Andrés G.R.C. 'CleanWiFi: the wireless network for air quality monitoring, community internet access and environmental education in smart cities'. 2016 ITU Kaleidoscope: ICTs for a Sustainable World (ITU WT); 2016. pp. 1–6.

[10] Kandhalu A., Rowe A., Rajkumar R., Huang C., Yeh C.-C. 'Real-time video surveillance over IEEE 802.11 mesh networks'. 15th IEEE Real-Time and Embedded Technology and Applications Symposium; 2009. pp. 205–14.

[11] Cilfone A., Davoli L., Belli L., Ferrari G. 'Wireless mesh networking: an IoT-oriented perspective survey on relevant technologies'. *Future Internet.* 2019;11(4):99.

[12] Raspberry Pi 3 Model B [online]. 2021. Available from https://www.raspberrypi.org/products/raspberry-pi-3-model-b/ [Accessed 25 Jun 2021].

[13] Davoli L., Cilfone A., Belli L., Ferrari G. 'Design and experimental performance analysis of a B.A.T.M.A.N.-based double Wi-Fi interface mesh network'. *Future Generation Computer Systems.* 2019;92(3):593–603.

[14] Neumann A., Aichele C., Lindner M., Wunderlich S. *Better approach to mobile ad-hoc networking (B.A.T.M.A.N.). Internet engineering task Force (IETF) [online].* 2008. Available from https://datatracker.ietf.org/doc/html/draft-openmesh-b-a-t-m-a-n [Accessed 25 Feb 2022].

[15] Clausen T., Jacquet P. *Optimized link state routing protocol (OLSR). Internet engineering task force (IETF) [online].* 2003.. Available from https://tools.ietf.org/rfc/rfc3626 [Accessed 25 Feb 2022].

[16] 'IEEE standard for information technology–telecommunications and information exchange between systems local and metropolitan area networks–specific requirements part 11: wireless LAN medium access control (MAC) and physical layer (PHY) specifications'. *IEEE Std 80211-2012.* 2012:1–2793.

[17] 'IEEE standard for information technology–telecommunications and information exchange between systems–local and metropolitan area networks–specific requirements part 11: wireless LAN medium access control (MAC) and physical layer (PHY) specifications Amendment 10: mesh networking'. *IEEE Std 80211s-2011*; 2011. pp. 1–372.

[18] Bari S.M.S., Anwar F., Masud M.H. 'Performance study of hybrid wireless mesh protocol (HWMP) for IEEE 802.11s WLAN mesh networks'. 2012 International Conference on Computer and Communication Engineering (ICCCE); 2012. pp. 712–16.

[19] Perkins C., Belding-Royer E., Das S. *Ad hoc on-demand distance vector (AODV) routing. Internet engineering task force (IETF) [online].* 2003. Available from https://tools.ietf.org/rfc/rfc3561.

[20] Quartulli A., Lo Cigno R. *Client announcement and fast roaming in a layer-2 mesh network [online].* DISI-11-472. University of Trento: Department of Information Engineering and Computer Science; 2011. Available from http://eprints.biblio.unitn.it/2269/1/report.pdf [Accessed 25 Feb 2022].

[21] Davoli L., Veltri L., Ventre P.L., Siracusano G., Salsano S. 'Traffic engineering with segment routing: SDN-based architectural design and open source implementation'. 2015 Fourth European Workshop on Software Defined Networks (EWSDN); IEEE; 2015. pp. 111–12.

[22] Seppänen K., Kilpi J., Suihko T. *'Integrating WMN based mobile backhaul with SDN control'* in Giaffreda R., Cagáňová D., Li Y., Riggio R., Voisard A.

(eds.). *Internet of Things. IoT Infrastructures*. Cham: Springer International Publishing; 2015. pp. 222–33.

[23]   Huang H., Li P., Guo S., Zhuang W. 'Software-defined wireless mesh networks: architecture and traffic orchestration'. *IEEE Network*. 2015;29(4):24–30.

[24]   Salsano S., Veltri L., Davoli L., Ventre P.L., Siracusano G. 'PMSR–poor man's segment routing, a minimalistic approach to segment routing and a traffic engineering use case'. 2016 IEEE/IFIP Network Operations and Management Symposium (NOMS); 2016. pp. 598–604.

[25]   Better Approach To Mobile Ad-hoc Networking (B.A.T.M.A.N.) IV [online]. 2021. Available from https://www.open-mesh.org/projects/batman-adv/wiki/BATMAN_IV [Accessed 25 Jun 2021].

[26]   De Couto D.S.J., Aguayo D., Bicket J., Morris R. 'A high-throughput path metric for multi-hop wireless routing'. Proceedings of the 9th Annual International Conference on Mobile Computing and Networking. MobiCom '03; New York, NY, USA: ACM; 2003. pp. 134–46.

[27]   Droms R.E., Lemon T. *The DHCP Handbook*. 2nd ed. Pearson Education; 2002.

[28]   Davoli L., Belli L., Cilfone A., Ferrari G. 'Integration of Wi-Fi mobile nodes in a web of things testbed'. *ICT Express*. 2016;2(3):96–9. Special Issue on ICT Convergence in the Internet of Things (IoT).

[29]   Codeluppi G., Cilfone A., Davoli L., Ferrari G. 'LoRaFarM: a LoRaWAN-based smart farming modular IoT architecture'. *Sensors*. 2020;20(7):2028.

[30]   Davoli L., Pagliari E., Ferrari G. 'Hybrid LoRa-IEEE 802.11s opportunistic mesh networking for flexible UAV swarming'. *Drones*. 2021;5(2):26.

[31]   Beuran R., Nguyen L.T., Miyachi T., *et al.* 'QOMB: a wireless network emulation testbed'. IEEE Global Telecommunications Conference; Honolulu, HI, USA, 30 Nov; 2009. pp. 1–6.

[32]   Davoli L., Protskaya Y., Veltri L. 'NEMO: a flexible Java-based network emulator'. 2018 26th International Conference on Software, Telecommunications and Computer Networks (SoftCOM); 2018. pp. 1–6.

[33]   Jovanović N., Zakić A., Veinović M. 'VirtualMeshLab: virtual laboratory for teaching wireless mesh network'. *Computer Applications in Engineering Education*. 2016;24(4):567–76.

[34]   Belli L., Cirani S., Davoli L. 'An open-source cloud architecture for big stream IoT applications' in Žarko I.P., Pripužić K., Serrano M. (eds.). *Interoperability and Open-Source Solutions for the Internet of Things: International Workshop, FP7 OpenIoT Project, Held in Conjunction with Soft- COM 2014, Split, Croatia*. Springer International Publishing; 2014. pp. 73–88.

[35]   Yi X., Liu F., Liu J., Jin H. 'Building a network highway for big data: architecture and challenges'. *IEEE Network*. 2014;28(4):5–13.

Chapter 5

Software-defined radio for wireless mesh networks

Rafik Zitouni[1], Laurent George[2], and Stefan Ataman[3]

In this chapter, we give the required background in order to properly introduce the software-defined radio (SDR) and discuss its usage in wireless mesh networks (WMN) applications such as wireless Internet of Things (IoT) and vehicular communications (V2X). These applications require a mesh interconnection of radio devices to ensure efficient data collection and forwarding. A number of standards have been specified for these networks' applications such as IEEE 802.15.4e and IEEE 802.11 p. These standards are rigid in their definition of physical layer (PHY) specifications. Indeed, their transceivers implement predefined modulators/demodulators, and they are not aware of new constraints of radio environments and applications, such as scarcity of the spectrum and the limited capacity of channels. We introduce SDR as a reconfigurable radio where transmitter and receiver functions are implemented in software rather than in hardware. For the IEEE 802.15.4e-based IoT mesh networks, we explore the feasibility to define in software the possible PHYs to deal with the scarcity of the spectrum. For the V2X, we expose the signal superposition in order to increase the channel capacity. We also highlight major challenges and recent developments in SDR.

5.1 Introduction

Over the past 20 years, many applications based on WMN have emerged such as the IoT and Intelligent Transportation Systems. According to the recent economic projection, the size of wireless connectivity market is estimated to grow from USD 69.0 billion in 2020 to USD 141.1 billion by 2025, at a Compound Annual Growth Rate of 15.4% [1]. The end-user expectations, industrial constraints, and market opportunities are the main driving forces of this significant progress. Diverse

[1]5GIC & 6GIC, Institute for Communication Systems (ICS), University of Surrey, Guildford, United Kingdom
[2]University of Gustave Eiffel, LIGM/ESIEE Paris, Marne-la-Vallée, France
[3]Extreme Light Infrastructure - Nuclear Physics (ELI-NP), Măgurele, Romania

applications can be supported, such as voice, video, and data communications. Each one needs an appropriate hardware air-interface supporting specific radio techniques and protocols. In fact, each application has its packet structures, data types, and signal processing techniques, and each radio has to communicate and decode signals using a dedicated circuitry, for example, GSM, 5G, SigFox, IEEE 802.15.4, LoRa, and IEEE 802.11 transceivers. In addition, hardware manufacturers wish to develop quickly and cheaper new wireless technologies. Moreover, radio modulation techniques are static, since they are implemented in hardware; neither radio designer nor the radio itself can change these techniques without replacing the hardware. The PHY of wireless networks uses these hardware radios and cope with their limitations. So with the high number of networks and technologies, the radio frequency spectrum is scarce and radio communications can interfere leading to the radio performance degradation. One main issue seen by a transceiver designer is how to ensure the scalability of radio communication capabilities. The evolution of wireless technologies requires the manufacturing of new hardware devices coupled with the withdrawal of the old ones. However, one possibly sustainable solution might be the use of universal radio hardware that is able to handle in software the technological specifications.

SDR is a set of technologies that defines radio transceiver parameters and functions in software running at top of a common hardware. It is a virtual definition of radio transceivers with the objective to substitute the signal processing objects and operation with software running on a computer machine. For WMN industrial and standard organizations, SDR is an ideal solution for prototyping/tests and reduces time to market of wireless technologies. SDRs are classified into two main classes: software communication architecture (SCA) and embedded SDR. Many research works focus on embedded SDRs, particularly by using Universal Software Radio Peripheral (USRP) devices [2]. Indeed, a number of test platforms exist such as OpenAirInterface [3], srsLTE [4], and OpenLTE [5]. However, they are available without an accurate characterization of their RF output power, particularly, when they are driven by GNU Radio flow graphs [6].

The IoT and the Internet of Vehicles are both based on the existing standards for physical and Medium Access Control (MAC) layers of network architecture, IEEE 802.15.4e and IEEE 802.11 p, respectively. Although these standards are mature, new issues arise with scarcity of the radio spectrum and the need to increase the channel capacity. IEEE 802.15.4e-based WMNs specify two ISM bands, 2,450 MHz and 868/915 MHz, for distinct hardware transceivers. Introducing an SDR to merge them both in software using one SDR hardware would allow the network to increase the available radio resources. On the other hand, only seven channels are specified for IEEE 802.11 p with a limited bandwidth of 10 MHz for 27 Mb/s. Implementing them in the new SDR radio access technologies such as non-orthogonal multiple access (NOMA) might be innovative and increase the channel capacity.

This chapter is structured as follows. The next section outlines the new challenges of WMN and motivates the SDR solution. In Section 5.3, we report the state of the art for SDR architecture, solutions, and our classification. Since USRP SDR devices are largely used by the research community, we focus on the analysis of their performances in Section 5.4. SDR solutions for IEEE 802.15.4e and IEEE 802.11 p are presented, respectively, in Sections 5.5 and 5.6. Finally, Section 5.7 gives some conclusions and future works.

5.2   Challenges for the wireless mesh networks

Surveys have been published on challenges, architectures, and protocols for WMN [7–9]. They are related to ad-hoc and wireless sensor networks (WSN) analyzing the design of transport, routing and MAC protocols, network configuration as well as deployment under mobility constraints. These challenges focus only on protocols and architecture without insights on how to experiment in real world a WMN. Moreover, the definition in software of PHY specifications has not been investigated. In this section, we report additional challenges such as cross layers, experiment of WMN, rigid specifications of standards, and scarcity of spectrum.

5.2.1   Cross-layer design

Parameters of PHY layer are essential and useful for the upper layers of the Open Systems Interconnection (OSI) model. For example, packet routing in WSNs under energy constrains requires a metric using transmitter output power and battery level as main parameters. Similarly, MAC protocols use carrier sense and measure the received energy in order to find out if a channel is idle or busy. The output power of a transmitted signal is managed by the PHY layer. Since it features the node's energy consumption, its capture by the network layer helps the design of efficient routing protocols. The PHY layer can also exploit information passed by the network layer. If the requested data rate is low, the PHY layer can choose an adapted digital modulation. The cross-layer concept has been proposed for WMN in order to make inter-layer communications [10, 11]. The idea behind the cross-layer design is to break down the rigid separation between OSI layers. The drawback of this design is the technology dependency, which means that the designed architecture is not portable to multiple technologies. Furthermore, software platforms are needed to test and implement this architecture. From a problem complexity and interoperability perspective, it is more efficient to isolate the problems by layers. But if the solution needs only a small modification to another layer, it would be more efficient to facilitate layer interactions. For example, if we aim to increase the network's data rate, one solution at the network layer might be improving queue management and optimizing packet routing. Whereas, this problem can be solved only by changing the digital modulation, for example, data rate can be improved by a factor of 6 by using 64-quadrature amplitude modulation (QAM) instead of binary phase-shift keying (BPSK) modulation. If these modulators are implemented in software, several possibilities of design, tests, and performances' evaluation become possible.

5.2.2   *Experiment of WMN communications*

Several software tools have been proposed to establish a model for PHY layer parameters of WMNs. Simulators are accessible and easy to configure in order to emulate the wireless hardware and transmission environment, for example, MATLAB with Simulink, NS-2, OPNET, and OMNET. Commonly, they are the base of tests and validation of proposed solutions for network research problems. For example, packet routing in WSNs under energy constrains require a metric using transmitter output power and battery level as main parameters. Similarly, MAC protocols use carrier sense and measure the received energy to know if the channel is idle or busy. Commonly, the performances of these protocols are measured through simulators as reported in Chen *et al.* [12]. The obtained simulation results depend on the ability to manipulate realistic PHY parameters, in particular, the hypothesis related to wireless channels. However, these solutions tested on a simulator can be neither reproduced nor verified without real-world experiments. In addition, several simulators are optimistic when they model network environment, such as the channel model between network's nodes. In simulation, we assume that the inter-node communications are ensured without possible changes of digital signal processing techniques. Furthermore, the sponsors of research and development projects often criticize simulations, mostly if its proposal is not realistic and without real-world tests. In Section 5.3, we analyze SDR that offers the possibility not only to simulate wireless networks but also to build real-world testbeds and prototypes.

5.2.3   *Rigid implementation of standards*

Standards help ensuring the product's functionality, compatibility, and interoperability. IEEE 802.15.4 [13] and IEEE 802.11 p [14] are the suitable standards for WSNs and Vehicular Ad-hoc NETwork (VANET), respectively. They define Physical and MAC layers of wireless networks. They deal with network requirements, such as energy and processing limitation of devices in IoT or low latency of V2X communications. However, device manufacturers must follow one standard instead of developing new proprietary techniques. But the applications of WSNs and VANET are diverse with specific requirements. To test a standard adaptation or exploration, the manufacturer requires passing through a manufacturing process. In addition, standards evolve gradually from a first version to an improved new one. For example, 802.15.4e is only one release of IEEE 802.15.4 standard. Other revisions have been published with new amendments such as 802.15.4 w in 2020 [13]. Similarly, IEEE 802.11 bd [15] is the new release of IEEE 802.11 p. Basically, these standards are implemented in hardware and their scalability is limited by the nature of hardware. However, software implementation via SDR is more suitable to adjust standard specifications adapted to applications' requirements.

5.2.4   *Scarcity of spectrum*

Another main problem of the WMNs is the scarcity of the radio spectrum. Industrial scientific medical (ISM) frequency bands are available for WMNs. They are shared with many other wireless communication standards and technologies, such as IEEE 802.15.1, IEEE 802.11b/g/n, LoRa, SigFox, etc. As we have explained above, standard organizations specify a carrier frequency in those frequency bands. A carrier frequency (or central frequency) defines a channel of communication between network nodes. The value in Hertz of that frequency is specified statically even if its selection can be done dynamically. However, these values should be chosen also dynamically to guarantee more robustness to spectrum perturbations. The crowded state of the spectrum can be addressed with dynamic spectrum access (DSA) or dynamic spectrum sharing. Instead of static spectrum access, spectrum users can adjust the carrier frequency dynamically. Cognitive radio (CR) [16] is a system which senses its electromagnetic environment and dynamically adjusts its radio parameters to improve radio performances. It is a conceptual layer over SDR or an abstraction layer to program the SDR satisfying application. Implementing CR at each device requires extra computing resources for processing and the virtualization of PHY [17].

5.3 Software-defined radio (SDR)

Software radio (SR) is a set of technologies that define radio transceiver parameters and functions in software, including carrier frequency, modulation, bandwidth, frequency, space, time, and code agility [18–20]. SDR is a realizable implementation of SR. It is a reconfigurable radio, in which the radio functionalities are defined as much as possible in software. The SDR allows network designers and researchers to prototype reconfigurable radio transceivers of WMN end devices and base stations. The objective is the substitution of circuitry signal processing chains by software processing chains [20]. For example, a transceiver designer implements modulators/demodulators, mixers, filters, and amplifiers in software with a higher flexibility level. However, a specific architecture is required to represent SDR where a combination of various software processing components is often called *waveforms* or radio applications.

5.3.1 Architecture

Figure 5.1 depicts the three main parts of the SDR architecture: the radio frequency (RF) front end (FE), intermediate frequency (IF), and software (or baseband processing). The digital-to-analog converter (DAC) and analog-to-digital converter (ADC)

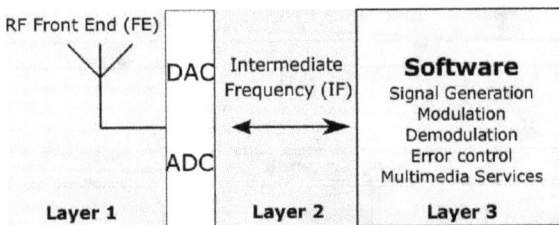

Figure 5.1   Architecture of SR

are, respectively, at the closest proximity to antennas of transmitter and receiver chains [21]. Fully SR architecture is hard to realize due to the limited power processing. An IF part is mandatory to ensure a pragmatic SDR. It is considered as a brake for the SDR domain, as it is limited by the sampling rate and the frequency bandwidth. It accomplishes a channelization, i.e., chooses a specific central frequency, which cannot be performed at a RF FE. In addition, the objective is to select a bandwidth and shifts it from RF to an intermediate range adapted to the processing power of baseband processing computers. This processing replaces analog functionalities by digital ones. Programmable-processing technologies perform this function using, for example, field-programmable gate array (FPGA), digital signal processors (DSP), general purpose processors (GPP), programmable System-on-Chip,) and other application-specific processing entities, for example, application-specific integrated circuit. According to these three parts of the architecture, i.e., baseband programming languages and front-end hardware, the SDRs are classified into two main classes: SCA and embedded SDR. We will briefly present the first class and focus on embedded SDR.

5.3.1.1 Software communication architecture (SCA)

The SCA was created by the US Department of Defense under the Joint Tactical Radio System (JTRS) program [22, 23]. Its objective is to hide radio hardware details via the encapsulation of hardware dependencies in driver software and Common Object Request Broker (CORBA) communications. It is based on an object-oriented approach with three main layers: operating environment, CORBA middleware, and Waveforms layer (see Figure 5.2). The SCA components are waveforms (or objects) prototyped by the users of SCA frameworks. Through the broker bus or the middleware, they interact with hardware devices.

Redhawk is an open source SCA framework [24, 25]. It has been designed and developed by the US National Security Agency and made public as an open source project. It provides tools to support the development, deployment, and management of real-time SR applications. Figure 5.2 depicts the position of Redhawk

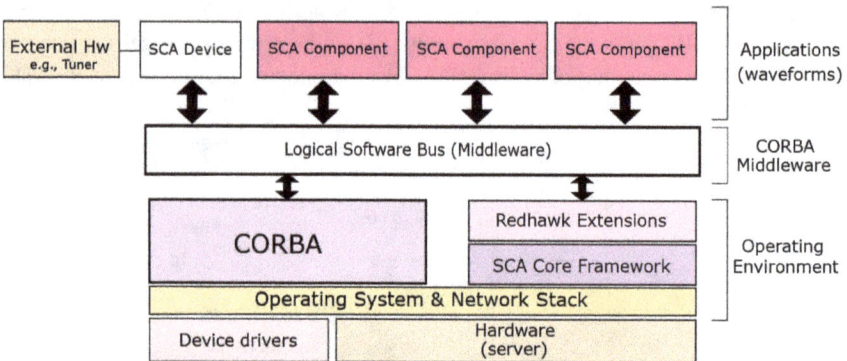

Figure 5.2   Software communication architecture [23]

extensions in an operating environment layer between the SCA core framework and the CORBA bus. Open Source SCA Implementation::Embedded (OSSIE) [26] is another open-source SCA SDR. It is implemented over a CORBA middleware working on a Linux operating system and carried out by Intel or Advanced Micro Devices-based computer. The core framework of OSSIE contains signal-processing components and software interfaces based on CORBA. Furthermore, for a proof of concepts using OSSIE, SDR prototyping can be faster compared to other software architecture.

The obtained results of research works on SCA have been brought by defense companies manufacturing a tactical SDR [27]. Many lessons were learned about the complexity and the scalability of the SCA. CORBA requires a lot of processing power and memory resources, and the objects communicate with fewer throughput. In fact, SCA performances depend on the object size, since several objects have to be exchanged within the SCA bus. Moreover, SCA remains limited since wideband waveform capability, reprogramability, cryptography, and multi-channel operations were never been implemented. For constrained end devices, it is hard to implement in software physical and MAC layers. A trade-off can be found for the base stations (or sink) of WMNs.

5.3.1.2 Embedded software-defined radio

In the literature, there is neither a general architecture nor a standard proposed for embedded systems for WMNs. The design of such an SDR is constrained by the application type and hardware capabilities. From our analysis of the state of the art, we can notice two main architectures: (i) reconfigurable hardware-based architecture and (ii) GPP-based architecture.

Reconfigurable hardware-based architecture
The objective is to benefit from the flexibility and the high performances of dedicated hardware [28, 29]. The entire radio communication stacks can be implemented on FPGA. Hardware synthesizer methods should be used, for example, register-transfer level. These methods allow a designer to implement a hardware conception but with an extra time needed to handle the associated programming languages, for example, Very High-Speed Integrated Circuit Hardware Description Language (VHDL). An SDR can be implemented via an Intellectual Propriety (IP) core, which is data or logic blocks. Ideally, an IP core should be entirely portable and easily inserted into marketed technologies. However, IPs are expensive and consume a lot of logic resources as well as require powerful hardware. Furthermore, coarse-grained reconfigurable architecture consists of a large number of function units interconnected through an embedded network. Comparing this architecture to an FPGA, the advantage is low in terms of power consumption and time needed to set up the platform. Therefore, the gate-level reconfigurability is limited but with a large increase in hardware efficiency.

Figure 5.3   USRP B210

General purpose processor (GPP)-based architecture

In this architecture, GPP performs signal processing operations implemented by SDR users in high-level programming languages. Usually, a computer represents GPP connected to an external device with an FPGA or DSP board. These devices allow an SDR to accomplish a lot of processing power such as Digital Down Conversion of radio receiver [30]. An interface controller is an intermediate adapter between these coprocessor boards and the GPP, for example, USB, Ethernet, and PCI. For example, USRPs [2] are external devices with an FPGA co-processor and RF FE (see Figure 5.3). USRP requires an RF FE, mixers, filters, oscillators, and amplifiers to translate the signal from the RF domain to the complex baseband or IF layer (see Figure 5.4). The baseband of IF signals are sampled by ADCs, and the obtained digital samples are clocked into the FPGA. The latter provides down conversion, which includes fine-frequency tuning and several filters for decimation [2]. The intermediate adapter between the GPP and USRP can be USB, Ethernet, or PCI interfaces. A survey on SDR platforms and standards has been presented by Zitouni [30].

GNU Radio is an open-source project toolkit for building SRs that run on PC [6]. The project was founded in early 2000 and revived by Thomas Rondeau from 2010 to 2015 [31]. It can be used with readily available low-cost external RF hardware to create SDRs (or waveforms), for example, USRP, or without hardware in a simulated environment. It is widely used in hobbyist, academic, and commercial environments to support both wireless communications research, real-world radio systems, and cyber security of wireless networks. GNU Radio provides signal processing blocks for modulation, demodulation, filtering, and various data processing operations. In addition, new blocks can be easily added to the toolkit. SR programs

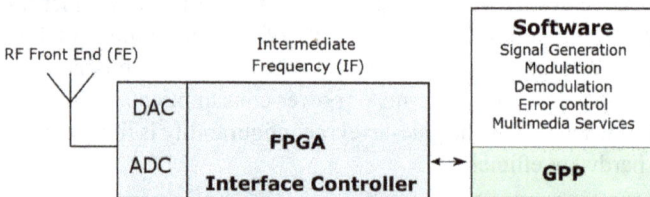

Figure 5.4   GPP-based SDR

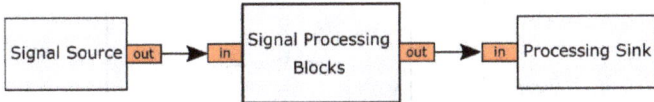

Figure 5.5   Logic of GNU Radio flow graphs

can be created by connecting these blocks to form *flow graphs*. Each block can be developed in oriented object Python or C++ programming languages. C++ is suitable for signal processing functions under real-time constraints. Python represents the logic or the flexibility of how to connect these blocks to build the software transmitter and receiver. Figure 5.5 depicts GNU Radio's flow graph logic. The data or signal samples are generated from a signal source and undergo processing through the signal processing block(s) until arriving at the last block, namely the sink. For example, the USRP FE can be a signal source when the flow graph demodulates the received signal and it can be a sink when the signal is modulated.

Recently, RF Network-on-Chip (RFNoC) has been developed by Ettus Research [32] for USRP hardware to mix between a reconfigurable-based architecture and a GPP one. The idea is to design GNU Radio flow graphs with some blocks executed by the FPGA and other ones by the GPP. The advantage of RFNoC compared to reconfigurable hardware-based architecture is the handling of non-processing tasks such as data movement between FPGA and GPP, exposing settings APIs, routing, etc. It allows the developer to spend more time in designing SDR rather than resolving setup problems. In addition, it speeds up signal processing functions such as filters and equalizers. However, RFNoC is available only for specific and expensive boards like USRP X Series which costs between 5k € and 6k € [2].

5.4   Performances analysis of SDR platform

The ideal SDR should allow researchers and engineers to program digital signal processing functions with high flexibility for baseband processing. One main desirable hardware quality of a SDR hardware is the capability to cover a wide range of both central frequencies and channel bandwidth. Besides, the software programming using high-level languages (C++ and Python) running on GPP is an adapted architecture. However, the available processing power is limited since GPP resources are shared with running applications such as operating systems, drivers, and graphical user interfaces. Consequently, the IF part of the SDR architecture (see Section 5.3.1) is unavoidable to get a working SDR given the processing power constraints on the SR. The boards from Ettus Research[TM] have shown high performance, as detailed in Table 5.1. The announced performances of USRP devices evolve with time, thanks to research and development in programmable logic devices and embedded systems.

In this section, we report a summary of our approach to assess performances of a number of USRP devices. We measured the RF output power of USRPs, and we highlight the most important results of our works [33–35].

Table 5.1   Performances of the SDR devices

	Name	Central frequency (f_c)	Baseband bandwidth (MHz)	Host connection
Bus USRP	USRP B200 and B210	70 MHz to 6 GHz with MIMO capabilities	50 MHz	USB 3.0
Networked USRP	USRP N200, N210, N300, and N310	N20x depends on daughter board placed in 1 slot. N310 covers a wide band up to 6 GHz	50 to 200 MHz	Gigabit Ethernet
Embedded USRP	USRP E313, E312, and E310	70 MHz to 6 GHz	Up to 56 MHz	Embedded interconnection
X series USRP	USRP X300 and X310	Defined by daughter boards placed in 2 slots	Up to 160 MHz	PCIe, dual 10 GigE, dual 1 GigE
LimeSDR	LimeSDR-USB	100 kHz to 3.8 GHz	Up to 80 MHz	USB 3.0
	LimeSDR-PCI	70 MHz to 6 GHz with MIMO capabilities	50 MHz	PCIe
WARP Radio	WARP Radio V3	2.4 to 5 GHz	40 MHz	Z Gigabit Ethernet

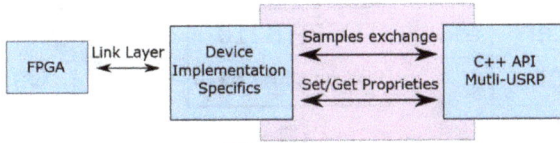

Figure 5.6   Universal hardware driver (UHD)

5.4.1   Analysis of USRP boards driven by GNU Radio

USRP B210, USRP N210, and SBX daughter boards are among the most popular SDR devices [2]. They are largely used in industrial and research prototyping without an accurate characterization of their RF output power, particularly, when they are driven by GNU Radio flow graphs. There is no typical hardware architecture of USRPs. USRPs require a RF FE, mixers, filters, oscillators, and amplifiers to translate the signal from the RF domain to the complex baseband or IF signals. Generally, the FE is implemented on daughter boards.

They are mapped by a host computer via the universal hardware driver (UHD) for cross-layer development like GNU Radio and LabView. Figure 5.6 illustrates the exchange between the high-level programming and USRP's hardware. The UHD is a single Application Programming Interface (API) developed in C++ for all USRP devices and daughter boards. The objectives of UHD are to recognize devices and instantiate device parameters. The UHD deals with input/output samples and sets/ gets USRP proprieties. The USRP source and sink blocks in a flow graph contain the transmission proprieties allowed by the UHD, such as gain, central frequency, sample rate, number of channels, and antenna selection. The UHD is specific to Ettus boards only. Other tools are currently under development to support a general type of SDR devices. SoapySDR [36] is an open-source project intended to be a generalized hardware abstraction library. It expects to support not only USRP but also general SDR devices like LimeSDR [37].

In this section, we analyze the impact of SR parameters on the performances of two SDR devices: the USRP B210 and the SBX daughter boards. USRP B210 [2] is a fully integrated or single-board USRP platform with a large frequency coverage running from 70 MHz to 6 GHz. It is able to stream up to 56 MHz of real-time RF bandwidth. The B210 board integrates both signal processing chains of the AD9361 transceiver [38], providing coherent Multiple Input Multiple Output (MIMO) capability (two transmitters and two receivers). The onboard signal processing and control of this transceiver are performed by a Spartan6 FPGA connected to a host PC using SuperSpeed USB 3.0. Similarly, the SBX daughter board [2] is plugged onto USRP N210. It is announced that the board covers a wide range of frequency from 400 to 4,400 MHz and features an output power up to 100 mW (20 dBm) with a typical noise figure equal to 5 dB. The advertised output gain flatness varies no more than 1 dB from 450 MHz to 3.5 GHz.

| Transmitter flow graph | Ethernet | USRP B210 / SBX on USRP N210 | Coaxial Cable | Digital Oscilloscope |

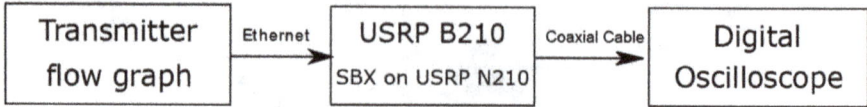

Figure 5.7   Experimental setup of RF measurements

5.4.1.1   Measurement approach

Figure 5.7 depicts the measurements setup. The GNU Radio flow graph is run by the host computer and instructed to transmit an unmodulated carrier, where the digital oscilloscope receives this signal and measures the output power of USRP B 210 and SBX daughter boards. The SBX daughter boards have been controlled by USRP N-210. Our measurement method was based on a frequency-domain analysis of a generated sinusoidal signal (FFT on multitone signals).

Figure 5.8 shows a simplified flow graph of sinusoidal signal source attached to the USRP sink. The central frequency can be modified from 350 MHz to 4.4 GHz for SBX boards and from 50 MHz to 6 GHz for USRP B210. The power of the output signal can be adjusted by two parameters: the DAC value (between 0 and 1) and the UHD gain value. The DAC value is an adjustable constant that multiplies all signal's samples (two components for complex samples), similar to an amplifier. The UHD gain, on the other hand, is an internal USRP sink parameter. In our case, it was built with GCC running on a Linux operating system. It is implemented through the parameter programmable gain amplifier (PGA) of the daughter board and is defined in a brief application note [39].

The output power of a daughter board, without additional transmission gain, is a function of the DAC value, and it is supposed to be deterministic (at least when statistically averaged). It is expected to follow the simple relation:

$$\langle P_{out} \rangle = \text{DAC}^2 \cdot \mathcal{P}_0 \tag{5.1}$$

where $\langle \cdot \rangle$ is the statistical average of the output signal, and $\mathcal{P}_0$ is a reference power level.

We used the spectrum analyzer of the digital oscilloscope LeCroy Waverunner 640 Zi [40]. The digital oscilloscope LeCroy 640 Zi features a 4 GHz bandwidth and sampling rates up to 40 GS/s. The oscilloscope was directly connected to the Tx output of the SDR devices through a high-frequency, low-loss, coaxial cable.

We focused on two parameters of the SBX daughter boards and USRP B210, namely the RF bandwidth and the average output power versus frequency. Parameters available to our disposal were the DAC and the UHD gain values.

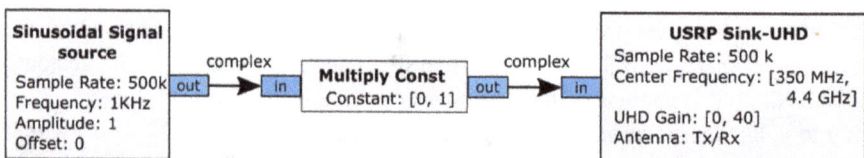

| Sinusoidal Signal source | complex | Multiply Const | complex | USRP Sink-UHD |
| Sample Rate: 500k / Frequency: 1KHz / Amplitude: 1 / Offset: 0 | out → in | Constant: [0, 1] | out → in | Sample Rate: 500 k / Center Frequency: [350 MHz, 4.4 GHz] / UHD Gain: [0, 40] / Antenna: Tx/Rx |

Figure 5.8   Transmitter flow graph (sinusoidal signal)

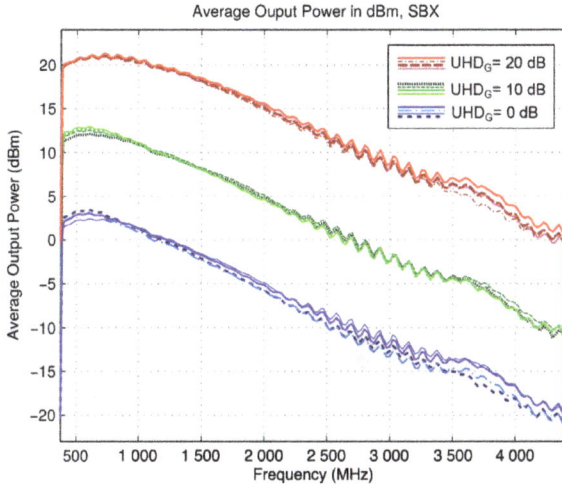

Figure 5.9   Average output power versus carrier frequency for three UHD$_G$ gains. Four SBX boards have been measured, and the four different shades of red/green/blue represent the individual result of each board.

A sinusoidal signal was generated at a given central frequency f_i, and all important peaks in the SBX's output spectral power were measured by the digital oscilloscope. The bandwidth measurements were performed by varying the central frequency f_i in steps of 50 MHz for the SBX board and 100 MHz for USRP B210. The output power of the sinusoidal signal was tuned by two parameters: the DAC value and the UHD gain (UHD_G). The first one is a software amplification, which we can continuously vary between 0 and 1. UHD_G takes integer values and is dependent on the SDR board specifications [2]. For the SBX daughter boards, UHD_G is allowed to vary between 0 and 20, whereas for USRP B210, UHD_G takes values in the interval [0, 89].

5.4.1.2   Measurement results

Among the obtained results, we note the unexpected behavior of the USRP daughter board's amplifier, compared to the advertised performances [2, 34, 35]. Figure 5.9 shows that the covered bandwidth size matches with the announced specifications for SBX daughter boards. Nonetheless, the gain throughout the passband decreases significantly as the frequency increases. These measured results have been obtained with four different boards and by imposing three different levels of UHD_G (0, 10, and 20 dB) while the software amplifier DAC value was set to 1. Furthermore, in Zitouni *et al.* [35], we completed our measurements with the analysis of the total harmonic distortion (THD) of the generated signal carrier. We found that increasing the output power of the signal causes important power emission on the unwanted higher harmonics. Thus, although the UHD gain is a valuable and easily modifiable

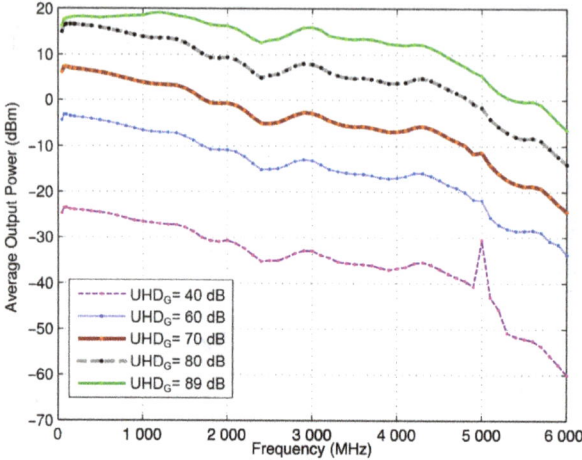

Figure 5.10   Average output power versus carrier frequency for UHD_G gains of 40, 60, 70, 80, and 89 dB for USRP B210

parameter, its increase has to be done with care. On the other hand, Figure 5.10 highlights the obtained bandwidth through different output power levels measured by increasing the UHD_G parameter, for values of 40, 60, 70, 80, and 89 dB. We observed the same behavior of the SDR amplifier at the two MIMO outputs. The output leakage is proportional to the growth of the output frequency. Indeed, at high central frequencies, a low output power is obtained.

These results are useful when experimenting with new protocols of mesh and cellular wireless networks for IoT and autonomous car applications. Experimenting new solutions for radio resources management goes through the accurate knowledge of the covered frequency band of the SDR hardware. Some platforms have been developed such as OpenAirInterface [3], srsLTE [4], and OpenLTE [5] using different versions of USRP hardware. For example in Nikitopoulos *et al.* [41], the authors have experienced limited output power with USRP X series using OpenAirInterface. The researchers can plan to use external power amplifiers to increase the coverage of their experiment scenarios. Similarly, research works on wireless network vulnerabilities [42] need an accurate estimate to perform benchmarks such as impress, Man-In-the-Middle, and Jamming attacks. Other projects have been launched to implement standards' specification of IoT networks such as IEEE 802.15.4 [43] and LoRa [44] on GNU Radio and USRP SDR.

5.5   SDR for IEEE 802.15.4e

After the assessments of a potential SDR platform, the implementation of software transceivers would be more effective. In Schmid [43] and Bloessl *et al.* [45], authors have been limited only to implement IEEE 802.15.4 PHY specifications on GNU Radio and USRP SDR. The scarcity of the spectrum has not been addressed in these works, although this network shares crowded unlicensed bands with other net-work technologies such as IEEE 802.11, IEEE 802.15.1, LoRa, and SigFox. IEEE 802.15.4e-based WMNs use two ISM bands 2,450 and 868/915 MHz. However, sharing the spectrum without coordination is the source of interference and unre-liable communications. Time-slotted channel hopping (TSCH) protocol has been proposed to handle multi-channels based on channel (frequency) hopping. In the 2,450 MHz band, the hopping among 16 channels is a function of time slots and the number of available channels. This allocation depends only on the time slots and not on the link quality. Link quality indicator (LQI) show an energy strength and qual-ity of received data frames in a selected channel. Even if the LQI is measured, the selected carrier frequency is predefined. Moreover, changing dynamically channels in TSCH is not expected without MAC protocol coordination. In Zitouni *et al.* [46], we suggested the design and the prototyping of a DSA on SDR for IEEE 802.15.4e, exploiting the two specified ISM bands.

5.5.1 Dynamic spectrum access

Our DSA is an open sharing model (or spectrum commons). It is a strategy where each network has equal rights in unlicensed frequency bands. We assume that IEEE 802.15.4 can use simultaneously 2,450 MHz and the 868/915 MHz bands, where this network is considered as secondary user (SU) and other unlicensed users are primary users (PUs). SU is an opportunistic network trying to access to the available spectrum. For each band, IEEE 802.15.4 Tx/Rx chains are implemented with GNU Radio and can be reused as black boxes. In addition, we dedicate the spectrum sens-ing and the frequency selection only to the SU receiver. A spectrum sensor measures the energy (power) strength in a given frequency band, and according to a threshold, a carrier frequency is selected. Notice that PUs could be an orthogonal frequency-division multiplexing (OFDM) transmitter in these two bands.

Figure 5.11 shows a simplified GNU Radio flowgraph of IEEE 802.15.4e SDR Receiver (Rx). We prototyped five chains. The two first ones are Rx in the two bands, 2,450 MHz and 868/815 MHz. The third and fourth chains are Transmitter (Tx) and Rx of Gaussian Minimum Shift Keying (GMSK) packets. Finally, the fifth chain is an energy detector Rx. It estimates the output of a time-averaged Power

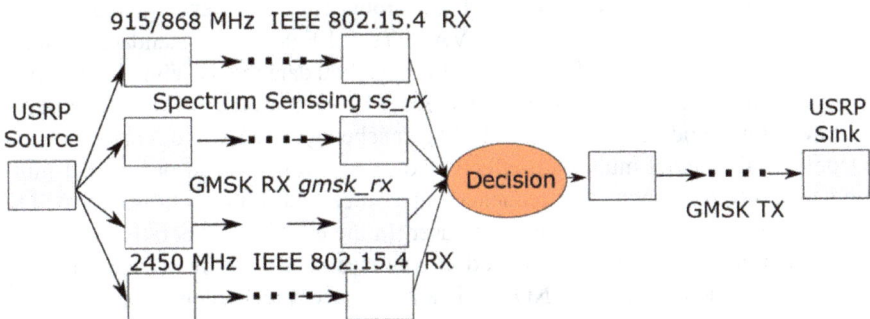

Figure 5.11   SU software-defined IEEE 802.15.4e receiver (Rx)

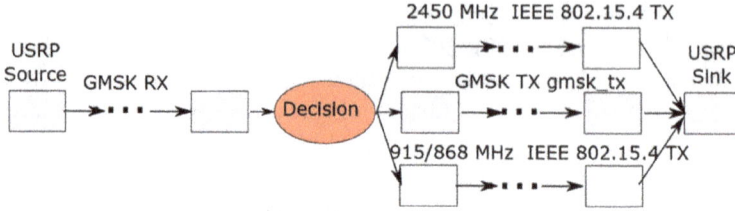

Figure 5.12   SU software-defined IEEE 802.15.4e transmitter (Tx)

Spectral Density. Each chain is selected to transmit or receive messages according to a synchronization algorithm. The particularity of our work is the capability of our Tx to send messages and reconfigure online the transmission chains. On the other side, simplified flowgraph of Transmitter (Tx) is highlighted in Figure 5.12, and two sub-transmitters are implemented for each frequency band. Similarly to the SU receiver, a frequency selection is coordinated via GMSK acknowledgment exchange. PU Transmitter might be any other modulator to emulate another technology like IEEE 802.11.

In this chapter, we are limited to presenting the general approach for DSA on SDR. It is the first step to build a real-world CR. Currently, only the sink of WSN can perform DSA, since the available SDR hardware is expensive and requires high-processing resources.

5.6 SDR for IEEE 802.11p

Increasing the channel capacity of the already specified transmitter is another issue when the standard is fully implemented in hardware. This is the case of IEEE 802.11 p for V2X. In Bloessl *et al.* [47], the authors have evaluated a SDR of IEEE 802.11 p PHY [48]. They compared the performances of SDR to the off-the-shelf boards (Cohda boards) [49] without proposing an approach to improve the obtained results. However, our contribution is to reuse this implementation and increase the channel capacity without changing the standard specifications [50].

Compared to WSN, V2X require high throughput, low latency, and availability of network services. The current VANETs IEEE 802.11 p standard answers partly these requirements. Theoretically, the specified data rate is defined to be up to 27 Mb/s for 10 MHz, depending on the used modulation/encoding mode over OFDM. This encoding is a suitable PHY layer scheme for V2X. It offers a moderate Doppler and channel multi-path effects. However, OFDM is an orthogonal multiple access (OMA) without a frequency reuse since each sub-carrier is allocated to one modulation mode and for only one user. In the case of collaborative vehicular perception and entertainment application, vehicles need high-throughput or high-frequency efficiency (b/s/Hz). NOMA is a radio access technology allowing vehicles to reuse frequency via the signal superposition in the power domain. Receivers perform Successive Interference Cancellation (SIC) to extract signals.

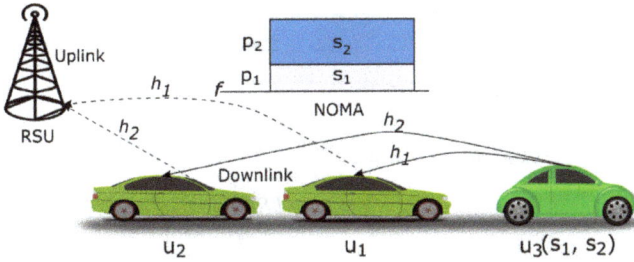

Figure 5.13   Utilization of NOMA and SIC in VANET

5.6.1   Non-orthogonal multiple access

NOMA and SIC have been intensively studied in the context of cellular networks and 5G technology [51, 52]. In the context of Mesh and VANETs, NOMA and SIC might be applied when vehicles are able to define in software the communication chains. Our NOMA/SIC scheme reuses the IEEE 802.11 p SDR encoding chains, i.e. BPSK, QPSK, 16QAM, and 64QAM, with NOMA at the transmitter (Tx) and SIC at the receiver (Rx). In the case of VANET (see Figure 5.13), a roadside unit (RSU) and vehicles would play at once the role of a base station and mobile users. For example, in downlink, the vehicles reuse a frequency through the superposition of signals, while in uplink, the RSU and vehicles perform an SIC function.

In the following section, we focus on the design of SDR of NOMA/SIC transceivers and outline their benefits.

5.6.1.1   Design of NOMA/SIC transceivers

The transmitter generates two superposed signals in the power domain on the same channel in the frequency band 5.9 (see Figure 5.14). Two parallel block branches (trunks) performing the same functions join at the adder. The two branches begin with the message generators or packet data units (PDUs), and they are sources for both users 1 and 2. They generate data messages periodically, similar to Cooperative Awareness Messages (CAM). Then, the OFDM Mapper selects the encoding modulation among the specified ones, i.e., BPSK, QPSK, 16 QAM, and 64 QAM. After the modulation, an amplifier multiplies the generated signal by a ratio $\sqrt{\alpha}$ (for p_1) and $\sqrt{1-\alpha}$ (for p_2). Finally, the amplified signals are superposed via the Add block.

Two variants of SIC techniques are possible: *Symbol level SIC* receiver and *Codeword level SIC*. The difference between them is the re-encoding – or not – of

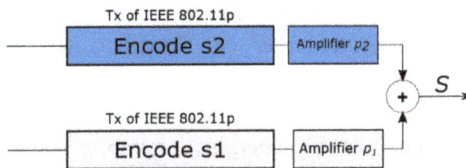

Figure 5.14   Simplified NOMA operations

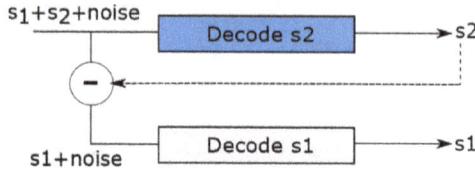

Figure 5.15   Simplified SIC operations

symbols. The *Codeword level SIC* performs channel decoding and re-encoding, while *Symbol level SIC* does not. Such a re-encoding technique would increase the probability of successful recovery of the signal at the receiver. Figure 5.15 shows a simplified schema of applied *Codeword level SIC*. In [50], GNU Radio flow graphs are described in detail.

5.6.1.2   Benefits of NOMA/SIC

Figure 5.13 shows the NOMA/SIC scenario in VANET. We consider three vehicles u_1, u_2, and u_3 and one RSU. Vehicle u_3 superposes and transmits two signals (S_1, S_2), which are amplified by the output powers p_1 and p_2, respectively. The two other vehicles u_1 and u_2 are the two downlink users. Obviously, the two signals are transmitted using one antenna on the same frequency (subcarrier). The received signal is represented as $y_i = h_i x + \omega_i$, where h_i is the complex channel coefficient between vehicles i. Note that ω_i denotes the received Gaussian noise including inter-vehicles interference. The power spectral densities or ω_i for each vehicle i is $N_{0,i}$.

We assume that the overall bandwidth is 1 Hertz ($B = 1$ Hz). One of the three vehicles transmits signals s_i for vehicle i ($i = 1, 2$), with an output power p_i. To deal with the variation of the output power level caused by the mobility, the vehicles u_1, u_2, and u_3 could form a platoon with a fixed inter-vehicle distance. Therefore, the sum of the two signals x is represented by:

$$x = \sqrt{p_1} s_1 + \sqrt{p_2} s_2. \tag{5.2}$$

The two receivers u_1 and u_2 decode the two signals in two steps. In the first step, the receiver decodes s_1, treating s_2 as Gaussian interference. After decoding s_1, the second step is to subtract the s_1 from the received signal and obtain s_2 (Figure 5.15). In this case, SIC is represented by the subtract operation used to extract s_2. In our case, only the user u_1 would apply the SIC approach because it is easy to implement noise cancellation on signal s_2 and thus extract s_1. It starts by decoding the signal with the highest output power, and after that, it extracts its appropriate signal. The receiver u_2 is able to decode signals without SIC. We limit our proposal to the benefit of NOMA downlink.

As discussed, downlink NOMA superposition improves the channel capacity or spectrum efficiency. If we consider the Shannon–Hartley theorem, we have for each vehicle i ($i = 1, 2$), the capacity C_{u_i} equal to:

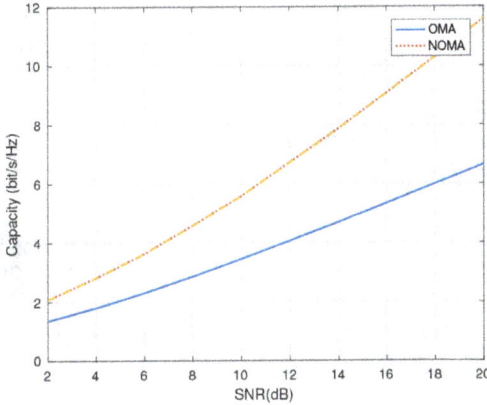

Figure 5.16   NOMA vs OMA channel capacity

$$\begin{cases} C_{u_1} = \log_2(1 + \dfrac{p_1|h_1|^2}{N_{0,1}}) \\ C_{u_2} = \log_2(1 + \dfrac{p_2|h_2|^2}{p_1|h_2|^2 + N_{0,2}}) \end{cases} \tag{5.3}$$

In the current OMA, the bandwidth is shared by all i vehicles. We consider β Hz for node u_1 and $(1 - \beta)$ Hz for u_2, where $0 \le \beta \le 1$ Hz. Then we obtain:

$$\begin{cases} C_{u_1} = \beta \log_2(1 + \dfrac{p_1|h_1|^2}{\beta N_{0,1}}) \\ C_{u_2} = (1 - \beta) \log_2(1 + \dfrac{p_2|h_2|^2}{(1 - \beta)N_{0,2}}) \end{cases} \tag{5.4}$$

If we compare the terms from (5.3) and (5.4), we notice that NOMA's capacity (or throughput) is better than that of the OMA scheme. Indeed, the spectrum efficiency (b/s/Hz) is multiplied by a factor of 2. Figure 5.16 shows a calculated channel capacity of NOMA and OMA for different power allocation ratios (or SNR) using MATLAB. We consider one transmitter and two receivers with a separated signal extraction. The ratio $\alpha \in [0, 1]$ between the two output powers p_1 and p_2 is taken equal to 0.9. We found it to be the best value in order to obtain a maximum throughput at user u_1, where $\alpha = \frac{p_1}{p_1 + p_2}$. In Zitouni and Tohmé [50], we reported how the channel capacity was increased by bringing down bit error rate related to the output power ratio between superposed signals.

5.7   Conclusion

In this chapter, we reviewed and summarized three main contributions: the output power analysis of USRP SDR boards and SDRs for the standards IEEE 802.15.4e and IEEE 802.11 p. We first presented challenges of WMNs related to the flexibility of radio transceivers and the possibilities to experiment radio solutions through

SDR. We suggested the classification of SDR architectures in two classes: SCA and embedded SDR. We focused on the GPP-based architecture. We then reported our analysis of two USRP boards' performances via an experimental approach. The importance to check the real output power of these USRPs driven by GNU Radio flow graphs was highlighted. We have also summarized our contributions addressed to alleviate issues of the two standards, IEEE 802.15.4e and IEEE 802.11 p. For the first one, we proposed merging the distinct specifications of the two frequency bands in an SDR. This SDR can perform DSA and then CR operations. For the second standard, we found that NOMA and SIC can be implemented on SDR to increase the channel capacity in V2X.

The security of WMN is an open issue, particularly when the SDR is used to perform attacks such as replay, jamming, and Man in the Middle (MinM). Our future works will be dedicated to discover and characterize new vulnerabilities and then develop countermeasures.

Acknowledgements

After the final editing of this work, sadly, our colleague and collaborator Prof. Laurent George passed away. He inspired us many times during our long-term collaboration and we were touched by his bravery and courage during his last months of life. We dedicate this chapter to his memory.

References

[1]   Wireless Connectivity Market [online]. 2022. Available from https://www.marketsandmarkets.com/Market-Reports/wireless-connectivity-market-192605963.html [Accessed 4 Apr 2022].

[2]   Ettus Research -- the leader in software defined radio (SDR) -- Ettus Research, a national instruments brand -- the leader in software defined radio (SDR) [online]. 2022. Available from https://www.ettus.com/ [Accessed 4 Apr 2022].

[3]   Kaltenberger F., Silva A.P., Gosain A., Wang L., Nguyen T.-T. 'OpenAirInterface: democratizing innovation in the 5G era'. *Computer Networks*. 2020;176(5):107284.

[4]   Gomez-Miguelez I., Garcia-Saavedra A., Sutton P.D., Serrano P., Cano C., Leith D.J. 'srsLTE: an open-source platform for LTE evolution and experimentation'. Proceedings of the Annual International Conference on Mobile Computing and Networking, MOBICOM; 2016. pp. 25–32.

[5]   OpenLTE Web [online]. 2020. Available from https://github.com/mgp25/OpenLTE [Accessed 4 April 2022].

[6]   Blossom E. 'GNU radio: tools for exploring the radio frequency spectrum'. *Linux Journal*. 2004;2004:4.

[7]   Gungor V.C., Natalizio E., Pace P., Avallone S. 'Challenges and issues in designing architectures and protocols for wireless mesh networks'. *Wireless Mesh Networks: Architectures and Protocols*. US: Springer;

2007. pp. 7–27. Available from https://link.springer.com/chapter/10.1007/978-0-387-68839-8{_}1.

[8]   Akyildiz I.F., Wang X., Wang W. 'Wireless mesh networks: a survey'. *Computer Networks*. 2005;47(4):445–87.

[9]   Cilfone A., Davoli L., Belli L., Ferrari G. 'Wireless mesh networking: an IoT-oriented perspective survey on relevant technologies'. *Future Internet*. 2019;11(4):99.

[10]   Akyildiz I.F., Wang X. 'Cross-layer design in wireless mesh networks'. *IEEE Transactions on Vehicular Technology*. 2008;57(2):1061–76.

[11]   Han C., Jornet J.M., Fadel E., Akyildiz I.F. 'A cross-layer communication module for the internet of things'. *Computer Networks*. 2013;57(3):622–33.

[12]   Chen Q., Zhang X.J., Lim W.L., Kwok Y.S., Sun S. 'High reliability, low latency and cost effective network planning for industrial wireless mesh networks'. *IEEE/ACM Transactions on Networking*. 2019;27(6):2354–62.

[13]   802.15.4-2020—IEEE standard for low-rate wireless networks—IEEE standard—IEEE Xplore [online]. 2020IEEE Standards Association. Available from https://ieeexplore.ieee.org/document/9144691 [Accessed 4 April 2022].

[14]   IEEE Standards Association. 802.11p-2010 IEEE standard for information technology – local and metropolitan area networks – specific requirements – Part 11: wireless LAN medium access control (MAC) and physical layer (PHY) specifications amendment 6: wireless access in vehicular [online]. 2010. Available from https://standards.ieee.org/standard/802_11p-2010.html [Accessed 4 Apr 2022].

[15]   Anwar W., Franchi N., Fettweis G. 'Physical layer evaluation of V2X communications technologies: 5G NR-V2X, LTE-V2X, IEEE 802.11bd, and IEEE 802.11p'. IEEE Vehicular Technology Conference. Institute of Electrical and Electronics Engineers Inc.; 2019.

[16]   Haykin S. 'Cognitive radio: brain-empowered wireless communications'. *IEEE Journal on Selected Areas in Communications*. 2005;23(2):201–20.

[17]   Battula R.B., Gopalani D., Gaur M.S. 'Path and link aware routing algorithm for cognitive radio wireless mesh network'. *Wireless Personal Communications*. 2017;96(3):3979–93.

[18]   Mitola J. 'Software radios: survey, critical evaluation and future directions'. *IEEE Aerospace and Electronic Systems Magazine*. 1993;8(4):25–36.

[19]   Mitola J. 'The software radio architecture'. *IEEE Communications Magazine*. 1995;33(5):26–38.

[20]   Mitola J. *Software Radio*. John Wiley & Sons, Inc.; 2003.

[21]   Akeela R., Dezfouli B. *Software-Defined Radios: Architecture, State-of-the-Art, and Challenges*. Elsevier B.V.; 2018.

[22]   *The joint tactical radio system: lessons learned and the way forward [online]*. 2012. Available from https://apps.dtic.mil/dtic/tr/fulltext/u2/a623331.pdf [Accessed 4 Apr 2022].

[23]   Rouphael T.J. 'High-level requirements and link budget analysis'. *Signal Processing for Software-Defined Radio*. Elsevier; 2009. pp. 87–122.

[24]   Robert M., Sun Y., Goodwin T., Turner H., Reed J.H., White J. 'Software frameworks for SDR'. *Proceedings of the IEEE*. 2015;103(3):452–75.

[25]   *The joint tactical radio system: lessons learned and the way forward [online].* 2021. Available from https://github.com/redhawksdr [Accessed 4 Apr 2022].

[26]   Snyder J. OSSIE: an open source software defined radio (SDR) toolset for education and research. Proceedings of the 43rd ACM technical symposium on Computer Science Education – SIGCSE '12; New York, New York, USA: Association for Computing Machinery (ACM); 2012. p. 672.

[27]   Shanton J.L. 'A software defined radio transformation'. Proceedings – IEEE Military Communications Conference MILCOM; 2009.

[28]   CCCP: Coarse-Grained Reconfigurable Architecture [online]. Available from http://cccp.eecs.umich.edu/research/cgra.php [Accessed 4 Apr 2022].

[29]   Rauwerda G.K., Heysters P.M., Smit G.J.M. 'Towards software defined radios using coarse-grained reconfigurable hardware'. *IEEE Transactions on Very Large Scale Integration (VLSI) Systems*. 2008;16(1):3–13.

[30]   Zitouni R. *Software Defined Radio for Cognitive Wireless Sensor Networks: A Reconfigurable IEEE 802.15.4*. France: University of Paris-Est; 2015.

[31]   Rondeau T.W., O'Shea T., Goergen N. 'Inspecting GNU radio applications with controlport and performance counters'. Proceedings of the Second Workshop on Software Radio Implementation Forum – SRIF '13; New York, New York, USA: ACM Press; 2013. p. 65.

[32]   Braun M., Pendlum J., Ettus M. 'RFNoC: RF network-on-chip'. *Proceedings of the GNU Radio Conference*. 2016;1(1).

[33]   Zitouni R., Ataman S., George L. 'RF measurements of the RFX 900 and RFX 2400 daughter boards with the USRP N210 driven by the GNU radio software'. Proceedings – 2013 International Conference on Cyber-Enabled Distributed Computing and Knowledge Discovery, CyberC 2013; IEEE Computer Society; 2013. pp. 490–4.

[34]   Zitouni R., George L. 'Output power analysis of a software defined radio device'. 2016 IEEE Radio and Antenna Days of the Indian Ocean, RADIO 2016; Reunion, France; 2016.

[35]   Zitouni R., Ataman S., Mathian M. 'Radio frequency measurements on a SBX daughter board using GNU radio and USRP N-210'. 2015 IEEE International Workshop on Measurements and Networking (M&N); 2015. pp. 31–5.

[36]   *Home – pothosware/SoapySDR Wiki – GitHub* [online]. Available from https://github.com/pothosware/SoapySDR/wiki [Accessed 4 Apr 2022].

[37]   LimeSDR – *Lime Microsystems* [online]. Available from https://limemicro. com/products/boards/limesdr/ [Accessed 4 Apr 2022].

[38]   *ANALOG DEVICES -RF Agile Transceiver AD9361* [online]. Available from http://www.analog.com/static/imported-files/data_sheets/AD9361.pdf [Accessed 4 Apr 2022].

[39]   *USRP Hardware Driver and USRP Manual* [online]. Available from https:// files.ettus.com/manual/classuhd_1_1gain__group.html [Accessed 4 Apr 2022].

[40] *LeCroy WaveRunner 640Zi Oscilloscope* [online]. Available from http://cdn.teledynelecroy.com/files/manuals/waverunner-6zi-operators-manual.pdf [Accessed 4 Apr 2022].

[41] Nikitopoulos K., Filo M., Jayawardena C., Tafazolli R. *Towards radio designs with non-linear processing for next generation mobile systems* [online]. 2020. Available from https://arxiv.org/pdf/2012.13371.pdf [Accessed 4 Apr 2022].

[42] Rupprecht D., Kohls K., Holz T., Poepper C. 'IMP4GT: IMPersonation attacks in 4G networks'. 27th Annual Network and Distributed System Security Symposium, NDSS 2020; San Diego, California, USA, February 23–26, 2020; 2020.

[43] Schmid T. *Technical report*. GNU radio 802.15.4 eEn- and decoding [online]; 2006. Available from http://citeseerx.ist.psu.edu/viewdoc/download? [Accessed 4 Apr 2022].

[44] Tapparel J., Afisiadis O., Mayoraz P., Balatsoukas-Stimming A., Burg A. 'An open-source LoRa physical layer prototype on GNU radio'. IEEE Workshop on Signal Processing Advances in Wireless Communications, SPAWC; Institute of Electrical and Electronics Engineers Inc; 2020.

[45] Bloessl B., Leitner C., Dressler F. A GNU Radio-based IEEE 802.15.4 Testbed. *12. GI/ITG KuVS Fachgespräch Drahtlose Sensornetze (FGSN 2013)*; Cottbus, Germany; 2013. pp. 37–40.

[46] Zitouni R., George L., Abouda Y. 'A dynamic spectrum access on SDR for IEEE 802.15.4 networks Wireless Innovation Forum SDR 2015'. Proceedings of WInnComm 2015; Cottbus, Germany; 2015.

[47] Bloessl B., Segata M., Sommer C., Dressler F. 'Performance assessment of IEEE 802.11p with an open source SDR-based prototype'. *IEEE Transactions on Mobile Computing*. 2018;17(5):1162–75.

[48] Society I.C. 'Part 11: wireless LAN medium access control (MAC) and physical layer (PHY) specifications'. IEEE Computer Society; 2012.

[49] WebSite C.W. *Cohda Wireless - Hardware* [online]. 2022. Available from https://cohdawireless.com/solutions/hardware/ [Accessed 4 Apr 2022].

[50] Zitouni R., Tohmé S. 'Non-orthogonal multiple access for vehicular networks based software-defined radio'. 2018 14th International Wireless Communications and Mobile Computing Conference, IWCMC 2018; Institute of Electrical and Electronics Engineers Inc.; 2018. pp. 1142–7.

[51] Wu Y., Qian L.P., Mao H., Yang X., Zhou H., Shen X. 'Optimal power allocation and scheduling for non-orthogonal multiple access relay-assisted networks'. *IEEE Transactions on Mobile Computing*. 2018;17(11):2591–606.

[52] Islam S.M.R., Avazov N., Dobre O.A., Kwak K.-sup. 'Power-domain non-orthogonal multiple access (noma) in 5G systems: potentials and challenges'. *IEEE Communications Surveys & Tutorials*. 2017;19(2):721–42.

Chapter 6

Backpressure and FlashLinQ-based algorithms for multi-hop flying ad-hoc networks

Benjamin Okolo[1], Chiara Buratti[1], and Roberto Verdone[1]

Recently unmanned aerial vehicles (UAVs) have been largely used for remote sensing and surveillance applications. Flying ad-hoc networks (FANETs) is an evolution of the mobile ad-hoc networks (MANETs) paradigm, where multiple UAVs, are capable of setting up a transmission network with minimal infrastructure available. When direct connection between an UAV and the infrastructure located on the ground is not possible, the UAV can relay on multi-hop communication using other UAVs as relays. The design of efficient distributed protocols, to manage both, the path definition and the access to the radio channel, is still an open issue in FANET. In this chapter, we propose a distributed joint routing and scheduling algorithm based on Backpressure and FlashLinQ. The solution includes features scenario-specific: UAVs trajectory-related information are used to optimize the selection of path. Numerical results, achieved via simulations, show that the proposed algorithm outperforms a benchmark solution, based on carrier sense multiple access protocol together with a neighbor-based routing decision algorithm. In addition, the impact of moving from a centralized to a more realistic distributed approach is investigated.

6.1 Introduction

Flying ad-hoc networks (FANETs) represent a new class and evolution of Mobile ad-hoc networks (MANETs), which are distributed and self-organizing networks composed of mobile nodes. FANETs can send information quickly and accurately in a situation where generic ad-hoc networks are not capable to do so. At the time of natural disaster like flooding, earthquakes and even in military battlefield, FANETs can perform better than other form of MANETs. One key difference between FANETs and MANETs, which contributes to their unique features, is the degree of mobility of the nodes. In FANETs, the high mobility of unmanned aerial vehicles (UAVs)

[1]DEI, University of Bologna, Bologna, Italy

contributes to several challenging networking problems, such as rapid topology variations, limited direct connectivity, etc. These networking challenges have made the existing routing algorithms for traditional MANETs, as for example, ad-hoc on-demand distance vector or optimized link-state routing, which are topology-based protocols, to be unsuitable for FANETs [1–5]. Hence, one of the open issues of FANETs is the design of a routing protocol able to create an effective route between UAVs as well as to adjust it to the promptly changing topology in real-time [6]. Given that topology-based protocols are unsuitable, other non-topology—based protocols could be considered and studied in the context of FANETs. As an example, the protocol could exploit the available geographical and/or trajectory information. This information could be provided either through Global Positioning System, Grid Location Service, or even through pre-defined drone trajectory (given that the movements of UAV could be governed by algorithms).

In this chapter, we consider a scenario where an UAV is considered as terminal device, in charge of exploring the environment taking pictures or videos during the flight and sending them to a final control center (CC). The UAV-CC communication can exploit long-range technologies, like 3G/4G/5G or LoRa, but also short-range technologies, as for example, Zigbee. In the latter case, a multi-hop communication is needed to overcome the limits of the direct link between the UAV and the CC. Assuming other UAVs are deployed in the environment, they can act as relays to alleviate the UAV-CC intermittent connectivity.

We propose a Predictive Trajectory-based Backpressure (PTBP) and FlashLinQ algorithm, being a distributed joint routing and scheduling algorithm, where decisions on path and radio resources to be used are made based on the trajectory followed by the UAV.

Backpressure (BP) has been largely studied in the context of multi-hop wireless networks and Internet of Things (IoT) [7]. The algorithm allows to stabilize any traffic input rates within the network capacity without the knowledge of channel state probabilities and arrival rates [7]. The algorithm only relies on queue sizes and does not require the knowledge of packet arrival rates. In BP, nodes monitor the queue size for each flow in their local queues and shares the information with the neighboring nodes. This information is locally used by each node at each transmission decision time, to determine the optimal flow based on the flow with the largest queue differential and to assign the transmission capacity of the link to the flow. The computation of the link capacity could be approximated through a chosen scheduler.

The FlashLinQ scheduler was first proposed by Qualcomm for peer-to-peer ad-hoc networks [8]. At each scheduling instance, the algorithm derives the maximal subset of links, which can be scheduled simultaneously in a given slot without causing harmful interference to any link. The performance of the algorithm depends on the accuracy of its knowledge of all potential links in the network. Hence, the performance is best for a centralized network scenario; while in a distributed network setting, it is best for a complete graph network scenario. Therefore, in order to optimize the algorithm for a general case scenario in a distributed network setting, an effective predictive method to locally obtain estimated knowledge of all potential links in the network is required.

The proposed protocol exploits the available trajectory information, such that route decision (or the choice on the next-hop) is made based on the trade-off with the queue backlog differential, and to solve forwarding decision problems, it relies on an efficient predictive method to locally obtain estimated knowledge of potential links in the network. The proposed solution is an extension of that presented in [9], where, however, static relays located on the ground were used to connect the UAV to the CC, and where only a centralized approach was considered.

The proposed solution is compared to a benchmark solution based on carrier-sense multiple access with collision avoidance (CSMA/CA) with request-to-send and clear-to-send (RTS/CTS) control packets and using a location-based routing scheme. CSMA/CA is implemented in several existing wireless communication technologies, and it is the default medium access control (MAC) mechanism used in Vehicular ad-hoc networks (VANET), described as the IEEE 802.11p MAC standard [10, 11]. According to CSMA/CA, all the nodes with packets to send will participate in the contention to gain access to the channel, by measuring the amount of energy in the channel. If the average amount of the channel energy over a known duration is above a given threshold, the channel is said to be busy, hence the node follows exponential backoff mechanism [12]. Under dense traffic scenario, the number of contentions and collisions occurrence increases enormously, resulting to increase in packet delay (i.e., high latency) and packet loss rate (i.e., low reliability). However, the reliability of the scheme can be improved using RTS/CTS control packets, without worsening packet delay overly. This improvement is made possible due to the capability of RTS/CTS in dealing with hidden terminal problems. In addition, it is also useful in forwarding decision-making. Location-based routing schemes are often greedy algorithms in which route decisions are made based on the path that minimizes the distance to the final destination node. For the case of this chapter, route-decisions are limited to the choice of the next-hop node and no knowledge of the full path to the final destination node is required.

Through simulations we characterize the performance of the proposed protocol and compared them to the CSMA-based scheme, to demonstrate its efficacy. Also the impact of having a distributed approach with respect to a centralized case is shown.

6.2 System model

6.2.1 Reference scenario

The considered scenario is reported in Figure 6.1, showing the UAV, in the following denoted as prime drone, used to gather data according to Poisson distribution with mean λ [packets/slot] from an environment of square area D, and to transmit the data to the CC. As reported in the figure, the prime drone uses other drones as relays to reach the CC. The other drones are successively admitted into the network at regular interval determined by the average transmission range of the drones, such that the connectivity between the last admitted drone and the CC is always maintained. Let m be the last admitted drone, then the $m + 1$ drone will be admitted into

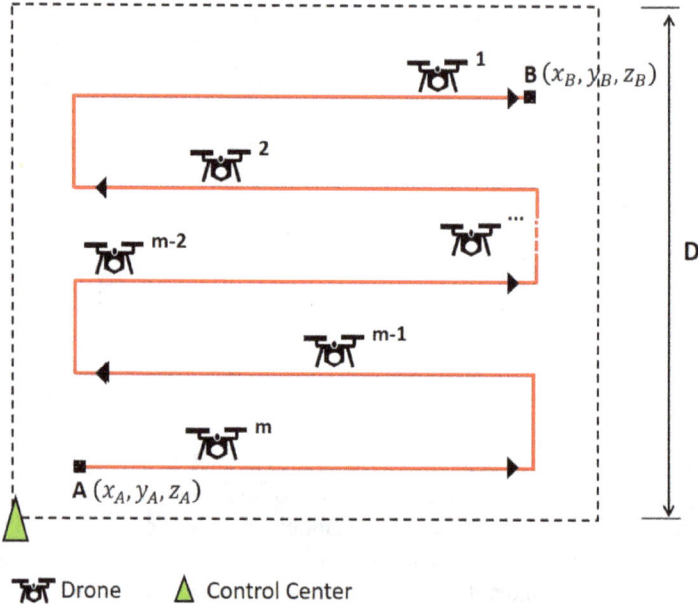

Figure 6.1   Reference scenario

the network, if $P_{rm,cc}/N_o \leq \gamma$, where $P_{rm,cc}$ is the average receive power from drone m to the CC and N_o is the noise power and γ is the signal-to-noise ratio (SNR) threshold. Each admitted drone flies through a pre-defined paparazzi path from point $A(x_A, y_A, z_A)$ to point $B(x_B, y_B, z_B)$, at a constant speed of x [m/s] and at an altitude of h [m]. In this work, we assume that all admitted drones are synchronized in time and that time is divided into slots.

We model the network at slot t as a connectivity graph, $\mathcal{G}(t) = (\mathcal{V}(t), \mathcal{L}(t))$, where $\mathcal{V}(t)$ is the set of vertices, corresponding to the N admitted drones at slot t and the CC, and $\mathcal{L}(t)$ is the set of links $l_{a,b}$ (with $a, b \in \mathcal{V}$) connected at slot t, which changes by passing time due to drones mobility. We assume a link is directly connected or that a is a neighbor to b, if $P_{r_{a,b}}/N_o \geq \gamma$. Therefore, $\mathcal{L}(t)$ represents the network topology state at slot t. In this work, we assume that $\mathcal{L}(t)$ is known to all nodes admitted into the network.

As far as drones mobility is concerned, we consider Paparazzi mobility model based on Paparazzi system [13]. According to this model drones scan the area defined by two points along the trip trajectories, from point A to point B in Figure 6.1.

We consider that our network is distributed, such that transmission decisions are done independently at each node. Let $x_a(t)$ be the set of transmission decision variables for node a and let $p_{a,b}(x_a(t)) \in \{0, 1\}$ denote the transmission decision of drone a at slot t with b as the intended receiver. Note that $p_{a,b}(x_a(t))$ does not only depend on x_a but also depends on the algorithm used by a. We may hence denote $\mathcal{S}(t)$ as the set of links for which $p_{a,b}(x_a(t)) = 1$. Clearly, $\mathcal{S}(t)$ is a subset of $\mathcal{L}(t)$, but unlike $\mathcal{L}(t)$, $\mathcal{S}(t)$ is not known by any node in the network, hence the possibility of having conflicting links in $\mathcal{S}(t)$. If $\mathcal{I}(t)$ is the optimal set at slot t for which no links conflict, then the discussions in this chapter (see below) are aimed at obtaining $\mathcal{S}(t)$ such that, $\mathcal{S}(t) \approx \mathcal{I}(t)$.

6.2.2   Channel model

We consider different path loss models depending on the specific link.

For the drone-to-CC communication we consider the air-to-ground probabilistic model for drones in urban environment presented in [14]. According to it, connections between drone and ground nodes can either be Line-Of-Sight (LOS) or Non-Line-Of-Sight (NLOS).

The LOS path loss model is given as:

$$L_{LOS}(dB) = 20 \log \left(\frac{4\pi f_c d}{c} \right) + \xi_{LOS} \tag{6.1}$$

where c is the speed of light, f_c is the center frequency, d is the node-drone distance and ξ_{LOS} is the shadowing coefficient which is set as described in [14]. In case of NLOS, the signals travel in LOS before interacting with objects located close to the ground, which results in shadowing effect. Therefore, $L_{NLOS}(dB)$ for the NLOS case is still given by (6.1), by substituting ξ_{LOS} with ξ_{NLOS}.

By denoting as p_{LOS} and $p_{NLOS} = 1 - p_{LOS}$ the probabilities of being in LOS or NLOS, respectively, the path loss in the air-to-ground case, L_{A2G}, is given by:

$$L_{A2G} = p_{LOS} \times L_{LOS} + p_{NLOS} \times L_{NLOS} \tag{6.2}$$

where p_{LOS} at a given elevation angle, θ, is computed according to the following (6.3)

$$P_{LOS} = \frac{1}{1 + \alpha \, \exp(-\beta [\frac{180}{\pi} \theta - \alpha])} \tag{6.3}$$

with α and β being environment-dependent constants, i.e., rural, urban, dense urban, etc. and adopted as given in [14].

For the drone-to-drone communication, we consider the following air-to-air path loss: $L_{A2A} = k_0 + 10 \, k_1 \log \frac{d}{d_0}$, where k_0 is the reference path-loss computed at a reference distance d_0 of 1 m, k_1 is the environment dependent path-loss exponent, and d is the distance of the receiver from the transmitter.

6.3   The proposed algorithms

We address a joint scheduling and routing problem, where, at each time slot, the following decisions should be taken: (1) Flow selection: identification of the data flow to be assigned to each link in the network; (2) Scheduling: identification of the maximal subset of non-conflicting links that can use the same radio resource (i.e., frequency channel) in a given slot. We aim at solving the two above problems jointly. The proposed solution relies on BP algorithm for the flow selection and on FlashLinQ for the scheduling decision. However, BP is modified in order to alleviate delays and to improve the network throughput. In particular, link flows are selected

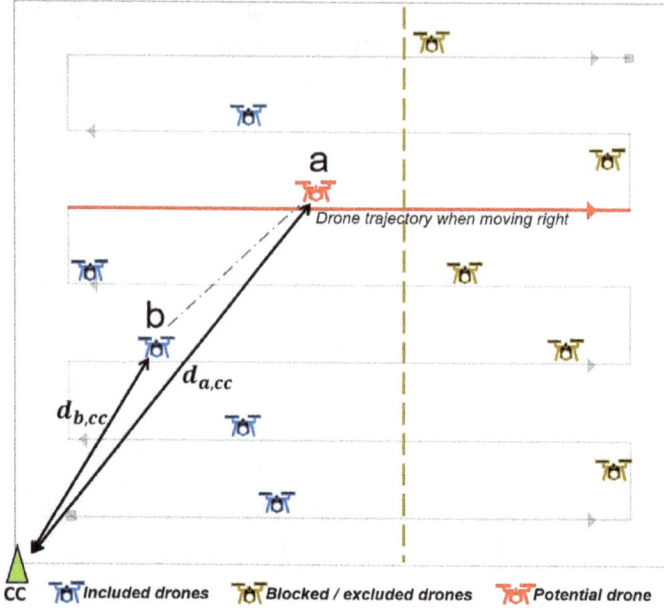

Figure 6.2   *Trajectory-based backpressure (T-BP)*

not only on the basis of the queue backlog differential (as largely done in BP algorithm) but also depending on the distance of the link receiver from the CC, such as on the direction towards which the drone is flying. In particular, in our solution we

- exclude relays whose distance from the CC is larger than the distance of the transmitter from the CC;
- exclude as potential relays those nodes whose position lies ahead of the current drone position, when the drone trajectory is in a direction away from the position of the CC. This is shown in Figure 6.2, where we are considering the drone moving towards right and where blocked relays are underlined in grey.

In the rest of this section, we present the proposed Trajectory-Based Joint BP and FlashLinQ algorithm, at first considering a centralized approach, and then reporting a distributed solution, more suitable for the considered scenario.

6.3.1   Trajectory-based joint backpressure and FlashLinQ

As stated above, flow selection problem is solved exploiting the BP algorithm, represented by the following weighted maximization problem:

$$
\begin{aligned}
\text{maximize} \quad & \sum_{a \in \mathcal{V}} \sum_{b \in \mathcal{V}} W_{ab}(t) * C_{ab}(\mathcal{I}(t), \mathcal{L}(t)) \\
\text{subject to} \quad & \mathcal{I}(t) \in \mathcal{I}_{\mathcal{L}}(t)
\end{aligned}
\tag{6.4}
$$

W_{ab} is the weight assigned to the link in the topology, during slot is computed according to the following two steps. First, the optimal flow of link among the available is derived as:

$$f_{ab}^*(t) = \arg\max_{f \in \mathcal{F}} \left[Q_a^f(t) - Q_b^f(t) \right] \tag{6.5}$$

Then, the corresponding optimal weight of the link at time t is computed as:

$$W_{ab}(t) = \max \left[P_{ab}(t) \cdot (Q_a^{f_{ab}^*}(t) - Q_b^{f_{ab}^*}(t)), 0 \right] \tag{6.6}$$

where $Q_a^f(t)$ is the queue size (in number of packets) at the beginning of slot t of flow f on node a.

$P_{ab}(t)$ is the function accounting for the drone trajectory and the distance from the CC, and it is given by:

$$P_{ab}(t) = \begin{cases} 0, & \text{if } b \text{ is blocked} \\ (1 - \frac{d_{b,cc}(t)}{d_{a,cc}(t)})^\rho & \text{otherwise} \end{cases} \tag{6.7}$$

where, $d_{a,cc}(t)$ and $d_{b,cc}(t)$ are the Euclidean distances from node a and node b to the CC, respectively, with a as the potential transmitter and b as the neighbour of a (see Figure 6.2), and where the blocking condition refers to condition two explained above for the drone-to-relay link, while it refers to condition one above for the relay-to-relay link. ρ is an arbitrary constant, $\rho \in \mathbb{Z}^+$; notice that for $\rho = 0$, $W_{ab}(t)$ reduces to the same function considered in the benchmark case.

$C_{ab}(\mathcal{I}(t), \mathcal{L}(t))$ is the transmission rate function for link l_{ab}, which depends on the network topology state at slot t, $\mathcal{L}(t)$, $C_{ab}(\mathcal{I}(t), \mathcal{L}(t))$ and on the scheduling decision taken at slot t, $\mathcal{I}(t)$ is approximated with the link capacity, given by: $C_{ab} = \log_2(1 + SINR_{ab})$ [packets/slot], where $SINR_{ab}$ is the Signal-to-Noise-plus-Interference Ratio (SINR) defined as follows:

$$SINR_{ab} = \frac{P_{r_{ab}}}{N_0 + \sum P_{r_{iab}}} \tag{6.8}$$

where $P_{r_{iab}}$ is the interference power on link l_{ab} and $P_{r_{ab}}$ is the useful received power, expressed as: $P_{r_{ab}}[dBm] = P_{t_{ab}}[dBm] - L_{ab}[dB]$, where $P_{t_{ab}}$ is the transmit power and L_{ab} is the path-loss.

Each queue in the network has the following dynamics:

$$Q_a(t+1) = \max[Q_a(t) - \mu_{ab}(t), 0] + A_a(t) \tag{6.9}$$

where $A_a(t)$ is the total number of arrivals at node a occurring at the end of slot t and $\mu_{ab}(t)$ is the transmission rate of the selected outgoing link, l_{ab}, at slot t, defined as the total number of transmitted packets during slot t. Since C_{ab} defines the maximum number of packets that can be transmitted on l_{ab} during slot t, the transmission rate, $\mu_{ab}(t)$, always satisfies $\mu_{ab}(t) \leq C_{ab}$. In particular, for each f_{ab}^*, if the $\max_{f \in \mathcal{F}} \left[Q_a^f(t) - Q_b^f(t) \right] > 0$ we have:

$$\mu_{ab}(t) = \begin{cases} C_{ab}(\mathcal{I}(t), \mathcal{L}(t)), & \text{if } Q_a^f(t) \geq C_{ab} \\ Q_a^f(t), & \text{otherwise} \end{cases} \qquad (6.10)$$

As far as the scheduling problem is concerned, we refer to the FlashLinQ algorithm, which relies on the SINR metric to derive a maximal subset of links which can be scheduled at the same time (i.e., time slot) on the same frequency channel, without causing damaging interference. The latter occurs if the SINR is above a given protection ratio (minimum desired SINR), denoted as η in the following.

The problem presented in (6.4) is Non-deterministic polynomial time (NP)-hard and, therefore, difficult to be solved. Indeed, the interference term at the denominator of (6.8) yields a difficult coupled problem since it depends on the scheduled set of links itself. This makes the link selection in the maximization problem reported in (6.4) non-convex. For this reason, we apply a greedy scheduling approach to obtain a sub-optimal solution, based on FlashLinQ heuristics.

Putting together the two above described problems of routing and scheduling, we can define our joint BP and FlashLinQ algorithm that is reported in Algorithm 1.

At each time slot t, $W_{ab}(t)$ is computed for each link of the topology $\mathcal{L}(t)$. Then, the scheduler works as follows. First, the algorithm orders all links in decreasing order with respect to the priority metric, given by $W_{ab}(t)$. Second, links are scheduled starting with those having higher priority. A lower priority link is scheduled if its inclusion in the set of scheduled links, $\mathcal{I}(t)$, does not violate the SINR constraint of any of the higher priority links and that of its own. Transmission rates are updated according to (6.10) for all links in $\mathcal{I}(t)$, at each new inclusion in the set.

Algorithm 1: Trajectory-Based Backpressure and FlashLinQ algorithm.

Input: $\mathcal{G}(0) = (\mathcal{V}, \mathcal{L}(0))$.
Output: Vectors $\mathcal{I}(t) \; \forall t$.
1 Set: $Q_a(0) = 0 \; \forall a; \; W_{ab} = 0 \; \forall a, b; \; t = 0$;
2 **for** *each t:* **do**
3   create link set $\mathcal{I}(t)$;
4   **for** *each l_{ab} in $\mathcal{L}(t)$:* **do**
5   compute $f_{ab}^*(t)$ according to eq. (5) ;
6   compute $W_{ab}(t)$ according to eq. (6) ;
7   **while** *$\mathcal{L}(t)$ is not empty:* **do**
8           $l_{ab} \leftarrow$ in $\mathcal{L}(t)$ with maximum $W_{ab}(t)$;
9   remove l_{ab} from $\mathcal{L}(t)$;
10   **if** *$SINR_{ab} \geq \eta$ given by eq. (9) with $i \in \mathcal{I}(t)$:* **then**
11   **for** *each l_{cd} in $\mathcal{I}(t)$, different from l_{ab}:* **do**
12   **if** *$SINR_{cd} \geq \eta$ given by eq. (9) with $i \in \mathcal{I}(t) \cup l_{ab}$:* **then**
13   add l_{ab} to $\mathcal{I}(t)$;
14   update $\mu_{ab}(t)$ according to eq. (10) ;
15   update $Q_a(t)$ according to eq. (9) ;
16   return $\mathcal{I}(t)$.

6.3.2 Predictive trajectory-based joint backpressure and FlashLinQ

The joint BP and FlashLinQ discussed in Section 6.3.1 is applicable in a centralized system, given that the determination of $I(t)$ (hence, $C_{ab}(\mathcal{I}(t), \mathcal{L}(t))$) in (6.4) and the FlashLinQ algorithm require the knowledge of all potential links in the entire network. However, to apply the scheme to a distributed system, we introduce a predictive capability to the scheme that allows each drone to locally estimate the set of potential links in the entire network. Note that in the following, our considerations are limited to a single flow network.

There are two major steps in PTBP: the first, is to update the local knowledge of the queue backlogs of all other drones admitted into the network and the second is to locally run either of the algorithms described above in Sections 6.3.1 and are based on the updated knowledge of queue backlogs. Each drone admitted into the network observes the first and second steps at the beginning of each slot. Thus, at the beginning of each slot t, each drone is able to locally predict the state of all other drones in the network and hence holds a rough (or estimated) knowledge of all potential links in the network.

Consider that M drones are admitted into the network at slot t, then each of the M drones maintains a local vector,

$$Q^a_{b_1:b_M}(t) := \{Q^a_{b_i}(t)\}^M_{i=1} \tag{6.11}$$

where $Q^a_{b_i}$ is the queue backlog of drone b_i local to drone a. In PTBP, we consider a mechanism for updating $Q^a_{b_1:b_M}$ through the help of local information sharing. We define $\lambda_{b_i}(t)$ as the local information shared by drone b_i to its neighbors, which provides the possibility to locally update $Q^a_{b_1:b_M}$ at each slot t. We consider $\lambda_{b_i}(t)$ as the union of three terms, given as:

$$\lambda_{b_i}(t) := Q_{b_i}(t) \cup \{Q^{b_i}_{b_j}(t-1)\}^N_{b_j \in \mathcal{N}_{b_i}} \cup \{\rho^{b_i}_{b_j}(t-1)\}^N_{b_j \in \mathcal{N}_{b_i}} \tag{6.12}$$

where the first term is the queue backlog of drone b_i at the beginning of slot t, the second term is the local vector of queue backlogs for all drones that are neighbors to drone b_i at the beginning of slot $t - 1$ and the third term is the local vector of predictions made for all drones that are neighbors to drone b_i at the beginning of slot $t - 1$ is the set of neighbors of drone b_i and N is the length of $\mathcal{N}_{b_i}$. We express $\rho^{b_i}_{b_j}(t)$ as the product of the two terms, $\rho^{b_i}_{b_j}(t) = n^{b_i}_{b_j}(t) * s^{b_i}_{b_j}(t)$. Where $n^{b_i}_{b_j}(t)$ is the number of packets estimated by b_i to be received or transmitted by b_j at slot t, and $s^{b_i}_{b_j}(t)$ is the state of drone b_j predicted by b_i at slot t. The state term $s^{b_i}_{b_j}(t)$ is defined below with three possibilities:

$$s^{b_i}_{b_j}(t) = \begin{cases} 1, & \text{if } b_i \text{ predicts } b_j \text{ as receiver} \\ -1, & \text{if } b_i \text{ predicts } b_j \text{ as transmitter} \\ 0, & \text{otherwise} \end{cases} \tag{6.13}$$

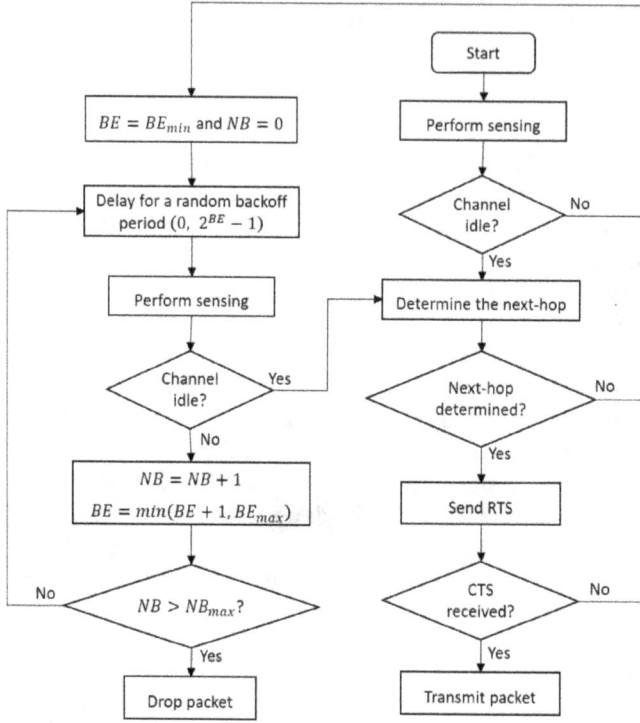

Figure 6.3   *Flow chart of the CSMA-based scheme*

In the scheme, each term in $Q^a_{b_1:b_M}$ (see (6.11)) is updated according on the following:

$$Q^a_{b_i}(t) = \begin{cases} Q_{b_i}(t), & \text{if } b_i \text{ is a neighbour to } a \\[2ex] e^a_{b_i}(t), & \text{otherwise} \end{cases} \tag{6.14}$$

where $Q_{b_i}(t)$ is the first term in $\lambda_{b_i}(t)$ and $e^a_{b_i}(t)$ is a local estimate derived by drone a based on the second and the third terms (see (6.12)), as expressed below:

$$e^a_{b_i}(t) = \begin{cases} Q^{b_k}_{b_i}(t-1) + \rho^{b_k}_{b_i}(t-1), & \text{if } b_k \text{ is neighbour} \\ & \text{to both } a \text{ and } b_i \\ Q^a_{b_i}(t-1) + \rho^a_{b_i}(t-1), & \text{otherwise} \end{cases} \tag{6.15}$$

6.4   The benchmark solution

Figure 6.3 describes the CSMA scheme considered in this work. Each drone with packet to transmit at any slot t implements the scheme. At the start of each implementation, if the channel is found busy after the initial channel sensing (or Clear Channel Assessment, CCA), the scheme initializes the Backoff Exponent (BE) and the Number of Backoff (NB) and then follows exponential backoff procedures. The channel sensing mechanism considered in this work is based on energy detection. The scheme determines the next-hop for drone a at slot t based on the following weight:

$$W_a(t) = \min_{b \in \mathcal{N}_a} \left(d_{b,cc}(t) \right) \tag{6.16}$$

where $d_{b,cc}(t)$ is the Euclidean distance from b to the CC and $\mathcal{N}_a$ is the set of neighbors of a.

6.5   Numerical results and discussions

6.5.1   Simulator setup

We discuss here the simulation results obtained from a custom simulator developed in C++. The simulated scenarios consist of a square field of length 1 000 m, multiple drones spaced at regular intervals, flying at a uniform speed of 15 m/s and height of 20 m above the ground, away from a single ground CC. The trajectory of the drones is as reported in Figure 6.1 and the channel model implemented is as described in Section 6.2. Parameters are set as reported in Table 6.1, if not otherwise specified.

Results are averaged over 10 similar scenarios with each scenario consisting of 6 000 frames. Each frame consists of 10 slots and each slot is equivalent to 10 ms. As performance metrics we consider:

1. Average Delay [s]: Computed as the mean of the end-to-end delays of all received packets.
2. Throughput [packets/frame]: Computed as Total packets received by the CC divided by the Total transmission time.
3. Packets Success Rate: Computed as Total packets successively received at each node divided by the Total packets transmitted.

6.5.2   Comparing protocols

We now report the performance of the CSMA-based scheme and the PTBP scheme applied to a distributed network setting. And for reference purposes on ideal case scenario, we also show the performance of joint T-BP and FlashLinQ applied to a centralized network setting. The aim is to compare the performance of the schemes when the input traffic rate is within the capacity region of the network.

Table 6.1   Default parameters

Parameter	Value
β	0.28
α	9.6
Drone speed, x	30 m/s
Drone height, h	20 m
(x_A, y_A, z_A)	(100,100,20) m
(x_B, y_B, z_B)	(900,900,20) m
Drone transmit power	2 dBm
Noise Power, N_0	−104 dB
SINR threshold, η	2.0 dB
SIR threshold, γ	1.3 dB
SNR threshold, γ	5 dB
Maximum queue size	∞
Square area, $D \times D$	1 000 × 1 000 m^2
k_0	40.7 dB
k_1	3
Centre frequency, f_c	2.4 GHz
Frame duration	10 ms
Air-to-air path-loss exponent	2.25
Average transmission range	≈ 360 m
Minimum backoff exponent, BE_{min}	3
Maximum backoff exponent, BE_{max}	5
Sensing sensitivity	−90 dBm

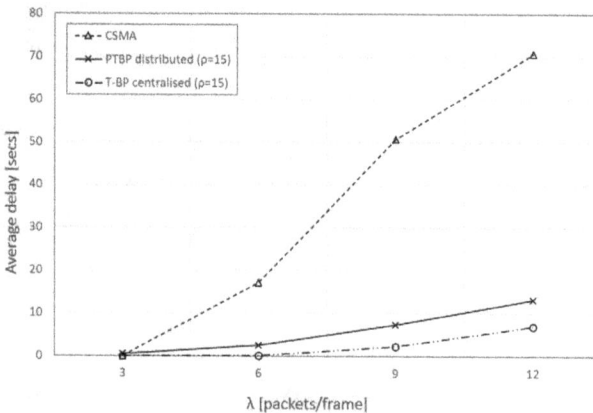

Figure 6.4   Delay performance by varying λ for 28 admitted drones covering an area of 1 000 m^2 and infinite queues size

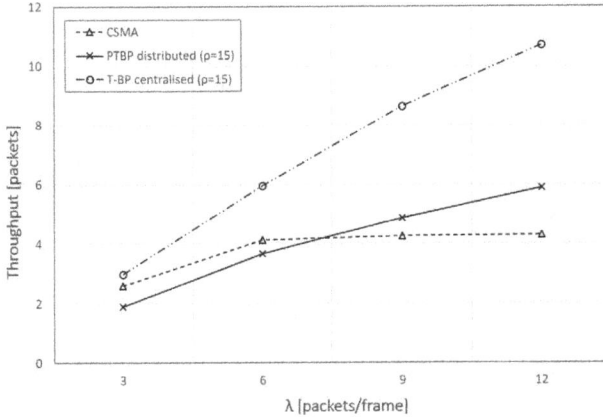

Figure 6.5   *Throughput performance by varying λ for 28 admitted drones covering an area of 1 000 m² and infinite queues size*

Figure 6.4 reports delay as a function of λ. As a general trend, the end-to-end delay increases with the input rate, because more packets have to wait in the queue for a longer time before they can be scheduled and transmitted. Comparing the CSMA and the PTBP algorithms which are applied to a distributed network setting, PTBP results into a significant reduction in delay for values of λ > 3 packets/frame. This performance is partly due to the fact that, while CSMA at higher traffic rate, employs more backoffs due to increased channel activity, PTBP does not employ backoffs, but consistently transmits packets based on its ability to estimate channel capacity.

Figure 6.5 shows throughput as a function of λ. The network through-put increases linearly with the input rate up to a maximum value where satura-tion occurs. The saturation for the CSMA occurred much earlier, from λ ≥ 6 packets/frame. While those of PTBP and T-BP could be determined for λ > 12 packets/frame. CSMA performed a bit better than the PTBP at lower input rates, but was quickly limited by its saturation point. The slow throughput saturation pace of the T-BP and PTBP algorithms is partly due to their ability to search for the channel capacity at each transmission. Hence if the maximum channel capacity is known, their saturation points could be determined.

Figure 6.6 shows the packet success rate as a function of λ. The difference in success rate between the T-BP centralized and the PTBP is relative to the esti-mation error incurred by the PTBP. The lesser the estimation error, the better the performance. Hence unlike the CSMA scheme, the success rate performance does not depend on the input rates but on the ability to minimize the estimation error. In order words, the cost of estimation in PTBP is reflected in Figures 6.4 and 6.5 as the difference between its delay and throughput performance with that of the T-BP centralized.

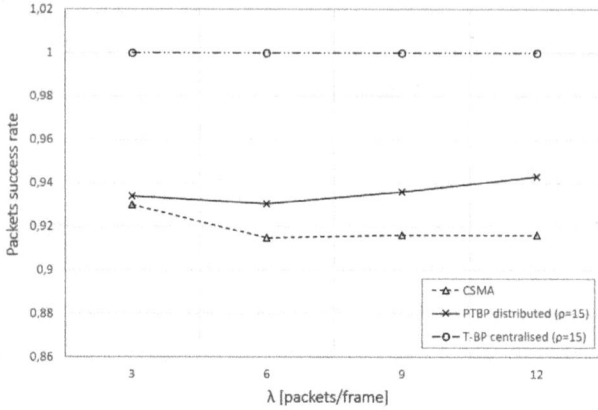

Figure 6.6   Packets success rate performance by varying λ for 28 admitted drones covering an area of 1 000 m² and infinite queues size

Finally, Figure 6.7 reports the effect of varying the transmit power on delay. We consider $\lambda = 6$ packets/frame, which is around the capacity region for the CSMA algorithm. From the figure, the delay performance of PTBP outweighs that of CSMA for all considered transmit power. In general, the delay performance for all three algorithms improves with increase in the transmit power. Table 6.2 also shows improvement in the energy consumed and the number of drones admitted, with increase in the transmit power.

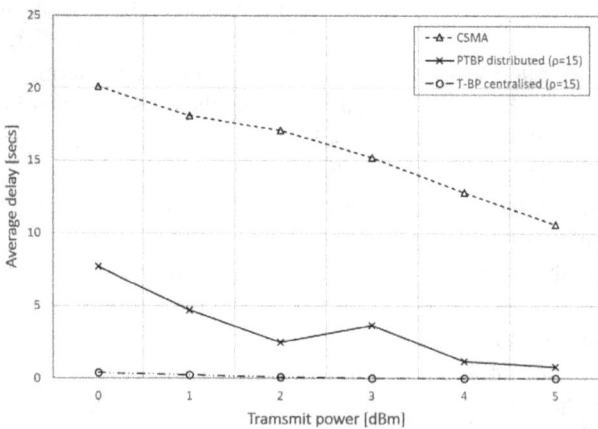

Figure 6.7   Delay performance by varying transmit power for λ = 6 packets/frame

Table 6.2   Upshots for varying transmit power for λ = 6 packets/frame

Transmit power [dBm]	Admitted drones	Energy spent [kJ]
0	35	552.1
1	31	490.6
2	28	436.5
3	25	388.6
4	22	346.5
5	20	309.1

6.6   Conclusion

This chapter considers a scenario where a mobile drone is used to monitor a large area on the ground and convey data to a remote CC in multi-hop, using other UAVs as relays. To achieve connectivity and high throughput gain and low delay, a modified BP algorithm was designed and evaluated through extensive simulations. The proposed algorithm takes the advantage of the position of the CC and the drone trajectory when scheduling the drone-relay link. The impact of moving from a centralized to a distributed approach is also reported, showing the feasibility of the proposed solution in a realistic scenario.

References

[1]   Singh K., Verma A.K. 'Applying OLSR routing in FANETs'. 2014 IEEE International Conference on Advanced Communications, Control and Computing Technologies; 2014. pp. 1212–15.

[2]   Leonov A.V., Litvinov G.A. 'Considering AODV and OLSR routing protocols to traffic monitoring scenario in FANET formed by mini-UAVs'. 2018 XIV International Scientific-Technical Conference on Actual Problems of Electronics Instrument Engineering (APEIE); 2018. pp. 229–37.

[3]   Rosati S., Krużelecki K., Traynard L., Mobile B.R. 'Speed-aware routing for UAV ad-hoc networks'. 2013 IEEE Globecom Workshops; 2013. pp. 1367–73.

[4]   Gupta L., Jain R., Vaszkun G. 'Survey of important issues in UAV communication networks'. *IEEE Communications Surveys & Tutorials*. 2015;18(2):1123–52.

[5]   Oubbati O.S., Lakas A., Zhou F., Güneş M., Yagoubi M.B. 'A survey on position-based routing protocols for flying ad-hoc networks (FANETs)'. *Vehicular Communications*. 2017;10(3):29–56.

[6]   Surzhik D.I., Vasilyev G.S., Kuzichkin O.R. 'Development of UAV trajectory approximation techniques for adaptive routing in FANET networks'. 2020 7th International Conference on Control, Decision and Information Technologies; 2020. pp. 1226–30.

[7]   Tassiulas L., Ephremides A. 'Stability properties of constrained queueing systems and scheduling policies for maximum throughput in multihop radio networks'. 9th IEEE Conference on Decision and Control; 1990. pp. 2130–2.

[8]   Wu X., Tavildar S., Shakkottai S., *et al.* 'FlashLinQ: a synchronous distributed scheduler for peer-to-peer ad-hoc networks'. *IEEE/ACM Transactions on Networking.* 2013;21(4):1215–28.

[9]   Katila C.J., Okolo B., Buratti C., Verdone R., Caire G. 'UAV-to-ground multi-hop communication using backpressure and FlashLinQ-based algorithms'. 2018 IEEE 29th Annual International Symposium on Personal, Indoor and Mobile Radio Communications (PIMRC); 2018. pp. 1179–84.

[10] Han C., Dianati M., Tafazolli R., Kernchen R., Shen X. 'Analytical study of the IEEE 802.11p MAC sublayer in vehicular networks'. *IEEE Transactions on Intelligent Transportation Systems.* 2012;13(2):873–86.

[11] Rajeswar Reddy G., Ramanathan R. 'An empirical study on MAC layer in IEEE 802.11p/WAVE based vehicular ad-hoc networks'. *Procedia Computer Science.* 2018;143(7):720–7.

[12] Bianchi G. 'Performance analysis of the IEEE 802.11 distributed coordination function'. *IEEE Journal on Selected Areas in Communications.* 2000;18(3):535–47.

[13] Bouachir O., Abrassart A., Garcia F., Larrieu N. 'A mobility model for UAV ad- hoc network'. *ICUAS.* 2014;2014:383–8.

[14] Al-Hourani A., Kandeepan S., Jamalipour A. 'Modeling air-to-ground path loss for low altitude platforms in urban environments'. 2014 IEEE Global Communications Conference; 2014. pp. 2898–904.

Chapter 7

Unmanned aerial vehicle relay networks

Evşen Yanmaz[1]

Unmanned aerial vehicle (UAV) or drone teams are deployed for a plethora of applications. Regardless of whether the drones work cooperatively in a distributed manner or assigned to individual tasks by a centralized controller, it is envisioned that some form of connectivity is required for the success of the mission. Connectivity of drones to ground control, ground users, and/or other drones can be maintained via cellular networks relying on an existing infrastructure or by incorporating connectivity needs into the mission plan, which depend on the application at hand. As an alternative, in this chapter, we consider a UAV relay network deployed to support mission-oriented UAV networks, where the mission and communication tasks are decoupled from each other. The relay UAVs are not capable of any mission tasks. Their purpose is to form a mesh network that connects the mission UAVs to a centralized controller and indirectly to each other. We employ a modular relay positioning and trajectory planning algorithm that guarantees connectivity needs of the UAV mission team with minimum number of relays and feasible trajectories, where the cost, network structure, and setup can be changed without relying on an existing infrastructure. We illustrate the usability of the algorithm for different applications, by studying scenarios that require all-time, periodic, or event-driven connectivity, and we analyze the performance in terms of connectivity and resource usage when the relay network is deployed.

7.1 Introduction

Drones or small unmanned aerial vehicles (UAVs) equipped with sensors, embedded processing, and wireless communication capabilities are deployed for various missions [1–4]. In these missions, a drone can be a flying base station (BS), a relay, or a user equipment; and hence it may require different levels of connectivity to a ground controller, ground users, and/or other drones.

[1]Department of Electrical and Electronics Engineering, Ozyegin University, Istanbul, Turkey

In this chapter, we focus on UAVs as relays or mesh nodes that help maintain connectivity of mission UAVs. Mission UAVs can be searching an area, delivering an item, monitoring a target, etc., and they are equipped with a communication interface. On the other hand, the relay UAVs are not capable of mission tasks, but can position themselves to create a multi-hop network between the UAVs and the controllers. Clearly, the mission nodes can be connected utilizing already existing infrastructure [2, 5] or the missions can be planned taking into communication constraints into account [6–12]. However, by deploying additional UAVs as relays, in this work, we decouple the mission and communication tasks and do not rely on any infrastructure, which might be limited or non-existent in some areas. Furthermore, issues including but not limited to network planning, resource allocation, handover, interference management are still open in cellular-assisted UAV networks [3].

Use of relay nodes is a traditional approach for maintaining connectivity of otherwise disconnected nodes. Relay nodes can be static [13], mobile on the ground [14], or UAVs [15–18]. A detailed overview of static or mobile ground relay node placement strategies can be found in [13], where criteria including minimizing the number of relay nodes, maximizing network lifetime, minimizing energy have been considered. Deployment of UAVs, on the other hand, has traditionally been proposed as relays or data collectors in ground wireless networks to connect sensor nodes, mobile users, or disconnected network parts [2, 3, 15, 16, 19–23]. In the recent years, UAVs are increasingly proposed to support connectivity of Internet of Things as well [24, 25]. The position and trajectory of UAVs can be planned such that the number of UAVs are minimized [26], total network coverage is maximized [27, 28], network throughput is maximized [29], among others.

Recently, we have proposed a modular offline relay positioning and assignment (RPA) algorithm such that the mission UAVs are continuously connected to the BS, and we have applied it to surveillance [30] and search and rescue (SAR) applications [31]. The RPA algorithm aims to minimize the total traveled distance and the number of relays for a given area and mission UAV paths, subject to velocity and transmission range constraints. To this end, we use the well-known Steiner tree problem with minimum number of Steiner points and bounded edge-length (STP-MSPBEL) [32] to find the minimum number of relay nodes and their positions such that a network of disjoint nodes is connected. Since the STP-MSPBEL algorithm does not consider the feasibility of the computed positions, i.e., whether any relay node can arrive at these positions on time, additional relay positions are inserted in the area to achieve a feasible trajectory, i.e., redundancy is added to guarantee connectivity. The paths are further optimized by removing any positions that do not contribute to the multi-hop network route between the UAV mission team and the BS. The modular nature of RPA allows us to use it for different types of missions without relying on existing infrastructure.

In this chapter, we deploy RPA for scenarios that require all-time, periodic, or event-driven connectivity, which represent surveillance, remote monitoring, and

SAR applications. We show how relaxed connectivity constraints affect the connectivity of the mission UAVs as well as the number of deployed relays and their velocities. We compare our results with an ideal scheme with no velocity constraint and a classic Voronoi-based relay positioning approach [33, 34]. Our results illustrate the trade-offs between the desired connectivity level and available resources.

The remainder of the chapter is organized as follows. In Section 7.2, we provide the system assumptions and performance metric definitions. The variations of the relay positioning algorithm and the multi-UAV path planners are presented in Section 7.3. Results are given in Section 7.4, and the chapter is concluded in Section 7.5.

7.2　System model

7.2.1　Assumptions

The search region, A, is a square area with sides a. The BS is located at the corner of the square region. The mission UAVs take off from the ground BS together to search/observe a given area, collect data, and deliver data to the BS according to a connectivity requirement. The drones are assumed to know their own positions (e.g., from onboard GPS) and are equipped with sensors (e.g., cameras) with range r_s suitable to their mission and a communication interface with a limited transmission range (r_c). Our experimental results have shown that aerial links can experience free-space path loss with a proper omni-directional antenna setup and in the absence of obstacles [35]. Therefore, we adopt disc model for the communication links; i.e., two nodes are connected to each other if they are within r_c of each other. There are a fixed number of N_s mission nodes that follow a pre-defined path at speeds V_s. The number of relay nodes is less than N_r. The relay nodes fly with speeds limited to V_r and adapt their trajectory based on the positions of the mission nodes according to RPA proposed in [30, 31] and the desired connectivity level. For applications that require continuous or periodic connectivity, the relay paths are pre-planned using the pre-defined mission paths. For scenarios with event-driven connectivity, the relays fly pre-planned paths until the event occurs, and then the paths of both mission and relay paths are computed online as information becomes available. All drones fly at the same altitude. Unless there is collision risk, the relay nodes fly to their next position on a straight line. The altitude of relay nodes can be adjusted for safety purposes, when needed. The mission nodes are connected to the BS if there is a multi-hop link between them, where the intermediate nodes can be a mission or a relay node. System parameters are defined in Table 7.1.

We analyze the performance for various use cases from resource utilization and networking viewpoints. The performance metrics of interest are as follows. *Percentage connected time* is the ratio of time the search drones are connected to BS over the coverage mission time. *Maximum and average number of relays* are the maximum and average number of drones in the air at any given timestep during the mission (including nodes idly hovering, nodes on their way to their next position, and nodes that are part of the network path between search drones and the BS).

Table 7.1   List of symbols

System parameters	Definition
N_s, N_r	Number of mission drones and maximum available number of relay drones
r_c, r_s	Transmission and sensing ranges
V_s, V_r	Maximum mission and relay drone speeds
T	Required connectivity period
Algorithm parameters	**Definition**
$\mathbf{P}_{BS}$	Ground station position (x_{BS}, y_{BS}, z_{BS})
$\mathbf{P}_S^i$	$N_s \times 3$ position matrix of mission nodes at timestep i, $(\mathbf{P}_S^i)_{j,*} = (x_{s_j}(i), y_{s_j}(i), z_{s_j}(i))$
$\mathbf{P}_R^i$	$N_r \times 3$ position matrix of mission nodes at timestep $i, (\mathbf{P}_R^i)_{j,*} = (x_{r_j}(i), y_{r_j}(i), z_{r_j}(i))$
$\mathbf{P}^{**}$	$N_r \times 3$ candidate relay position matrix at timestep $i+1$, $(\mathbf{P}^{**})_{j,*} = (x^{**}, y^{**}, z^{**})$, where ** can be **init** or **temp**
IdleDuration$_R^i$	$N_r \times 1$ vector of total number of consecutive timesteps each relay node has been idle at timestep i

Average velocity is used as a measure of energy consumption and is computed using the distance between consecutive relay positions and the timestep duration for each relay during the whole mission.

7.2.2   Pre-defined mission paths

RPA is applicable to any pre-defined mission plan. In this work, we assume that the pre-defined mission paths are designed to optimize coverage of an area. To this end, we assume the mission UAVs follow a multiple traveling salesmen problem (mTSP) based paths.

For a given sensing range r_s, we represent the square search region A by $(a/r_s)^2$ equal-sized, disjoint cells, where the goal is to determine a set of routes for N_s drones such that the total traveled distance is minimized. mTSP can be solved using several methods including dynamic programming, ant colony optimization, particle swarm optimization, to name a few [36]. Since we are only interested in maintaining connectivity after the paths are generated, we do not have any presumptions on computation limitations. Therefore, we set up a Genetic Algorithm to solve mTSP[11]. We use Joseph Kirk's implementation in MATLAB. The mTSP path planner takes the cell locations, the number of mission UAVs (N_s), minimum tour length for each UAV, population size, and number of iterations as input. It returns the paths to travel from a start location (e.g., BS) to a unique set of cells and back to the original starting place for each UAV such that all cells are visited exactly once. The plans are generated at the BS prior flight and are known to the relay positioning algorithm.

The design of the online portion of the mission paths (e.g., for event-based applications) will be explained in the next section.

7.3 Path planning for UAV relay networks

In this section, we first summarize the RPA algorithm proposed in [30, 31]. Then, we illustrate how the algorithm can be applied to different use cases, namely, with all-time, periodic, and event-driven connectivity.

7.3.1 Relay positioning and assignment algorithm

The aim of RPA is to determine the relay positions and the feasible relay trajectories such that a given set of N_s drones are connected to a root (i.e., the BS) at all times or until an event occurs. The mission nodes follow their mTSP paths and their positions are known during the pre-planned phase of the relay positioning. If and when an event occurs, the paths of the mission and relay nodes are continuously re-planned based on the available information and open tasks. The relay positioning algorithm is depicted in Figure 7.1 and will be briefly explained in the following. For the details of the algorithm, as well as the pseudocodes, the readers are referred to [30, 31]. RPA first determines the relay positions that would guarantee connectivity according to different criteria, which might lead to redundant positions. Then, it eliminates any redundant positions and assigns the relays to open positions such that total distance is minimized. The positions are represented by matrices denoted by P, where each row contains the (x, y, z) coordinates, and number of rows correspond to the number of nodes (see Table 7.1 for definitions). In the following, we describe the procedures in the algorithm within the context of pre-planned and event-driven application use cases.

7.3.1.1 Pre-planned relay paths

Pre-planned relay path (PPRP) planner determines the trajectory of a minimum number of relay nodes such that the mission nodes have connectivity to the BS either until the mission ends or an event occurs. If there are not enough relay nodes to guarantee that, then it uses some candidate position selection strategies leading to relay node positions such that the broken links are repaired quickly.

At each timestep, PPRP checks whether the mission nodes will be connected in the next timestep, given that the current relay nodes keep their positions (I in Figure 7.1). The timestep duration depends on the time interval (or period) T, i.e., when the connectivity of the mission nodes should be guaranteed. At each timestep i, network graph is formed, where the vertices are the locations of the BS, and all UAVs ($\mathbf{P_{BS}} \cup \mathbf{P_S^{i+1}} \cup \mathbf{P_R^i}$), and N is the corresponding set of mission and relay UAVs and the BS (i.e., $|N| = N_s + N_r + 1$). Node 1 is assumed to be the BS. A mission node is connected to the BS at a given timestep i if there exists a multi-hop path between the node and the BS. If all mission nodes are determined to be connected, the relay positions are kept as is. If at least one mission node will not be connected with the

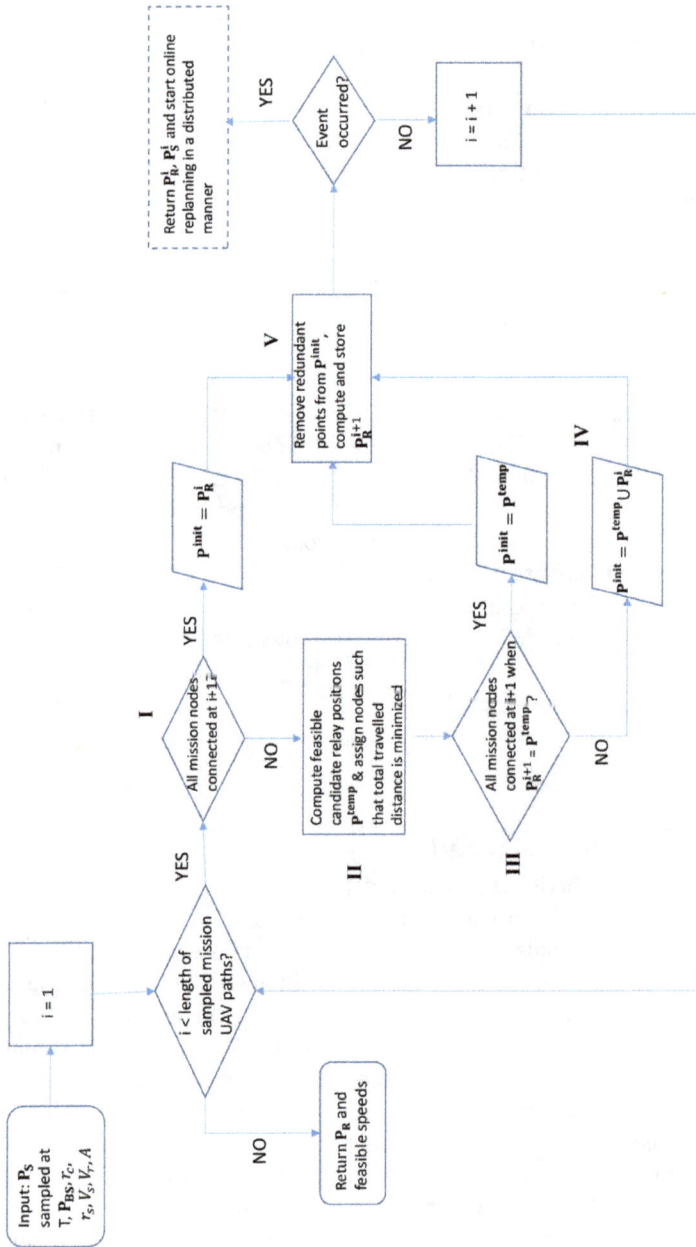

Figure 7.1   Overview of the RPA algorithm with its pre-planned and online phases

current relay positions, then the new relay positions and trajectories are computed given the mission node positions at the next timestep. The goal is to obtain an initial set of feasible relay positions, $\mathbf{P}^{init}$, such that in the next timestep all mission nodes are connected.

The optimal solution to connecting N_s disjoint nodes with minimum number of relays is found using STP-MSPBEL [32, 33]. First, minimum spanning tree over a set of nodes denoted by $\mathbf{P}^{i+1}_{MST}$ is computed and then equidistant Steiner points are inserted on edges longer than r_c. In [30], several $\mathbf{P}^{i+1}_{MST}$ choices were proposed, which can include all nodes, all busy nodes, or only mission nodes. In this chapter, we choose mixed-Voronoi and busy-Voronoi methods, which are defined as follows:

- **Busy-Voronoi:** $\mathbf{P}_{MST}$ corresponds to the vertices of the Voronoi diagram for the set of points given by $\mathbf{P}_{BS} \cup \mathbf{P}^{i+1}_S \cup \mathbf{P}^i_B$, where $\mathbf{B} \in \mathbf{R}$ is the set of relay nodes whose **IdleDuration**$^i_R = \mathbf{0}$.
- **Mixed-Voronoi:** This hybrid method determines the number of required additional relay nodes in the next step if $\mathbf{P}_{MST} = \mathbf{P}_{BS} \cup \mathbf{P}^{i+1}_S$ or if $\mathbf{P}_{MST} = \mathbf{P}_{BS} \cup \mathbf{P}^{i+1}_S \cup \mathbf{P}^i_B$. Then, it chooses the $\mathbf{P}_{MST}$ that leads to the minimum number of additional relay nodes in the next timestep.

Several methods for relay positioning in wireless ad hoc and sensor networks use Voronoi diagram or its dual Delaunay triangulation [33, 34]. The nodes to be connected (e.g., sensor nodes) are used to divide the area into Voronoi cells and the vertices of the Voronoi diagram (or the circumcenters of the Delaunay triangles) form the potential positions for the relays. Busy-Voronoi is chosen to represent these approaches as a benchmark. Mixed-Voronoi scheme is chosen due to its performance.

The next step is the assignment of relay nodes to the Steiner points (II in Figure 7.1). First, we check if a feasible trajectory exists to the Steiner points. STP-MSPBEL aims to minimize the number of relays on an otherwise broken link; however, it might generate positions that cannot be reached by relay nodes already in the air in a single timestep due to speed limitations. Therefore, if there are positions that cannot be reached by any relay node with feasible speeds, RPA first injects additional relay positions ($\mathbf{P}^{temp}$) to the Steiner positions such that the trajectory is feasible. Once the positions are determined, the relay assignment commences. Relay assignment takes as input the candidate positions, current relay positions, and the idle durations of the relay nodes. A linear matching method is followed to assign relays to candidate positions such that the total cost in terms of traveled distance is minimized. Once the assignment is done, the connectivity of the mission nodes at the next timestep is checked, given the computed position and assignment (III in Figure 7.1). If they are not all connected, the relay positions are amended to $\mathbf{P}^{init} = \mathbf{P}^{temp} \cup \mathbf{P}^i_R$ (IV in Figure 7.1). Final step is the removal of redundant positions from the initial relay position set and reassignment of relay nodes (V in Figure 7.1). The assigned relay nodes move toward their new positions, and unassigned nodes either hover in their current positions or move toward the BS.

7.3.1.2   Online path planning due to an event

The pre-planned path planning procedure is suitable for applications that require all-time or periodic connectivity, where the mission nodes fly mostly fixed paths. However, for applications such as SAR or event-coverage, connectivity needs might be triggered or altered after an event happens, e.g., a target is detected. Therefore, PPRP is assumed to be flown till an event occurs (or till the mission is over). Once the event occurs, the online phase commences (shown by dashed rectangle in Figure 7.1.). As opposed to the PPRP phase, in the online path planning, there is no centralized controller, and planning is done at each node in a distributed way depending on their state; i.e., depending on the information available at a given node. It is assumed that when a node comes into contact with another node, they exchange information and decide their next position cooperatively. If there are open tasks from the viewpoint of a clique (i.e., a node set connected to each other), nodes are assigned to the open tasks such that total traveled distance is minimized. The application of the RPA to an SAR mission is detailed in [31]. In the following, we define the potential states a node can be in:

- *Searching*: A mission node is considered to be in search state if an event has not yet occurred or the node has not yet been informed about it.
- *Detecting*: A mission node moves into detecting node state once it detects an event.
- *Uninformed*: A mission or relay node is in uninformed state, if it has not yet been informed of the event.
- *Informed*: A mission or relay node moves into informed state, if it becomes aware that an event has occurred.

During the online phase, nodes are involved in different actions and tasks in different states. In the following, we summarize the actions in each state:

- *Searching state*: The mission nodes at this state follow their pre-planned paths.
- *Detecting state*: If a mission node detects an event, it first checks if there are any other nodes within its transmission range. If it has no neighboring nodes, it moves toward BS until another node or BS is informed. Then, it returns to event location.
- *Uninformed state*: The mission and relay nodes at this state follow their pre-planned paths.
- *Informed state*: The nodes at this state aim to reach a desired final configuration as soon as possible. For instance, in [31], we have assumed that the final configuration is a relay chain formed from the target to the BS with a minimum number of relay nodes. Once a node is informed of the event location, they compute the node positions at the final configuration, e.g., equidistant relay chain positions from the event to BS, where the positions are r_c apart. As nodes meet, the nodes exchange information such as event location or open positions in the final configuration (if they know it). The informed nodes then cooperatively and

continuously assign the best node (in terms of traveled distance) to open relay locations until all locations are occupied. If a node is occupying a relay position, it hovers there until otherwise tasked. If a node is idle, it moves toward the BS, continuously checking for open tasks. Both relay and mission nodes can be at informed state and participate in the task allocation.

Due to the distributed nature of the algorithm and the limited transmission ranges, it is possible that multiple disjoint groups of nodes become informed of the open tasks and assign nodes to the same open task simultaneously. As soon as two or more nodes assigned to the same task come within contact, they decide jointly who is the best for the task. The online phase is concluded, once the final configuration is reached.

7.3.2   Illustration of different P_{MST} methods

The mixed-Voronoi method aims to minimize the overall relay nodes that connect the mission nodes to the BS, whereas busy-Voronoi aims to minimize the number of newly injected nodes in the network in each timestep by fixing only the "broken" links. Therefore, mixed-Voronoi follows a global approach, whereas busy-Voronoi focuses on local state. A snapshot of relay positions corresponding to both methods for the pre-planned phase at different time periods is shown in Figure 7.2 at different time instants. The red and blue lines correspond to the mTSP paths of the two mission nodes (S1, S2). Rxs represent the relay nodes. Gray lines show the network graph. First, observe that at the required connectivity periods (i.e., $T = 1, 30, 60$ s), the relay nodes are positioned such that the mission nodes are connected. This does not imply that in between the periods the mission nodes are still connected, as shown at the network graph of busy-Voronoi at 150 seconds when, $T = 60$ s. As the period increases, we observe that fewer number of relays are required in the air, since the relay nodes have more time to fly to an open relay position. Mixed-Voronoi scheme consistently uses less number of relays than busy-Voronoi scheme due to its global view. As the period increases, the relay positions of the mixed-Voronoi method converges to the Steiner points, whereas busy-Voronoi method suffers from fixing disconnections locally and keeps injecting new drones in the network. However, these snapshots do not say anything about the utilization of the nodes in the air. It has been shown in [30] that busy-Voronoi scheme leads to fewer idly hovering nodes than mixed-Voronoi scheme when $T = 1$ s, since the links and node positions are kept as they are as long as possible.

7.4   Results and discussions

In this section, we analyze the performance of the RPA for establishing connectivity in drone networks for different applications. The length of the coverage region is $a = 400$ m. The mission UAVs fly at speeds of 2.5 m/s, and the relay nodes can fly up to four times faster than the mission nodes. The number of relay nodes, N_r, range from 10 to 100. The number of mission nodes, N_s, range from 2 to 20. The sensing

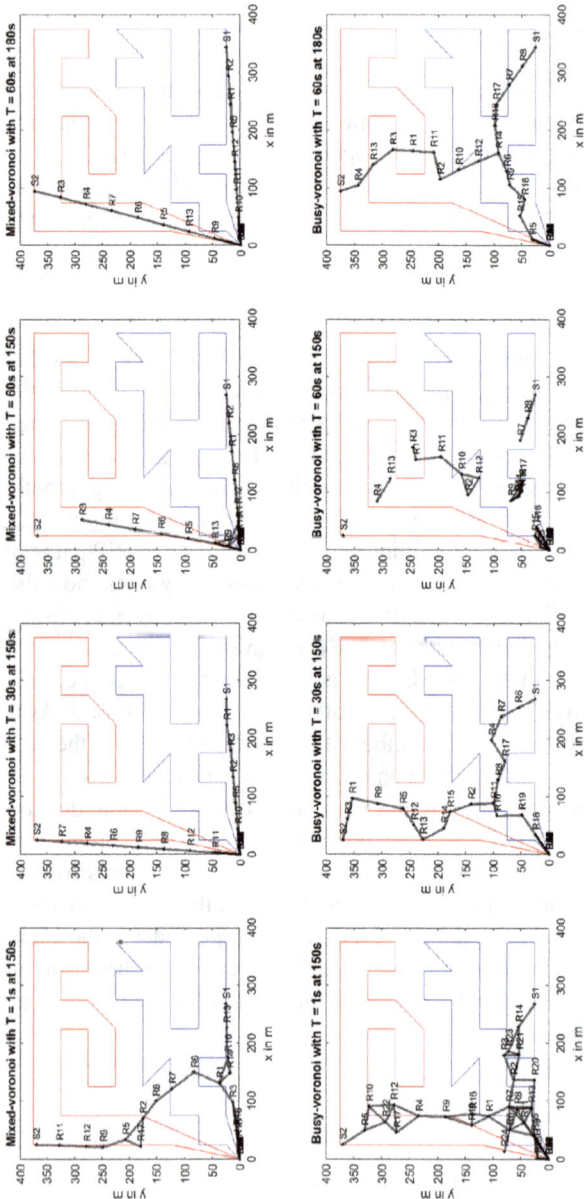

Figure 7.2   The red and blue lines correspond to the mTSP paths of the two mission nodes (S1, S2). Rxs represent the relay nodes. Gray lines show the network graph. Top row corresponds to network graph for mixed-Voronoi method, whereas bottom row shows busy-Voronoi layout, for T = 1, 30, 60 s and at 150 s and 180 s into the mission.

range is $r_s = 50$ m and the transmission range r_c is {50, 100} m. For mTSP, the area is divided into 64 cells that correspond to full coverage of the search region and the population size and number of iterations are set to 80 and 5 000, respectively.

The mission drones monitor a given area and provide information to the BS at time periods of $T = \{1, 30, 60\}$ s, to represent applications with all-time and periodic connectivity needs. To represent event-driven connectivity scenario, we consider an SAR mission, where in the online phase, the relay and search drones need to re-plan their positions to form a communication chain from the target to the BS as soon as the target is detected. The performance metrics of interest are (i) percentage connected time; (ii) maximum and average number of relays to guarantee connectivity to BS at the given time period; and (iii) average velocity of the relay nodes. These metrics are chosen to illustrate the trade-offs between required resources and networking.

7.4.1   Percentage connected time

Figure 7.3 shows the average percentage connected time of the mission nodes to the BS during their coverage and SAR missions. The connectivity of the mission nodes is analyzed for different N_s, r_c, and N_r values for different required connectivity periods. As expected, increasing T results in intermittently connected mission nodes. This reduction in percentage connected time affects both busy-Voronoi and mixed-Voronoi schemes. When N_s is low, even with a high number of available relays the connectivity is low for high periods, since the relay nodes are not injected into the network unless it is data dissemination time. As a comparison, an ideal case where there is no speed limit on the relay nodes, when $T = 1$ s is also analyzed. Comparing with this ideal case, we observe that mixed-Voronoi scheme comes closer to the ideal scheme than the busy-Voronoi scheme. As N_s increases, a high connected time is achieved with less N_r. When r_c is increased, the connectivity of all schemes increase above 70% for all scenarios. Note that our earlier studies have shown that mTSP paths while optimizing coverage have almost non-existent connectivity between the nodes even with a high N_s, especially, when r_c is low [30]; i.e., adding more and more mission nodes do not significantly decrease coverage time or improve connectivity. Therefore, achieving connected times that are high even with low N_r and low r_c with more efficient assignment of available drones is encouraging.

7.4.2   Required number of relays

In this section, we determine the number of relays required for providing connectivity in different scenarios.

Figures 7.4 and 7.5 show the maximum and average number of relays in the air for different number of mission and relay nodes, respectively, when $r_c = \{50, 100\}$ m. The no-speed-limit case is also included as a reference for minimum possible relay nodes. When the relay speed is limited to 10 m/s, mixed-Voronoi scheme is closer to the minimum than busy-Voronoi scheme, but still almost double the minimum number is required to guarantee continuous connectivity for some N_s and N_r. Compared to the benchmark busy-Voronoi scheme, on the other

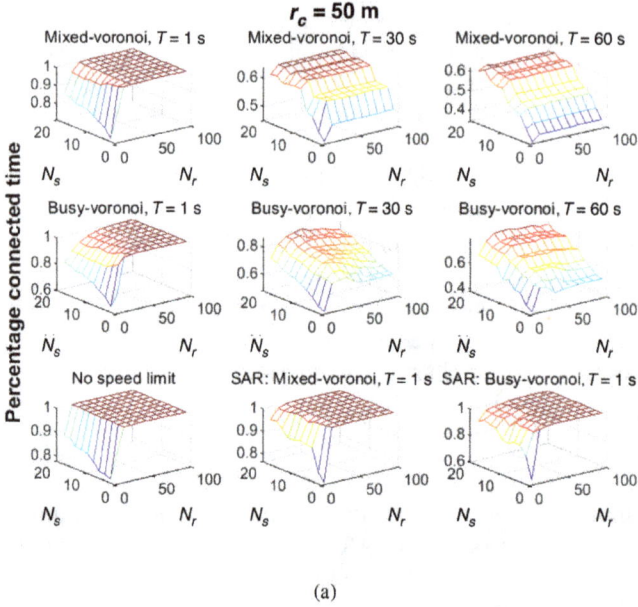

(a)

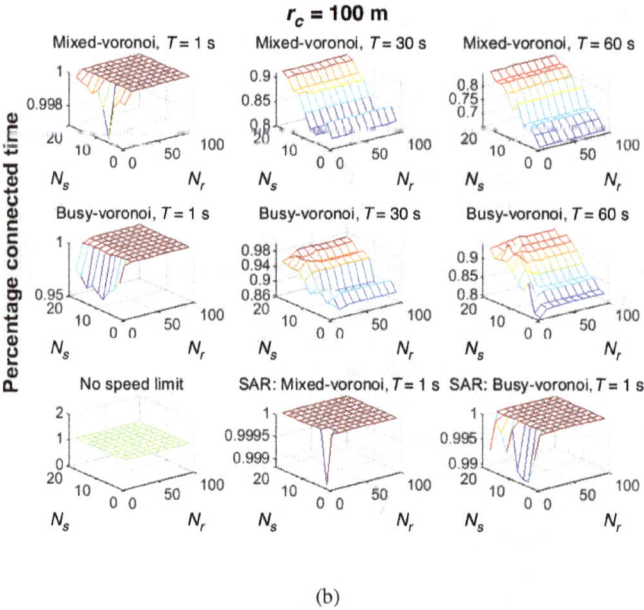

(b)

Figure 7.3   *Percentage connected time of mission nodes versus* N_r *and* N_s *for all-time (T = 1 s), periodic (T = 30, 60 s), and event-driven (SAR) connectivity, when (A)* r_c = *50 m, (B)* r_c = *100 m*

(a)

(b)

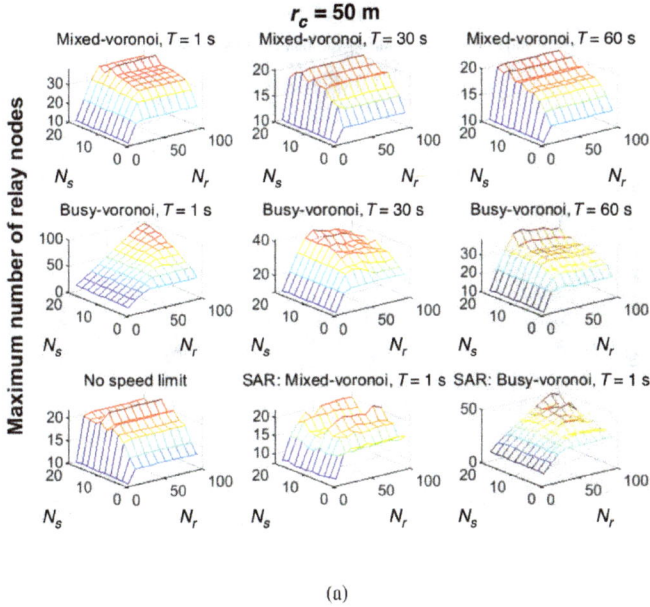

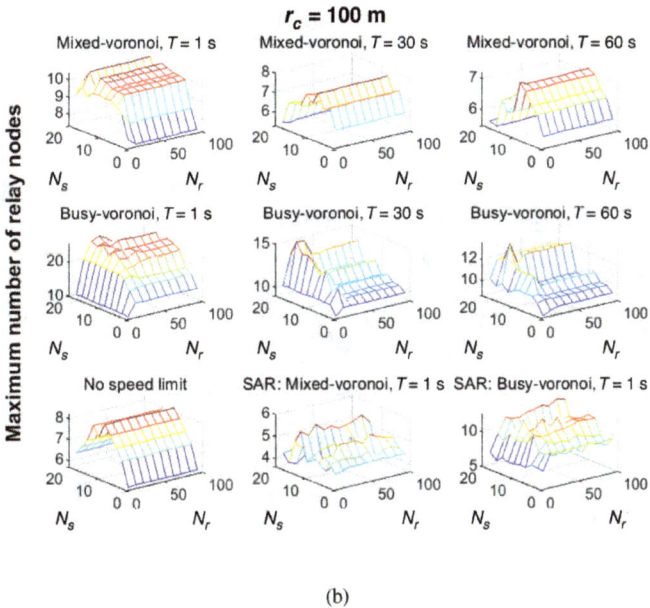

Figure 7.4   *Maximum number of relay nodes in the air versus* N_r *and* N_s *for all-time (*$T = 1$ *s), periodic (*$T = 30, 60$ *s), and event-driven (SAR) connectivity, when (A)* $r_c = 50$ *m, (B)* $r_c = 100$ *m*

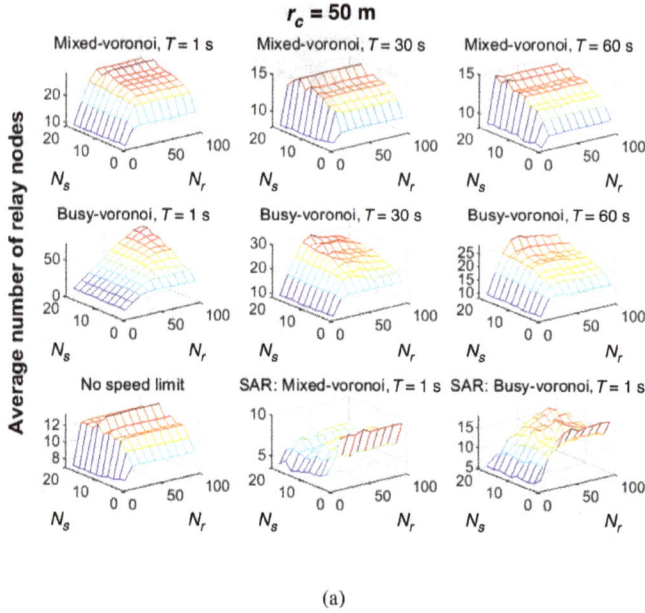

(a)

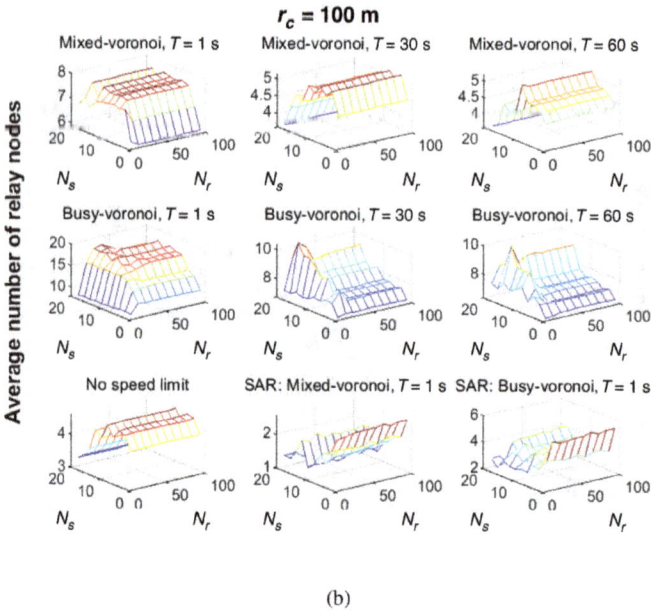

(b)

Figure 7.5   *Average number of relay nodes in the air versus N_r and N_s for all-time (T = 1 s), periodic (T = 30, 60 s), and event-driven (SAR) connectivity, when (A) r_c = 50 m, (B) r_c = 100 m*

hand, there is a significant reduction of around 50%, for all Ts. Both schemes suffer from speed limitations. Similar trends are observed when the transmission range is higher, with the only difference that the required number of relays drops significantly. This can be seen in part (b)s of Figures 7.4 and 7.5. Busy-Voronoi scheme requires and uses consistently a higher number of nodes than the other schemes. While the busy-Voronoi scheme requires a higher number of nodes, since it aims to fix "only the broken links"; there aren't significant changes in the node positions or velocities, and therefore, the links between the nodes are also expected to be more stable, which is relevant to the network performance. We have also analyzed the average number of relay nodes required for each scheme and have observed that the performance trends do not change, where mixed-Voronoi scheme requires lower number of relays, ranging between 10 and 25 depending on N_s when $r_c = 50$ m, compared to busy-Voronoi's 10–50 relay nodes. These results show that with careful selection of nodes to be connected at a given time, the relay paths can be planned more efficiently. However, the difference between the required number of nodes and the ideal case implies that the utilization of the nodes in the air can be further optimized. As T is increased, mixed-Voronoi performance approaches the ideal case. With these results, we observe that as the updates are required more frequently, the more resources are required to guarantee connectivity. The requirements are much less stringent for the event-driven scenarios. Since connectivity is essential only after an event occurs, and desirable before the event so that the final configuration can be reached as fast as possible, the number of required relay nodes are also lower for the SAR case with mixed-Voronoi scheme.

7.4.3   Average relay node velocity

Finally, we analyze the performance of the relay positioning schemes from energy viewpoint. To this end, we determine the average velocity of the relay nodes.

Figure 7.6 shows the average velocity of the relay nodes throughout the mission for different N_s, N_r, and T values for $r_c = \{50, 100\}$ m. In the ideal no speed-limit case, with increasing r_c, when T is small, the required velocity increases since the distance between relay positions in consecutive time intervals increase. Observe from Figure 7.6 that when there is no speed limit, relay nodes require average speeds less than 20 m/s when r_c is small, but grows fast to values above 70 m/s when r_c is 100 m. Recall that the RPA algorithm is forced to return velocities less than $V_r = 10$ m/s. Since the minimum relay positions from Steiner points do not guarantee this, RPA injects intermediate points that can be reached with a velocity less than V_r. This approach caused the number of required relays to increase, as shown in the previous subsection. In the all-time connectivity case ($T = 1$ s), most of the scenarios lead to an average velocity close to the maximum allowed. This is due to the fact that minimizing number of relays is achieved by spreading the nodes; i.e., by relay positions that are as far as possible from each other, which can be achieved if the relay UAVs are forced to fly at maximum speed. As T is increased, the time that the relay UAVs have to move between consecutive relay positions is also longer; therefore, lower speeds can be sufficient to enable connectivity. Since busy-Voronoi

(a)

(b)

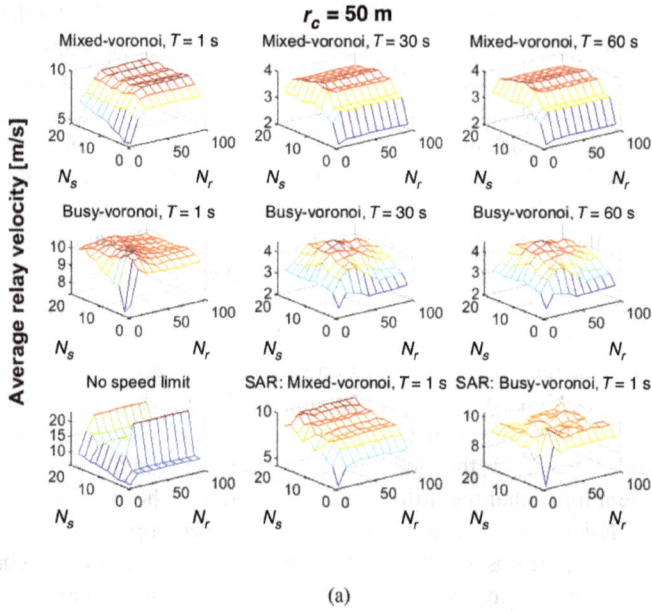

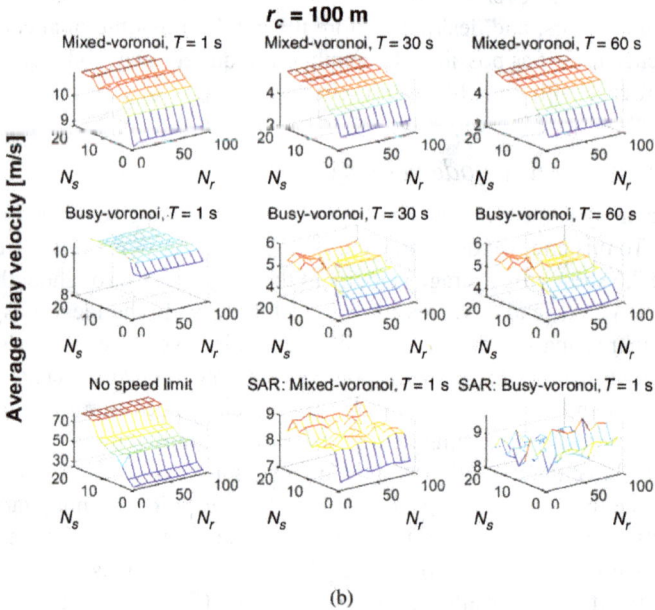

Figure 7.6   *Average relay node velocity in m/s versus* N_r *and* N_s *for all-time* $(T = 1\ s)$, *periodic* $(T = 30, 60\ s)$, *and event-driven (SAR) connectivity, when (A)* $r_c = 50\ m$, *(B)* $r_c = 100\ m$

places relay positions almost at the edge of transmission range, the required velocity is slightly higher than the case for mixed-Voronoi. When the data dissemination period requirement is relaxed, both schemes perform similarly. For SAR, the nodes also fly at higher speeds when $T = 1$ s.

The velocity can be interpreted as a feasibility parameter; i.e., if the required velocity is not achievable (or desirable) by the UAVs, connectivity of the mission nodes might not be guaranteed with a minimum number of drones. It can also be seen as an indicator for energy consumption. Measurements in [37] show that increasing velocity beyond a certain point will increase power consumption, which will decrease flight time and in turn will result in more drones required to replace the drones that would run out of battery during the mission. The schemes under investigation that are bounded by the maximum relay speed of 10 m/s all have average speeds higher than 6 m/s.

These results show that there is a delicate balance between the number of relay nodes and system requirements from energy and network viewpoints. While mixed-Voronoi scheme results in a lower number of nodes, busy-Voronoi may lead to better connectivity if the data dissemination requirements are relaxed. Therefore, depending on the application and the trajectory of the underlying mission nodes, the suitable relay positioning scheme is expected to change.

7.5   Conclusions

In this chapter, we analyzed a multi-UAV mission plan approach that decouples the mission and communication tasks and uses a UAV relay network to support UAV missions. The modular nature of the relay positioning and trajectory planning algorithm, RPA, allows its use for different types of missions, without relying on infrastructure. To this end, we employed RPA for scenarios that require all-time, periodic, or event-driven connectivity to represent common UAV applications such as surveillance, monitoring, SAR, respectively. Depending on the connectivity requirements, the relay positions can be computed offline for coverage-based applications or might need to be re-planned for event-driven applications. To this end, we introduced both pre-planned relay positioning and online distributed positioning phases in the algorithm. The algorithm computes relay positions based on Steiner points and determines feasible trajectories for relay UAVs to maintain connectivity of mission UAVs. We analyzed different strategies and illustrated that there was a delicate balance between the number of relay nodes and system requirements from energy and network viewpoints. Results show that with mixed-Voronoi scheme, the maximum number of relays in the air can be reduced up to 50% in comparison to the benchmark scheme, with less average velocities for the relay nodes. We also observe that the more stringent the communication requirements, the more resources need to be injected into the system. While decoupling the mission and communication networks enables application of RPA to different mission scenarios; further investigation is required to determine whether joint optimization of mission and networking requirements could perform better in terms of required resources. Moreover, further work is necessary for moving the algorithm to scenarios with unknown or partially known mission plans or for scenarios where limited infrastructure is available to assist UAV communications.

References

[1]   Hayat S., Yanmaz E., Muzaffar R. 'Survey on unmanned aerial vehicle networks for civil applications: a communications viewpoint'. *IEEE Communications Surveys & Tutorials*;18(4):2624–61.

[2]   Zeng Y., Wu Q., Zhang R. 'Accessing from the sky: a tutorial on UAV communications for 5G and beyond'. *Proceedings of the IEEE*. 2019;107(12):2327–75.

[3]   Mozaffari M., Saad W., Bennis M., Nam Y.-H., Debbah M. 'A tutorial on UAVs for wireless networks: applications, challenges, and open problems'. *IEEE Communications Surveys & Tutorials*. 2019;21(3):2334–60.

[4]   Shakhatreh H., Sawalmeh A.H., Al-Fuqaha A., *et al*. 'Unmanned aerial vehicles (UAVs): a survey on civil applications and key research challenges'. *IEEE Access*. 2019;7:48572–634.

[5]   Zeng Y., Zhang R., Lim T.J. 'Wireless communications with unmanned aerial vehicles: opportunities and challenges'. *IEEE Communications Magazine*. 2016;54(5):36–42.

[6]   Yanmaz E. 'Connectivity versus area coverage in unmanned aerial vehicle networks'. Proceedings of IEEE International Conference on Communications; 2012.

[7]   Hayat S., Yanmaz E., Bettstetter C., Brown T.X. 'Multi-objective drone path planning for search and rescue with quality-of-service requirements'. *Autonomous Robots*. 2020;44(7):1183–98.

[8]   Messous M., Senouci S., Sedjelmaci H. 'Network connectivity and area coverage for UAV Fleet mobility model with energy constraint'. Proceedings of IEEE Wireless Communcation and Networking Conference; 2016. pp. 1–6.

[9]   Danoy G., Brust M.R., Bouvry P. 'Connectivity stability in autonomous multi-level UAV swarms for wide area monitoring'. Proceedings of ACM Symposium on Development and Analysis of Intelligent Vehicular Networks and Applications; New York, NY, USA; 2015. pp. 1–8.

[10]   Flushing E.F., Kudelski M., Gambardella L.M., Di Caro G.A. 'Connectivity-aware planning of search and rescue missions'. Proceedings of IEEE International Symposium on Safety, Security, and Rescue Robotics; 2013. pp. 1–8.

[11]   Scherer J., Rinner B. 'Multi-robot persistent surveillance with connectivity constraints'. *IEEE Access*. 2020;8:15093–109.

[12]   Yanmaz E., Yahyanejad S., Rinner B. 'Drone networks: communications, coordination, and sensing. advances in wireless communication and networking for cooperating autonomous systems'. *Ad Hoc Networks*. 2018;68:1–15.

[13]   Younis M., Akkaya K. 'Strategies and techniques for node placement in wireless sensor networks: a survey'. *Ad Hoc Networks*. 2008;6(4):621–55.

[14] Magán-Carrión R., Camacho J., García-Teodoro P., Flushing E.F., Di Caro G.A. 'A dynamical relay node placement solution for MANETs'. *Computer Communications*. 2017;114(4):36–50.

[15] DL G., Pei G., Ly H., Gerla M. 'UAV aided intelligent routing for ad hoc wireless network in single-area theater'. Proceedings of IEEE Conference on Wireless Communications and Networking (WCNC). IEEE Vol. 3; 2000. pp. 1220–5.

[16] Palat R.C., Annamalai A., Reed J.H. 'Cooperative relaying for ad hoc ground networks using swarms'. Proceedings IEEE Military Communications Conference (MILCOM). Vol. 3; 2005. pp. 1588–94.

[17] Chen Y., Liu X., Zhao N., VTCVTC. 'Using multiple UAVs as relays for reliable communications'. 2018 IEEE 87th Vehicular Technology Conference; 2018. pp. 1–5.

[18] Li L., Chang T., Cai S., SPAWC. 'UAV positioning and power control for wireless two-way relaying'. 2019 IEEE 20th International Workshop on Signal Processing Advances in Wireless Communications; 2019. pp. 1–5.

[19] Hauert S., Zufferey J.-C., Floreano D. 'Evolved swarming without positioning information: an application in aerial communication relay'. *Autonomous Robots*. 2009;26(1):21–32.

[20] Zhan P., Yu K., Swindlehurst A.L. 'Wireless relay communications with unmanned aerial vehicles: performance and optimization'. *IEEE Transactions on Aerospace and Electronic Systems*. 2011;47(3):2068–85.

[21] Zhu M., Liu F., Cai Z., Xu M. 'Maintaining connectivity of MANETs through multiple unmanned aerial vehicles'. *Mathematical Problems in Engineering*. 2015;2015:1–14.

[22] Ladosz P., Oh H., Chen W.-H. 'Trajectory planning for communication relay unmanned aerial vehicles in urban dynamic environments'. *Journal of Intelligent & Robotic Systems*. 2018;89(1-2):7–25.

[23] Mardani A., Chiaberge M., Giaccone P. 'Communication-aware UAV path planning'. *IEEE Access*. 2019;7:52609–21.

[24] Liu Y., Dai H.-N., Wang Q., Shukla M.K., Imran M. 'Unmanned aerial vehicle for Internet of everything: opportunities and challenges'. *Computer Communications*. 2020;155(3):66–83.

[25] Qi F., Zhu X., Mang G., Kadoch M., Li W. 'UAV network and IoT in the sky for future smart cities'. *IEEE Network*. 2019;33(2):96–101.

[26] Kalantari E., Yanikomeroglu H., Yongacoglu A. 'On the number and 3D placement of drone base stations in wireless cellular networks'. 2016 IEEE 84th Vehicular Technology Conference (VTC-Fall); 2016. pp. 1–6.

[27] Mozaffari M., Saad W., Bennis M., Debbah M. 'Efficient deployment of multiple Unmanned aerial vehicles for optimal wireless coverage'. *IEEE Communications Letters*. 2016;20(8):1647–50.

[28] Bor-Yaliniz I., Yanikomeroglu H. 'The new frontier in RAN heterogeneity: multi-tier drone-cells'. *IEEE Communications Magazine*. 2016;54(11):48–55.

[29] Zeng Y., Zhang R., Lim T.J. 'Throughput maximization for UAV-enabled mobile relaying systems'. *IEEE Transactions on Communications*. 2016;64(12):4983–96.

[30] Yanmaz E. 'Positioning aerial relays to maintain connectivity during drone team missions'. *Ad Hoc Networks*. 2022;128(4):102800.

[31] Yanmaz E. 'Dynamic relay selection and positioning for cooperative UAV networks'. *IEEE Networking Letters*. 2021;3:114–8.

[32] Lin G.-H., Xue G. 'Steiner tree problem with minimum number of Steiner points and bounded edge-length'. *Information Processing Letters*. 1999;69(2):53–7.

[33] Senel F., Younis M. 'Optimized relay node placement for establishing connectivity in sensor networks'. Proceedings of IEEE Global Communications Conference (GLOBECOM); 2012. pp. 512–7.

[34] Li J.-S., Kao H.-C., Ke J.-D. 'Voronoi-based relay placement scheme for wireless sensor networks'. *IET Communications*. 2009;3(4):530–8.

[35] Yanmaz E., Kuschnig R., Bettstetter C. 'Achieving air-ground communications in 802.11 networks with three-dimensional aerial mobility'. Proceedings of IEEE International Conference on Computer Communications (INFOCOM), Mini conference; 2013.

[36] Bektas T. 'The multiple traveling salesman problem: an overview of formulations and solution procedures'. *Omega*. 2006;34(3):209–19.

[37] Zeng Y., Xu J., Zhang R. 'Energy minimization for wireless communication with rotary-wing UAV'. *IEEE Transactions on Wireless Communications*. 2019;18(4):2329–45.

Chapter 8

Multimedia content delivery in wireless mesh networking

Ting Bi[1] and Shengyang Chen[2]

8.1 Introduction

Over the past decades, the demand for supporting data communications has increased significantly. With the advances in wireless network technologies, usage of wireless devices has also increased rapidly, accompanied by growth in the data traffic associated with rich network services, such as video-related application services on mobile devices. High user quality of service/experience (QoS/QoE) for such services is considered essential for their further development.

Wireless mesh networks (WMNs) currently appear to be one of the most widely accepted last-mile connectivity approaches for their flexibility, usability, low cost, and rapid deployment. However, it is challenging to provide high-quality video wireless services, as the network resources involved are often constrained. Limited bandwidth, scarcity of wireless channels, and multi-hop connections coupled with a highly dynamic topology pose a severe challenge to the quality and interactivity levels of multimedia communications [1].

This chapter will first introduce the background of the multimedia content delivery with QoS requirements and the quality evaluation. The next section summarizes recent advances at the transport level of networking, focusing on proposals with significant impacts aiming at maximizing user-experience-based quality of content delivery. The protocols reviewed would be evaluated with respect to the QoS guarantees that can be achieved for multimedia applications. The last section presents analytics of essential architectural requirements for deploying innovative multimedia services in real-life use-case scenarios involving WMNs, with particular attention given to open problems and challenges in existing related academic research projects and industrial solutions.

[1]Maynooth University, Maynooth, Kildare, Ireland
[2]BlackBox Network Services Ltd, Lawrence, PA, USA

8.2   Multimedia content delivery and quality evaluation

8.2.1   Overview

Over the years, multimedia has evolved and become an essential part of modern life. Youtube*, Netflix[†], and Disney Plus[‡] are just some of the most popular video content providers that are available on the Internet. The main motivation that drives the development of communication technology is this explosion of media-rich content with a non-stop increase in bandwidth demand. However, transporting multimedia content, especially videos is not easy, and the quality of the content can be significantly degraded depending on network conditions. Even if the network throughput is sufficient for transporting the video, network-related parameters such as delay, jitter, and packet loss could highly affect the quality of the received video.

In this section, multimedia delivery protocols, some of the most important network-related parameters that could affect the quality of received videos will be discussed, the general streaming standard, and the metrics for evaluating user perceived quality for videos are also described.

8.2.2   Quality of service requirements

Network parameters were first introduced in the context of QoS by ITU-T E.800 in 1994 [2] in which QoS was described as "the totality of characteristics of a telecommunications service that bear on its ability to satisfy stated and implied needs of the user of the service." Figure 8.1 illustrated an end-to-end QoS scenario that shows how distributed multimedia systems perform active monitoring and maintenance of the delivered QoS by various components working in unison [3]. As Figure 8.1 shows, QoS comprises both network performance and non-network-related performance components. Regarding the network performance, network QoS parameters refer to a set of metrics that include delay, jitter, Bit Error Rate (BER), packet loss, and throughput.

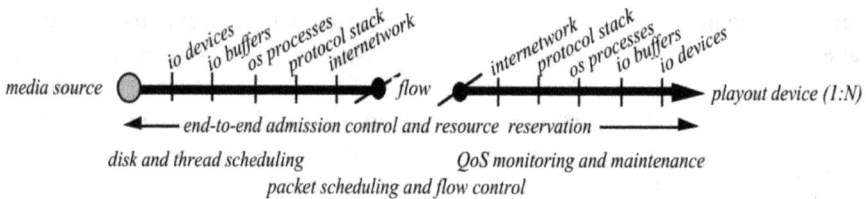

Figure 8.1   *End-to-end QoS scenario [3]*

**Youtube: https://www.youtube.com/.
[†]Netflix: https://www.netflix.com/.
[‡]Disney Plus: https://www.disneyplus.com/.

8.2.2.1   Delay

Generally, the delay in multimedia communication comprises application delay and end-to-end delay. Application delay represents the time difference between the data arrival time and drain time of the media content. Most of the time the delay is dependent on the hardware/software performance, which is influenced by the CPU/GPU performance and multimedia encoding/decoding solutions. End-to-end delay refers to the duration a packet travels across a network from the source to the destination. End-to-end delay is a summation of the four components as shown in (8.1): processing delay, queueing delay, transmission delay, and propagation delay.

$$end-to-end\ delay = processing\ delay + queueing\ delay \\ + transmission\ delay + propagation\ delay \tag{8.1}$$

where the processing delay is the summation of the time taken to process a packet to determine output link and check the bit errors, queueing delay refers to the time waiting on the output link for transmission, transmission delay is the time taken to transmit a packet on a link, and propagation delay refers to the time to deliver a bit over the transmission medium.

8.2.2.2   Jitter

Jitter is the difference between the current packet delay and the delay of the reference packet that represents the delay variation caused by network condition dynamics. In multimedia transmission, high jitter may result in distorted or jerky videos, which seriously affect user perceived quality. To avoid being affected by high jitter, a buffer is implemented at the receiver for video applications.

8.2.2.3   BER

BER is a key characteristic of the network channel condition and counts the number of bit errors per unit time. BER is affected by channel noise, interference, distortion, wireless multipath fading, or other transmission-related reasons.

8.2.2.4   Packet loss

Packet loss is the ratio of the packet dropped to the number of packet transmitted during the transmission through the network. During a transmission session, a packet might be lost due to various factors such as network congestion, buffer overflow, network connection failure, channel contentions, and collisions. For multimedia content delivery, packet loss can greatly impair user-perceived multimedia quality.

8.2.2.5   Throughput

Throughput is the rate of successful information delivery over a communication channel. Throughput is usually measured in bits per second or packets per second/time slots. Throughput is one of the most important parameters that can determine the network performance on the user side. Low throughput can cause long transmission times and low user-perceived quality, especially for real-time services.

Different applications have different demands on the above-discussed network parameters to achieve a good QoS. The International Telecommunication Union (ITU) released recommendation Y.1541 [4] on network performance objectives for Internet Protocol (IP)-based services. Table 8.1 illustrates the associated network performance requirements for various applications as described in ITU-T Rec Y.1541.

8.3   Video content delivery quality measurement

For measuring the user's experience on the service quality levels, two important concepts have been introduced: QoS and QoE. QoS is the overall performance of a telecommunication service that bears on its ability to satisfy the stated and implied needs of the user of the service. QoE represents the overall acceptability of service as subjectively perceived by an end user [5]. The concepts, scopes of application, and differences between QoS and QoE in the context of network services are illustrated in Figure 8.2. QoE includes the complete QoS-based effects and also can be influenced by additional psychological factors of end-user perception in a different environment with different types of application services.

QoE is one of the most important factors when measuring the quality of a service and is focused on understanding the overall human quality requirements. It involves various research fields such as social psychology, cognitive science, economics, and engineering science.

QoS and QoE can be measured in the context of a certain service application. This thesis focuses on assessing video perceived quality level following its network delivery.

8.3.1   Subjective and objective quality assessment

In order to assess end-user perceived quality regarding the delivered video content, various methodologies were developed to quantify the received video quality. Two major approaches exist: subjective methods and objective methods.

8.3.1.1   Subjective methods

Subjective methods require direct human exposure and interactions via pre-designed subjective tests, with feedback from test participants to score the perceived video quality within the tests. This involves large monetary and time costs and does not work for prediction or real-time quality assessment.

8.3.1.2   Objective methods

Objective methods are based on the use of metrics, and the calculation processes are performed by algorithms. These algorithms can be classified into three main subgroups [7].

Table 8.1 Y.1541 IP network performance requirements for different applications [4]

Parameters	Class 0	Class 1	Class 2	Class 3	Class 4	Class 5
Delay (IPTD)	100 ms	400 ms	100 ms	400 ms	1s	U
Jitter (IPDV)	50 ms	50 ms	U	U	U	U
Packet loss ratio (IPLR)	1×10^{-3}	1×10^{-3}	1×10^{-3}	1×10^{-3}	1×10^{-3}	U
Applications	Real-time, Highly interactive, Delay variation sensitive (VoIP, video conference)	Real-time, Interactive, Delay variation sensitive (VoIP)	Transaction data, highly interactive (signaling)	Transaction data, interactive	Low loss only (short transactions, bulk data)	Traditional application of default IP networks

"U" means "unspecified" or "unbounded."

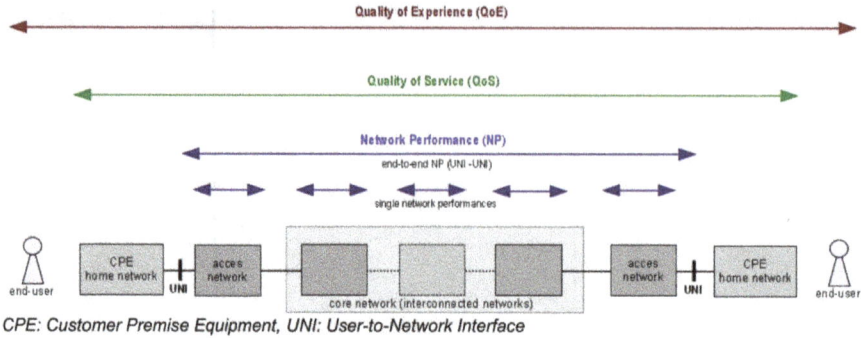

CPE: Customer Premise Equipment, UNI: User-to-Network Interface

Figure 8.2　Network performance, QoS and QoE [6]

A. *Full reference methods*: They are based on the comparison between the original video (before transmission) and the received video.
B. *Reduced reference methods*: They require a feature vector derived from the statistical model of the reference video for quality evaluation.
C. *No reference methods*: They use no-reference models that are based on the network-related or application-specific characteristics information and the received video only.

In this work, the Mean Opinion Score (MOS) [8] is one of the metrics adopted for subjective video quality estimation and comparison. MOS is one of the most commonly used metrics in assessing video quality. Five quality levels are defined in the MOS to measure the human quality of the video, from one representing "bad" quality to five representing "excellent" quality, as shown in Table 8.2.

Another metric is the peak signal-to-noise ratio (PSNR). PSNR is a full reference objective metric commonly and widely used for assessing video quality. The formula for PSNR is shown in (8.2):

$$PSNR_{dB} = 20 \log_{10} \frac{MAX_I}{\sqrt{MSE}} \tag{8.2}$$

MAX_I is the maximum possible pixel value of the image (e.g., 255), mean squared error (MSE) represents the difference between the original video and the received one and can be calculated by (8.3):

Table 8.2　ITU MOS quality and impairment scale [8]

MOS scale	Quality	Impairment
1	Excellent	Imperceptible
2	Good	Perceptible but not annoying
3	Fair	Slightly annoying
4	Poor	Annoying
5	Bad	Very annoying

Table 8.3   PSNR to MOS conversion [10]

PSNR [dB]	MOS scale	Quality	Impairment
>37	1	Excellent	Imperceptible
31–37	2	Good	Perceptible but not annoying
25–31	3	Fair	Slightly annoying
20–25	4	Poor	Annoying
<20	5	Bad	Very annoying

$$\text{MSE} = \frac{1}{MN} \sum_{m=1}^{M} \sum_{n=1}^{N} \left[x\left(n, m\right) - y\left(n, m\right) \right]^2 \tag{8.3}$$

$x\left(n, m\right)$ is the sample of the original source signal, and $y(n, m)$ is the distorted video signal. Various different approaches in defining PSNR for different purposes appear in the literature. In this work, we use the PSNR metric proposed by Lee *et al.* [9], which estimates the video quality subject to its network transmission. The computation is shown in (8.4):

$$\text{PSNR} = 20 \log_{10} \left(\frac{\text{MAX\_Bitrate}}{\sqrt{\left(\text{EXP}_{\text{Thr}} - \text{CRT}_{\text{Thr}}\right)^2}} \right) \tag{8.4}$$

MAX_Bitrate represents the bit rate of the multimedia stream after the encoding process. EXP_{Thr} represents the average throughput expected to be obtained. CRT_{Thr} is the actual measured throughput during the transmission.

The PSNR metric values are mapped to the MOS scale in Ke *et al.* [10] by using conversion as shown in Table 8.3.

Structural Similarity (SSIM) metric [11] is another known full reference objective metrics that aim at being more consistent with the human vision by comparing the similarities between two frames based on luminance, contrast, and structural similarity. The SSIM metric sits between [0, 1], where 0 means there is no correlation with the original video frame (totally different frames) and 1 means full correlation (identical frames). SSIM metric is highly correlated with subjective quality assessment methods; however, the computational complexity of SSIM is also much higher than that of PSNR.

Video Quality Model (VQM) metric [12] is a novel objective video assessment metric developed by the Institute of Telecommunication Science. In VQM, the perceptual effects of video impairments such as blurring, noise, and distortion are assessed and combined. The computational complexity of VQM is even higher than SSIM, yet the correlation with the subjective quality assessment method is better.

8.4   Energy consumption issues during content delivery

As end-user terminal devices in wireless networks are deployed with limited-amount battery-based power resources, the energy conservation of wireless devices has become a critical issue to extend the life spans of devices. Therefore, pursuing high energy efficiency will be the trend for the design of future wireless communications. Energy efficiency is generally defined as information bits per unit of transmission energy. A typical function of energy efficiency calculation for an additive white Gaussian noise channel is shown in (8.5) [13]:

$$\eta_{EE} = \frac{2R}{N_0(2^{2R}-1)} \tag{8.5}$$

in which the channel capacity R is defined as [13]:

$$R = \frac{1}{2} \log_2 \left(1 + \frac{P}{N_0 B}\right) \tag{8.6}$$

In equation (8.6), P represents the transmit power, N_0 represents the noise power spectral density, and B represents the system bandwidth. It is obvious that with this energy consumption model, the energy efficiency decreases monotonically with the channel capacity.

The energy consumption model implied by the equations (8.5) and (8.6) is based on the assumption that the information block size is infinite. However, in real-life cases where the information block size is always finite, the situation is different. Energy conservation on wireless devices results from various aspects such as data transmission, traffic encoding/decoding, content playback, and radio access interface switching. Among these sources, most of the energy is conserved during data transmission over wireless devices when their radio access interfaces are active, yet the state transition of these interfaces also consumes a considerable amount of energy. For example, a typical commercial 802.11g network consumes 990 mW energy at the idle state and 1980 mW at the transmitting state [14]. On the other hand, the energy consumption from circuits on the devices differs according to the type of radio access technology used. For example, telephony and data over the WCDMA network are more power-intensive on many types of mobile devices with widely known manufacturers, such as HTC and Samsung smartphones, in comparison with comparable GSM network. HSPA+ delivers significant improvement on the battery life of wireless devices by maintaining a long-term active connection that allows a shorter time for the device to transfer from wake to idle.

Apart from the effect of circuit energy consumption, recent research efforts have indicated that the amount of energy consumed by the devices in wireless networks is closely related to the amount of data transferred [15]. Usually, the higher network traffic indicates the larger energy consumption. This makes it more necessary to increase energy efficiency for content delivery services in wireless networks as most of these services, especially when involved with rich-media content, are with high traffic load so that the QoS/QoE requirements of these services can be achieved.

8.5   Protocols, schemes, and algorithms

8.5.1   Transport layer protocols

The multimedia content delivery services are built at the application layer with support from various transport layer protocols. Transport layer protocols enable end-to-end data transmission between the source and destination hosts. Two fundamental transport layer protocols were designed and are widely deployed in the network environment: Transport Control Protocol (TCP)[§] and User Datagram Protocol (UDP)[¶]. TCP supports congestion control, retransmission, and flow control functions to provide reliable and in-sequence data delivery. UDP does not support reliable transmission, is message-oriented, and is preferred for multimedia delivery. Some other transport layer protocols will be discussed in detail.

8.5.1.1   Datagram Congestion Control Protocol (DCCP)

The DCCP [16] is a message-oriented transport layer protocol that provides unreliable data delivery to achieve timely transmission. It involves two optional congestion control mechanisms: a TCP-like congestion control [17] and a TCP-Friendly Rate Control [18]. Compared to UDP, DCCP has session and congestion control. Compared to TCP, DCCP does not provide reliability and retransmission.

8.5.1.2   Stream Control Transmission Protocol (SCTP)

The SCTP[**] is a reliable, message-oriented transport protocol for IP network data communications. Compared to UDP, SCTP provides reliability and congestion control. Compared to TCP, SCTP provides "multi-streaming" and "multi-homing." SCTP supports multi-homing by exchanging and maintaining lists of IP addresses for each SCTP endpoints that are associated with each other and support multi-streaming by managing several separated data streams that can transmit data via independent sequence deliveries. For each SCTP association, the end host has one primary path and one or more backup paths. The end hosts with multiple network interfaces can connect to several separate networks concurrently using STCP, which makes it suitable for mobility. Moreover, the end-hosts can change their primary communication path to a new path before the breakdown of the current path is triggered by a handover decision. This provides the possibility for seamless handover under SCTP. A TCP-like congestion control mechanism is employed in SCTP at the association level.

8.5.1.3   Multipath Transmission Control Protocol (MPTCP)

The MPTCP represents an extension of the classic legacy TCP [19, 20] that was designed to be transparent to both applications and network. It allows multiple

---

[§]Transmission Control Protocol: https://tools.ietf.org/html/rfc793.
[¶]User Datagram Protocol: https://www.ietf.org/rfc/rfc768.txt.
[**]Stream Control Transmission Protocol: https://tools.ietf.org/html/rfc4960.

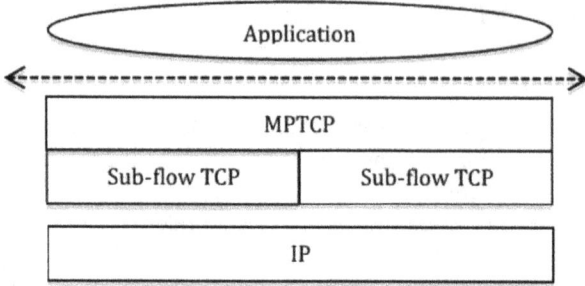

Figure 8.3   MPTCP architecture [22]

sub-flows to be set up for a single connection session between two hosts [21]. MPTCP is connection-oriented and allows mobile devices to concurrently utilize multiple interfaces and network access technologies to improve both network delivery performance and QoS, especially in heterogeneous wireless network environments.

Figure 8.3 illustrates MPTCP architecture that consists of two levels [22]: the "MPTCP level" is an application-oriented level, which gathers the semantic from the application. The "sub-flow TCP level" is the network-oriented level and helps with the protocol's reuse of the TCP architecture, which ensures the MPTCP packets are not blocked by middle-boxes.

A summary of the transport-layer protocols described in the sections above is listed in Table 8.4.

8.6   MAC-layer schemes

8.6.1   *QoS-related wireless mesh MAC-layer schemes*

To solve some existing QoS performance issues specifically for WMNs using the IEEE 802.11 MAC protocol, several novel MAC protocols have been proposed in the research literature.

One of the first MAC solutions for improving the QoS performance of single-channel WMNs is introduced in Benveniste [23] that includes three steps: increase the contention window size, express forwarding, and express retransmission. Increasing the size of the contention window decreases the possibility of a collision occurring at hidden nodes. Express forwarding involves increasing the duration field value of the current data frame being sent so that the neighbor nodes stop being active and the receiver node is able to access the channel, and the ACK (Acknowledges) packet of this data frame derives the extended duration field value from it, adjusted for elapsed time. Express retransmission allows immediate retransmission of the packet once the ACK timer expires. In this way, the end-to-end delay occurred in regular retransmission of a multi-hop flow is shortened. These three schemes are all based on existing mechanisms, such as CSMA/CA (Carrier-Sense Multiple Access with Collision Avoidance) [24] proposed in the IEEE 802.11 MAC protocol and TXOP proposed in the IEEE 802.11e protocol [25], so the complexity

Table 8.4   Features of transfer protocols

Feature name	TCP	UDP	DCCP	SCTP	MPTCP
Packet Header Size (Bytes)	20–60	8	12 or 16	12	50–90
Transport Layer Packet Entity	Segment	Datagram	Datagram	Datagram	Segment
Connection Oriented	Yes	No	Yes	Yes	Yes
Reliable Transport	Yes	No	No	Yes	Yes
Unreliable Transport	No	Yes	Yes	Yes	No
Preserve Message Boundary	No	Yes	Yes	Yes	No
Ordered Delivery	Yes	No	No	Yes	Yes
Unordered Delivery	No	Yes	Yes	Yes	No
Data Checksum	Yes	Optional	Yes	Yes	Yes
Checksum Size (Bits)	16	16	16	32	16
Partial Checksum	No	No	Yes	No	No
Path Mtu	Yes	No	Yes	Yes	Yes
Flow Control	Yes	No	No	Yes	Yes
Congestion Control	Yes	No	Yes	Yes	Yes
Explicit Congestion Notification	Yes	No	Yes	Yes	Yes
Multiple Streams	No	No	No	Yes	Yes
Multi-Homing	No	No	No	Yes	Yes
Bundling/Nagle	Yes	No	No	Yes	Yes

Comparison of transport layer protocols: https://en.wikipedia.org/wiki/Transport_layer.

of modification is low. Performance evaluation of the three schemes is done with a WMN structure together with three WLAN structures, using VoIP applications and high-/low-resolution videos as traffic. Results have proven that the proposed schemes are able to reduce end-to-end retransmission delay at an obvious level and to decrease the occurrence of dropped data frames.

2P [26] aims at improving the performance of the IEEE 802.11 MAC protocol by redesigning the CSMA/CA mechanism to enable data transmission and reception along with all links of a node simultaneously, making full use of the available channel capacity and supporting low-cost rural connectivity. The basic principle of 2P is based on the synchronous transmitting/receiving operation state (SynTx/SynRx) [24] of the links at a node switching between these two states, which are not enabled in the IEEE 802.11 MAC protocol. 2P modifies the immediate ACK and the carrier-sense-based back-off mechanism in the IEEE 802.11 MAC protocol to enable SynTx/SynRx. Meanwhile, 2P introduces a timeout mechanism to handle temporary synchronization loss and link recovery on each individual link and a SynTx notification mechanism to coordinate the state switching of multiple nodes. In terms of performance, 2P offers significant throughput improvement within a WMN topology containing 25–30 nodes and the throughput benefit is robust against losses, in comparison with the IEEE 802.11 MAC protocol with CSMA/CA.

Another contention-based MAC solution is the DBTMA (Dual Busy Tone Multiple Access) [27] protocol, which aims at solving the hidden/exposed terminal problem. In DBTMA, two narrow-bandwidth and out-of-band busy tones are used together with the RTS (Request to Send) packet at the receiver node and RTS transmitter node of data packet transmission. The busy tones are used to provide protection for the RTS packets so that the chance of successful reception of RTS packets increases and the throughput increases consequently. As a result, the exposed terminal nodes can initiate data packet transmission, and the hidden terminal nodes are able to reply to RTS packets and initiate data packet reception. DBTMA is proven to have better performance in terms of channel throughput in comparison with traditional RTS/CTS-based MAC protocols, such as ALOHA [24]. However, the usage of multiple busy tones requires the extra incorporation of circuits into wireless devices and also increases bandwidth consumption.

An advanced MAC-layer protocol used for parallelism packet transmission in WMNs is proposed in Hasan *et al.* [28] as an extension of the Medium Access via Collision Avoidance with Enhanced Parallelism (MACA-P) [29] protocol. The MACA-P protocol supports the concurrent transmission on any mesh node by scheduling transmitting data packets and receiving ACK packets in the outgoing RTS packets to the neighbor mesh nodes. The neighbor mesh nodes schedule their data transmissions and ACK receptions according to the schedule included in the RTS packets received from that node. The main modifications made by the extended solution based on MACA-P include consideration of whether the remaining time is long enough for transmitting frames and the usage of a pair of query/query reply signals and an associated two-way handshake mechanism for connection setup. The information of neighbor mesh nodes is collected, and the network allocation vector table is created after connection establishment. The control gap between RTS/CTS packet exchanges is calculated with the prediction of packet transmission time based on the send-out time included in the query signal and the arrival time of the query reply signal. The RTS sender mesh node retransmits its RTS packets with the value of the control gap, causing its neighbor mesh nodes to reschedule the reception time accordingly. By continuously adjusting the value of the control gap, the solution tries to synchronize the maximum number of mesh nodes in a parallel transmission without the unnecessary growth of wait time for control packets. However, the protocol has some performance issues when detecting different numbers of neighbors at different nodes.

The MAC-layer scheme proposed in Shankar *et al.* [30] works as a part of a cross-layer solution that contains an overlay network structure for improving delay-constrained video streaming service quality in multi-hop WMNs. The scheme uses an algorithm to optimize different control parameters at mesh nodes, especially transmission retry limits, together with an IEEE 802.11e HCCA-based [25] traffic scheduling mechanism. The optimized parameters are exchanged between different layers in the overlay network structure to be further used as essential information for pre-building the mesh topology before the initialization of video streaming. As a result, negative network reconfiguration effects such as additional delays and link failures are minimized for the application service. Experimental results based on simulations have shown that the scheme provides a significant improvement on received video quality as frequent feedback via the overlay network and end-to-end optimizations with different set of utility functions are used. The results also indicate that (1) the balancing between overhead in the overlay network and improvement brought by the feedback and (2) the prioritization of video packets with respect to distortion reduction and the received video quality.

The Time-Division Multiple Access (TDMA)-mini-slot-based scheme proposed in Jiao and Li [31] improves the performance of the IEEE 802.11 MAC protocol. To solve the issues of packet loss, bandwidth decrease, and transmission delay caused by contention in traditional MAC protocols, this scheme allocates channel resources in mini-slots synchronized within two-hop neighborhood mesh nodes to avoid packet transmission corruption. Each individual mesh node is assigned a unique mini-slot within a two-hop neighborhood, and all the slots are synchronized so that data packets from different nodes do not corrupt each other during transmission. Moreover, it provides priority access to extend cooperative communication to multi-hop cases, which is always reserved by a mini-slot as needed. This is achieved by using an instantaneous Signal-to-Noise Ratio (SNR)-based helper selection algorithm in the case of faulty wireless channels. The proposed scheme improves the end-to-end throughput in a multi-hop manner in comparison with the IEEE 802.11 MAC protocol and offers generally steady performance when packet size varies during transmission.

MAC-ASA [32] is another MAC-layer scheme aiming at the following improvements based on the IEEE 802.11 MAC protocol: the enhancement of end-to-end transmission throughput for both single-hop and multi-hop WMNs, scalability, and collision avoidance capability. MAC-ASA combines a distributed link scheduling algorithm and a power control mechanism for router-to-router communication together. It also uses a modified version of the IEEE 802.11 CSMA/CA contention handling procedure within a coordinate dynamic TDMA-like protocol to reduce collision and to enable a data aggregation pipeline. In the case of communication between mesh clients and routers, MAC-ASA provides good throughput improvement and robustness to increasing packet size. However, MAC-ASA does not increase performance significantly when there is only mesh traffic in the network (only mesh routers are involved in data transmission).

8.6.2.   *Energy-related wireless mesh MAC-layer schemes*

The MAC-layer solutions discussed above mainly focus on transmission performance in WMNs and take little consideration of the energy consumption of the mesh devices, especially energy consumption during data packet transmission. Factors such as access delay and packet collision have an impact on the energy consumption of wireless devices during data packet transmission, which cannot be ignored. In this context, several energy-related MAC solutions for WMNs are also presented and discussed in this section.

DSMA-S [33] is proposed to solve the control packet collision problem. DSMA-S is based on the contention-based scheme in DBTMA and uses two out-of-band busy tone signals at the receiver node to indicate successful transmission and packet collision to other nodes. Meanwhile, control packets and data packets in DSMA-S are transmitted separately on the common wireless channel, which is divided into time-synchronized slots, to maximize the channel efficiency. The busy tone signals are sensed only at the beginning of the time slots so that the idle listening state of mesh devices is reduced to save energy. As no busy tones are broadcasted by the sender nodes, losing normal functionality, the improvement on throughput performance of DSMA-S in comparison with DBTMA is more significant in the case of hidden terminals in the network than no hidden terminals. In terms of energy consumption, DSMA-S offers better performance than DBTMA with a larger ratio of data packet duration and control packet duration.

The energy-efficient wireless mesh node monitoring scheme proposed in Hassanzadeh *et al.* [34] tries to ensure continuous and complete detection coverage for WMN intrusion and to save energy for the non-monitoring mesh nodes via duty cycle management. In this scheme, each node in the WMN is integrated with an intrusion detection engine for wireless link monitoring. The intrusion detection engines on nodes selected for wireless link monitoring consume more energy than others, as they detect traffic in the entire network while others detect only traffic from/to the mesh client. A centralized and a distributed monitoring mesh node selection algorithm are proposed at the mesh gateway with a trade-off between the time and message complexity for intrusion detection rate, respectively. Meanwhile, duty cycle management schemes are used on the mesh nodes that are not selected as monitoring nodes, without affecting the intrusion detection process. Simulation-based results show that the mesh node monitoring scheme proves full wireless link coverage regardless of the network capacity, while the level of residual energy of nodes in the WMN is increased.

8.7   Routing protocols and algorithms

Apart from MAC-layer solutions, routing algorithms during data transmission also play an important role in data packet transmission performance. In wireless networks, nodes between the sender node and the receiver node are responsible for data forwarding, working as routers. The principle of routers on how to generate data forwarding path determines the overall performance and resource cost of the wireless network, as each individual router has different states in terms of power supply, network load, mobility, and duty cycle. It is essential that routing algorithms considering both energy efficiency and performance should be employed for wireless networks. Reviews on some of the state-of-the-art routing algorithms are presented in this section.

8.7.1   Routing protocols

Many classic routing protocols and algorithms have been widely used in various network scenarios with significant benefits. One of the earliest routing schemes is the classic Dijkstra's algorithm [35]. It is originally designed for solving the shortest path problem in graph searching: finding the path with the lowest cost between any endpoint to others in a graph structure. In the Dijkstra's algorithm, the cost calculation starts from one end node of the structure and recursively updates for each unvisited neighbor node of the current node in the structure until the destination node is visited. In a wireless sensor/mesh network, the cost of finding the shortest path is determined by multiple network-condition-related factors such as network latency, traffic load, and energy consumption. The principle of Dijkstra's algorithm is commonly used in many more advanced path-searching algorithms for generating traffic routes with the lowest cost in wireless networks. These protocols and algorithms can be divided into two main categories: **global** and **distributed**, according to whether any node in the network has the information of all the other nodes or just the nodes that are adjacent.

One of the common examples of global routing algorithms is the link state (LS) routing protocol [36] used in packet-switching networks. In LS protocol, all the switching nodes in the network, i.e., the routers, construct a connectivity graph structure showing connection states between nodes. Each individual node in the graph structure has the knowledge of the connectivity of the entire network. Based on this, each node in the LS protocol calculates the optimal path to other possible destination nodes independently, similar to the principle of Dijkstra's algorithm. This is done by broadcasting short messages called link-state advertisements throughout the network in order to update routing information such as sequence numbers. The global routing table is updated when the shortest path calculations of nodes are complete.

The Optimized Link State Routing Protocol (OLSR) [37] is an IP-based extension of the classic LS routing protocol and is mainly designed for mobile ad-hoc networks but also useful for other wireless ad-hoc networks. It is a typical proactive routing algorithm that exchanges control messages periodically between nodes in the network, optimizing the classic LS protocol in terms of control packet size and control traffic load from the following two aspects.

- Only a subset of links is declared in OLSR instead of all the links with neighbor nodes in the classic LS routing protocol. The neighbor nodes of the subset of links are multipoint relay selectors.
- Only the multipoint relays (the selected nodes) are used for retransmission in OLSR.

In OLSR, the multiple relays for any nodes are chosen from the range of its one-hop neighbor nodes when they have a bi-directional link toward any existing multipoint relays. Next at each node, the routes to all known destinations through its multipoint relays are calculated by periodically exchanging information of the relays with other nodes. Once the routes are built, unicasting packets are forwarded hop by hop along the optimal route from the source node to the destination node. Broadcasting packets, on the other hand, are forwarded by the multipoint relays based on the rule that retransmission occurs only when the first copy from any multipoint relay selector node is received.

For distributed routing algorithms, one common example is the distance vector (DV) routing protocol defined in RFC 1058[††]. It uses distance and direction as the two main parts of the route calculation information exchanged between adjacent nodes. Each node in the network keeps a routing database with one entry for every possible destination. Information options such as an address, gateway, and timer are included in the database that is updated according to information received from neighboring gateways. The update of information is done in each node by periodically broadcasting a set of routing table messages to its neighbor nodes according to its own clock, so the update of information in the entire network is asynchronous. The optimal route from the source node to the destination node is determined by recursively judging the sum of the cost for individual hops along the route. The cost could be (but not limited to) message delay, sending power, loss rate, etc.

Although a small amount of drops of updated information is tolerant to the DV routing protocol, which moves to a new equilibrium when the drops of updated information cause changes of network topology, rules such as the split horizon and triggered update are applied in the real implementation of the DV routing protocol to ensure the stability of route calculation. It is also worth to mention that the determination of the maximum cost value in the DV routing protocol is a trade-off between the speed convergence and capacity of the network, as the value has to be larger than any possible real route but not exceeding the information update time limit.

Most classic routing algorithms derived from Dijkstra's algorithm, including OLSR and the DV routing protocol just discussed, determine the shortest path by iteratively considering every node with the lowest cost. This involves a complicated iterative calculation process resulting in slowness and ineffectiveness for large-scale wireless network topologies. Another issue caused by such algorithms is the count-to-infinity problem, in which parts of the network become completely inaccessible before the network realizes from the update of route information, as the time spent on the update is not short enough due to the definition of large "infinite" route cost values on certain nodes in the network.

Ad-hoc On-Demand Distance Vector (AODV) [38] is one of the most commonly used protocols in wireless networks, which is an efficient reactive routing protocol designed for wireless multi-hop networks. In AODV, each node maintains a neighbor table of all the directly connected neighbors in order to provide quick response for new route establishment requests and for routing maintenance. For route discovery, AODV uses a broadcast-based route discovery mechanism in which a route request is broadcast from the source node across the network. Hybrid Wireless Mesh Protocol (HWMP) is based on AODV, which is the default routing protocol defined from the IEEE 802.11 second standard [39]. HWMP has a configurable extension for proactive routing, which uses MAC address with layer two routing and uses radio-aware as routing metric.

[††††]Routing Information Protocol: http://tools.ietf.org/html/rfc1058.

8.7.2   Routing algorithms

Many advanced routing algorithms have been proposed lately to endure the issues of classic routing algorithms based on Dijkstra's algorithm. As an example, the Recursive Best-First Search (RBFS) algorithm [36] uses linear space in the maximum search depth to expand nodes in the best-first order. The RBFS algorithm considers all the nodes in the network as a tree structure, in which every node has an upper bound value on the cost defined as the minimal cost value among its parent and sibling nodes. The principle of the RBFS algorithm is briefly described as follows.

The expansion starts from the sender node by searching all its neighboring nodes for the one with the minimal cost as the parent node of the sub-tree in the next recursive call. When the neighboring node with the minimal cost is found, the upper bound value is defined as the minimal cost value among its parent (the sender node, in this case) and all sibling nodes (all the other neighbors of the sender node, in this case).

In every recursive call, the sub-tree below the current parent node is explored when it contains frontier nodes with cost not exceeding the upper bound value, which is updated according to the cost value of the current parent node and all of its children nodes.

To avoid inefficiency caused by the redundant expansion in the recursive process, every child node in each recursive call inherits the cost value of its parent node as its own if this value is greater than its original own cost value. In order to be expanded, the upper bound of a node must be at least as large as its updated cost value. If a node has been previously expanded, its updated cost value will be higher than its original cost value. In this case, its updated cost value is the minimum of the last updated cost values of its child nodes. In general, the updated cost value of a parent node is passed down to its child nodes, which inherit this value only if it exceeds both the original cost value of the parent node and itself.

Research efforts have proved that the RBFS algorithm has better performance than the classic routing algorithms based on Dijkstra's algorithm with shortest-path problems for direct graph network topologies. However, the performance benefit is limited when it comes to other issues related to routing and optimization, such as the traveling salesman problem [40], the graph coloring and scheduling problem [41], and the knapsack problem [42].

The simulated annealing arithmetic (SAA) algorithm [43] has been proposed as a better solution for the combinatorial optimization problems which determine the principle of the cost function in routing issues. The SAA algorithm uses a discrete-time inhomogeneous Markov chain [44] with the updated states of each node in the network whose neighbor states (new states of the problem produced after altering a given state in an optimal way) are considered. At each iteration step, the algorithm probabilistically makes a decision whether to move to the neighbor state. Moving to the neighbor state results in minimal alterations of the last state in order to help the algorithm keep the better results and change the worse results. After a tour of successive neighbor states comes the final solution considering all the neighbors of one state to explore the state space. States that are more optimal during this process are preferred when the algorithm is making decisions for the final solution, yet other states are also possibly accepted in certain circumstances in order to avoid the local optimum problem from simple heuristic algorithms.

In the SSA algorithm, the probability for a node to move to neighbor states is determined by an acceptance probability function that depends on the state results and a global parameter that varies according to time. States with smaller results are considered more optimal. If the difference between the results of moving/not moving to the neighbor state becomes more significant after passing a period of time, the probability of moving decreases accordingly. This process ends with a global optimal solution as the annealing schedule is extended.

It is obvious that for the SSA algorithm, the choice of parameters such as the state space, the cost function, the acceptance probability function, and the annealing schedule has a significant impact on the benefit of the algorithm. However, as there is no general way to define a generally optimal choice of the parameters for different situations, in real-life usage the choice depends on the given problem.

8.7.3   *Routing mechanisms in wireless mesh networks*

This section investigates some research efforts that have been made specifically for advanced routing mechanisms in WMNs, apart from the general routing algorithms introduced in the previous sections, which are widely developed in the research area and used in industry.

Simple Opportunistic Adaptive Routing protocol (SOAR) [45] is a proactive link-state-based routing protocol proposed for explicitly supporting multi-flow in WMNs. It attempts to improve the network throughput and fairness by introducing the following mechanisms: adaptively selecting forwarding paths to leverage path diversity and reducing duplicate transmissions, determining optimal forwarding nodes in terms of priority timer, local loss recovery to handle dropped packet detection and retransmission and adaptively controlling data sending rate according to network conditions. With these mechanisms, SOAR offers better tolerance on the instability of wireless network medium with the hop-by-hop data forwarding in comparison with traditional shortest-path routing protocols. The performance of SOAR is evaluated through simulations and real test-bed experiments for single-flow and multi-flow scenarios with various network topologies. Results show the SOAR achieves higher improvement on the network throughput under symmetric losses than asymmetric losses with single-flow scenarios and a significant improvement on the flow index fairness with multi-flow scenarios.

The multi-flow joint optimization routing algorithm proposed in Krishnaswamy *et al.* [46] works as the key part of a cross-layer cross-overlay architecture. It provides fast information exchange during cross-layer parameter update in order to enable proactive traffic performance optimization using a mesh inter-networking system with network-centric computing. The routing algorithm gathers link-state information of multiple traffic flows from a global database deployed in the mesh internetworking system and makes a joint optimization to meet the constraint of every flow. Factors utilized in the joint optimization for route decision differ according to constraints of different flows. Examples of such factors include the end trip time over the link (for applications with strict end-to-end delay constraints) and the effective throughput of the flow (for applications with significant bandwidth demand constraints). The preferred routing choice is decided independently by each flow based on the result of the joint optimization using extensions of Dijkstra's algorithm.

In Liu *et al.* [47], a QoS-aware backup routing algorithm is proposed to work with an available bandwidth estimation mechanism to accommodate stable QoS for multimedia flows in mobile WMNs. The bandwidth estimation of any node in a network is based on the effective channel capacity and the total occupied bandwidth of this node and its neighbor nodes that share a common channel. The backup routing algorithm includes such information of the node into route calculation information packets to be broadcasted to neighboring nodes sharing the same channel for bandwidth estimation. Meanwhile, to reduce the overhead caused by frequent route discovery in mobile WMNs with unstable link quality, a backup piggybacked path (whose available bandwidth is the second maximal among all the paths) is selected apart from the primary path when a route is established after an exchange of route request/reply messages between nodes, in order to provide more reliable connectivity. Multimedia streams are transmitted via the primary path by default unless it is disconnected, in which case the backup path is activated for transmission. Simulation-based results have proven that the backup routing algorithm successfully shortened the route establishment and re-built time, which is beneficial for real-time multimedia communications.

8.8 Multimedia content delivery services

8.8.1 Overview

Content delivery network (CDN)-based network services and the associated technologies offer low-latency access to remote contents for end users, including but not limited to common web context, HD-quality video/audio stream, applications and system updates, across multiple business areas with high demand on transmission quality, such as manufacturing, retailing, finance, and health care.

Technology development related to CDN services has been in process for many years by various service providers, structuring the unseen backbone of the Internet, while confronting the challenges of multiple-type content delivery, device-detection-based adaptive delivery, data security and standardization, especially on mesh networks. This section gives an introduction to streaming service, specific on HTTP-based adaptive streaming service, and then summarizes state-of-art research outcomes and industrial cases of CDN services based on mesh networks.

8.8.2 Streaming service

Traditional adaptive bitrate streaming protocols are associated with classic standards such as the Real-Time Transport Protocol (RTP)‡‡, the Real-Time Streaming Protocol (RTSP)§§, and Hypertext Transfer Protocol (HTTP)¶. RTP is designed for real-time streaming between end-to-end devices and was mostly performed over UDP, which has no QoS guarantee, transmission control and data protection. RTSP is used to provide remote multimedia playback control support such as play/pause commands from the end-user devices. Synchronization between media streams is handled by the control protocol Real-Time Control Protocol (RTCP)*** which also includes network QoS-related information such as loss, delay, and jitter. Most RTSP servers use the RTP in conjunction with RTCP for media stream delivery. There are also other protocols such as RTMP, etc., but they are not widely deployed.

The main disadvantage of using RTP/UDP is that it cannot traverse Internet firewalls and NAT devices as most of them are configured to restrict UDP traffic. Lately, in order to overcome this problem, HTTP is widely used, as it is allowed by the majority of firewalls. HTTP uses TCP or MPTCP as underlying transport protocols. This is the main reason for which the majority of the deployed adaptive multimedia solutions are based on HTTP, and hence either TCP or MPTCP.

Adaptive bitrate (ABR) streaming uses either a source video format that is encoded at multiple bitrates or performs transcoding from an original rate to the desired one on the fly. ABR works by detecting the delivery conditions (e.g., network bandwidth at the end device and/or CPU capacity, energy level in real time) and by adjusting the quality of the transmitted video stream accordingly.

HTTP-based adaptive bitrate streaming is client-driven and the adaptation logic resides at the client slide, which can reduce the requirement of persistent connections between client and server. This architecture also increases scalability by removing the session maintenance from the server side and was seamlessly adopted by the existing HTTP delivery infrastructure (e.g., HTTP caches and servers).

8.8.2.1 HTTP-based adaptive streaming standards

The first international standard on the adaptive bit-rate HTTP-based streaming solution is the Dynamic Adaptive Streaming over HTTP (DASH) [48] referred to as MPEG-DASH, which started in 2010, had a draft in January 2011 and was released as a final Standard in November 2011. DASH is based on adaptive HTTP streaming [49] in 3GPP Release nine and on HTTP Adaptive Streaming (HAS) [50] in Open IPTV Forum Release 2.

MPEG-DASH is an adaptive bitrate streaming technique that enables high-quality streaming of media content over the Internet. The video content is partitioned into one or more segments and delivered from conventional HTTP web

---

‡‡Real-time Transport Protocol: https://www.ietf.org/rfc/rfc3550.txt.
§§Real-time Streaming Protocol: https://www.ietf.org/rfc/rfc2326.txt.
¶Hypertext Transport Protocol: http://www.ietf.org/rfc/rfc2616.txt.
***RTP Control Protocol: http://tools.ietf.org/html/rfc4961.

servers to the client using HTTP. DASH consists of two main components [51]: Media presentation and Media presentation description (MPD). Media Presentation is a sequence of one or more segments that incorporate periods, adaptation sets, and representations, which break up the video from start to finish. MPD is like a manifest file and is an eXtensible Markup Language (XML) document that identifies the various content components and the location of all alternative segments, providing the relationship between them.

In addition to the standards, some other HTTP-based adaptive streaming solutions are adopted by the key industry players (e.g., Microsoft, Apple, and Adobe).

Microsoft Smooth Streaming (MSS)[†††], referred to as Smooth Streaming, was introduced from a patent "Seamless Switching of Scalable Video Bitstreams" [52] from Microsoft, and is an Internet Information Services Media Services extension. MSS switches between streams of different quality levels according to the network's available bandwidth.

Apple HTTP Live Streaming (HLV)[‡‡‡] is a client-side adaptive HTTP streaming solution as part of QuickTime X and iOS, which supports both live and video on-demand content. HLV uses its own segmenter to divide the stream/video content into small MPEG2-TS files as video chunks with different duration and bitrate.

Adobe HTTP dynamic streaming (HDS)[§§§] enables on-demand and live adaptive bitrate video delivery of standard-based MP4 media over regular HTTP connections. HDS is deployed on the Adobe Flash media delivery platform [53], which means it is available on any device running a browser with an Adobe Flash plug-in.

Compared to MSS, HLS, and HDS, MEPG-DASH not only supports all the features from the other three solutions but also has some special features such as HTML5 support, the definition of quality metrics, multiple video views, and so on. 3GPP Release 10 has adopted MEPG-DASH for use over wireless networks [54].

8.8.2.2   Streaming service over WMNs

Video streaming applications as one of the most experienced rapidly developed service have certain requirements such as sufficient bandwidth and low end-to-end delay to provide QoS performance. The mass-market adoption of high-end mobile devices and an increasing amount of video traffic has led the mobile operators to adopt various solutions to help them cope with the explosion of mobile broadband data traffic while ensuring high QoS levels to their services. WMNs offer a low capital expenditure solution for such services by providing a multi-hop relay backbone composed of low-cost mesh gateways or routers with fixed power supplies [55].

With the improvement of video quality from Standard Definition (SD) to High Definition (HD) and Ultrahigh Definition (UHD), the video streaming applications require higher bandwidth, which is difficult to find a path to support that requirement in the single-radio single-channel WMNs. Fortunately, IEEE 802.11 standards

---

[†††]Smooth Streaming: https://www.iis.net/downloads/microsoft/smooth-streaming.
[‡‡‡]HTTP Live Streaming: https://tools.ietf.org/html/draft-pantos-http-live-streaming-19.
[§§§]HTTP Dynamic Streaming: http://www.adobe.com/products/hds-dynamic-streaming.html.

provide multiple non-overlapping frequency channels, which allow wireless nodes equipped with multiple network interface cards to transmit packets over several non-overlapping channels simultaneously, such bandwidth aggregation ability of multi-radio multi-channel (MRMC) WMNs can significantly increase the network capacity [56].

8.9   Research-related works

Major strategies adopted on multimedia content delivery via WMNs focus on efficient wireless resource utilization to create more transmission opportunities, including the use of channel diversity to provide non-interfering transmission capacity, the scheduling of transmission rates or video multicasts to make the most of wireless connection capacity, the collaboration of WMNs peer nodes to provide alternative paths to avoid bottlenecks, and the exploration of cognitive radio networks to seek extra wireless bandwidth [55].

Lee *et al.* [57] proposed a vEB-tree (van Emde Boas tree)-based architecture for interactive VoD services in peer-to-peer networks, including dividing video stream data into multiple segments distributed among participating peers and searching segment based on a pre-defined distribution scheme that reduces network server stress. The proposed architecture introduced performance improvement in terms of lower jump latency, server load, and searching topology maintenance cost.

A novel virtual content-aware network layer architecture has been introduced in Borcoci *et al.* [58], with the main architecture consisting of service providers, network operators, and end-users for multimedia distribution with QoS assurance. The main concept of the virtual network layer architecture and its responsibility in collaboration with all the included interfaces among overall system layers were discussed, as part of the European FP7 ICT research project, ALICANTE [59].

The energy consumption features of pervasive video CDN have been investigated in Boskovic *et al.* [60], with mathematical analysis on the energy efficiency of video content placement among access points of mesh networks as video servers. Mixed integer linear programming model has been used to derive optimal placement/caching strategies that achieve low level of energy consumption of the mesh routers while maintaining the video delivery quality evaluated by storage capacity and latency.

Performance evaluation has been done in Lertsinsrubtavee and Mekbungwan [61] for community WMN with limited-quality links against CDN using a push-based content delivery method where multimedia contents were pushed closest to mesh routers with high-priority cache storage, in order to ensure delivery quality for HD video content to communities within mesh networks that consist of a small number of users. The evaluation was done in a laboratory environment where the content distribution and request patterns were based on the log files collected from a real-life community mesh network environment in a rural area.

An energy-aware content distribution scheme has been introduced in Mandal *et al.* [62] as a hybrid CDN-P2P network system, which raised and evaluated the possibility of exploiting the P2P system to reduce CDN load while maintaining and decreasing the energy consumption of network nodes. Two network topologies with and without the energy-aware content distribution selection mechanism were deployed with the same network service providers on simulations, and the result proved that the energy-aware scheme achieved considerable level of system-wide energy efficiency improvement while significantly reducing the server load on traditional CDN.

CMPVoD [63] proposed a cluster mesh-based scheme for hybrid video-on-demand (VoD) streaming with VCR functions, addressing the issues related to overhead and sub-optimal neighbor selection by organizing peers into clusters based on the associated video playback features, and ensuring peers with higher bandwidth in the cluster were assigned with closer locations to the video servers. An adaptive buffer map exchange was also introduced to reduce the impact from bandwidth overhead. The features in CMPVoD efficiently reduced video delivery overhead and significantly improved delivery QoS under various network conditions.

A two-step server selection algorithm for VoD streaming in P2P-based CDN was proposed in [64] to address the combined increased delay and bandwidth consumption by neighbor selection of both P2P and CDN. The algorithm assigns the closest video servers within the network based on the number of hops to the network topologies central node, followed by video servers hosting groups of peers closest to the incoming video client in average distance, so that the selected hybrid CDN-P2P overlay is both closest located and consists of nearest located neighbors. The performance of the algorithm has been evaluated with dynamic simulation in terms of startup delay, discontinuity, distortion, server direct connections, server data bandwidth consumption and peer data bandwidth contribution.

PPISM [65] was introduced as a hybrid framework to solve the control overhead, stability and scalability problems for interactive multimedia streaming over P2P-based CDN. It combines traditional CDN and P2P structures into a tree-mesh hierarchical structure, with an associated mechanism of node selection and tree construction. Multimedia stream data was pushed to the tree with push-pull strategies in parallel by edge servers and node clients requesting data. Simulation results showed that the proposed framework significantly decreased the end-to-end delay, data distortion and control overhead, in comparison with the original P2P and traditional CDN-P2P architectures.

The research work in [66] presented a cloud-based CDN with device-to-device and unmanned aerial vehicle (UAV)-enabled caching. Content delivery was achieved within the network that shared the content from servers to clients and between clients, whose payoff depends on a combination of expected delivery cost and delay for the content subscriber clients. Cooperative game strategies have been introduced to form coalition to content exchange between clients to reduce overall delivery cost and delay. Lyapunov optimization has also been used for active allocation of contents from content servers to UAVs. Analytical and numerical evaluation proved that the clients in the proposed network achieved balanced coalitional state on the condition of rational servers and clients.

The FastTrack algorithm [67] presented a performance improvement scheme to solve the formulated optimization challenges existing in the network model based on centralized server and multiple CDN sites, including limited caching spaces at the CDN sites, allocation of CDN for video requests, choice of different ports from the CDN and the central storage and bandwidth allocation. The performance of the algorithm was evaluated in terms of the stall duration tail probability (SDTP) metric, whose theoretical bounds were analysed via simulation results in comparison with the existing baseline strategies used by CDN service providers.

An extension approach based on classic OLSR with content-awareness was presented in [68], which was integrated into a Fog-based CDN architecture together with a popularity-based caching strategy to maintain overall delivery performance. The performance of the approach has been compared with the standard Fog-based CDN framework with the default caching strategy in OLSR, whose result proved that the proposed approach archived higher data delivery ratio, higher caching hit rate, and lower delay.

8.10   Industrial solutions and products

Lumen has provided a CDN mesh delivery solution [69] for business-level scalable and reliable video delivery, including high-quality live streaming and VoD broadcasting.

The solution consists of the following components:

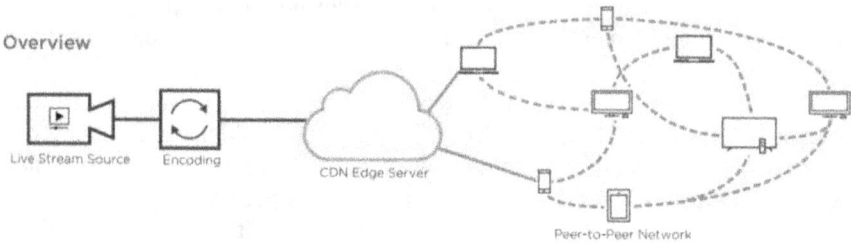

Features provided by the solution include:

- Improve QoS with higher bit rates and less rebuffering
- Optimize delivery by leveraging variables such as user location and ISP
- Easy plug-and-play integration with a broad range of HTML5 and mobile players
- CDN and DRM agnostic for easy integration into existing workflows
- A feature-rich dashboard provides teams with a comprehensive view of the platform
- Effective resource management that accounts for device limitations
- Works with monetization and ad insertion workflow for uninterrupted personalized ads

Linear Technologies (now part of Analog Devices) has proposed an industrial IoT application solution named SmartMesh [70] as part of its Dust Networks product. It is a wireless sensor network with chips and PCB modules managed by mesh networking software that enables sensors to communicate with each other in complex industrial IoT environments.

The SmartMesh network architecture consists of:

- a highly scalable self-forming multi-hop mesh network, whose nodes (referred to as motes) collect and relay data
- a network manager that monitors and manages network performance and security, and exchanges data with a host application

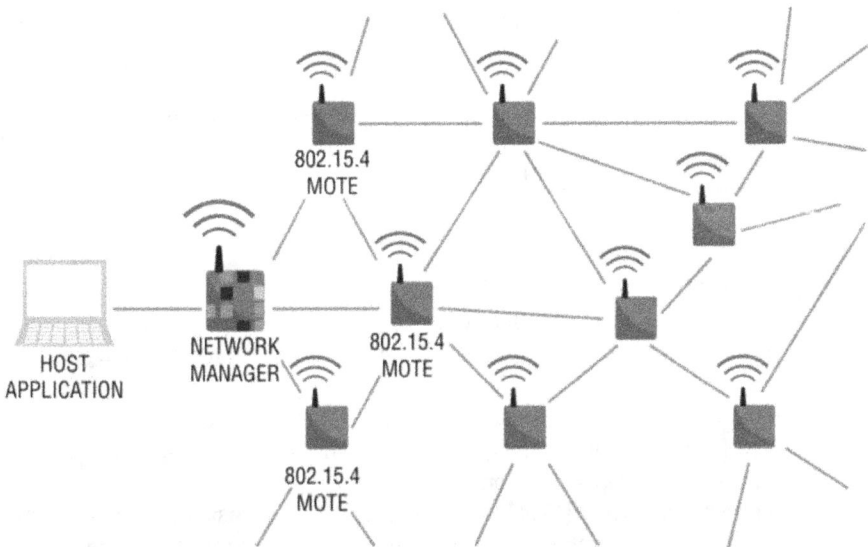

Each mote node in the network maintains redundant path for failover/retry during content delivery caused by interference, to assure transmission quality and reduce resource-consuming path recovery. Each node tracks and exchanges delivery statistics and reports to the manager application to proactively operate load balancing between delivery paths to optimize network performance.

SmartMesh also collaborates with a time-synchronized, channel-hopping link layer framework based on mesh network standards such as WirelessHART and IEEE 802.15.4e, to provide a complete wireless networking solution with data reliability, device lifespan, and data encryption/authentication.

8.11 Challenging multimedia content

Currently, the most commonly used video codecs in multimedia content delivery are H.264/AVC‡‡‡ and H.265/HEVC‡‡‡‡. The most commonly used audio codec is AAC††††. From SD to HD and UHD, the MEPG-DASH works well to adapt and stream video content on WMNs. However, for some challenging multimedia content, there is still need for more investigations. This content includes 3D video, Virtual Reality (VR), and mulsemedia content and discussed next.

8.11.1 3D video

Currently, there are two categories of 3D video technologies: stereoscopic 3D, as the first-generation 3D video technology and multi-view 3D video as the second generation. Table 8.5 shows the comparison of the two technologies from different points of view.

8.11.2 VR, AR, 360-degree videos, and mulsemedia content

The 360-degree video is not VR. These are videos in which a view in every direction is recorded at the same time using either a special rig of multiple cameras (e.g., omnidirectional camera) or a dedicated VR camera. The viewer can control the viewing direction during playback.

Virtual reality (VR) is a realistic and immersive simulation of a three-dimensional environment. VR video used 360-degree video [71] and relies on the mechanism that our brain achieves stereo vision i.e., by fusing two images from our eyes, in which nearby objects have greater disparity than far away objects. The recommendation about the minimum resolution for VR video from YouTube is 5120 × 5120 [72], which is much higher than the 4K requirements. The VR video needs other features to support the omnidirectional stereo vision.

By contrast, Augmented Reality (AR) essentially inserts virtual objects into the real-world view, which the virtual object elements are augmented by computer-generated sensory input such as sound, video, graphics data.

Mulsemedia [73] or multi-sensorial media consists of other media types from human senses (i.e., haptic, olfaction, taste) in addition to audio and visual content. Unlike traditional multimedia, mulsemedia aims to provide immersive communications and enhances user QoE. Mulsemedia services may include any combination of traditional media objects such as text, graphical images, and video, as well as non-traditional media such as olfactory, haptic, and skin-sensorial data. As mulsemedia is essentially about using these multiple media objects to communicate information to users, achieving synchronization between the component media objects that make up the mulsemedia is essential to the success of these systems.

---

‡‡‡H.264: Advanced video coding: https://www.itu.int/rec/T-REC-H.264.
‡‡‡‡H.265: High-efficiency video coding: https://www.itu.int/rec/T-REC-H.265.
††††Advanced Audio Coding (AAC): http://www.iso.org/iso/iso_catalogue/catalogue_ics/catalogue_detail_ics.htm?csnumber=43345.

Table 8.5 Comparison of two 3D video technologies [5]

	Stereoscopic 3D	Multi-view 3D
Idea	(1) Creates or enhances the illusion of depth in an image and presents two offset images separately to the left and right eyes. The two images are perceived by humans as 3D depth enhanced. (2) Uses different input layouts: side by side, top/down, alternating rows, etc.	(1) Simultaneously encodes sequences captured by multiple cameras using a single video stream. (2) Uses as input layout: multiple view streams
Strength	(1) Compatible with conventional 2D video. (2) Saves bandwidth and storage in comparison to Multi-view 3D. (3) Good for broadcasting	(1) Experiences natural depth perception. (2) No glasses. (3) Multiple angles
Weakness	(1) Resolution of individual view is lower compared to 2D. (2) Glasses needed in most cases. (3) Lenticular sheet technology can avoid using glasses but currently provides narrow spots. (4) Fixed viewing angle, no free-view capability	It is challenging for broadcasting due to limited bitrate channel
Adaptation	Bitrate scaling, e.g., (1) Assign lower bitrate for chrominance than for luminance component; (2) Reduce bitrate by discarding enhancement layers for either/both left and right eye(s)	View scaling, e.g., (1) Discard certain views which might be outside of the user's field of view. (2) Depth-based rendering is always adopted to enhance the experience with low added bitrate
Codec	MPEG4/H.264 AVC for 2D+MPEG4/H.264 for depth; MPEG4/H.264 AVC for 2D+MVC for depth as enhancement; Multi-view Video Coding (MVC)	Multi-view Video Coding (MVC)
Delivery	MPEG-2 transport stream, e.g., Blue-ray disc IETF RTP, e.g., real-time transport via IP ISO base media file format, e.g., progressive download in video-on-demand, HTTP streaming	

References

[1]   Zhang Y., Luo J., Hu H. (eds.). *Wireless Mesh Networking: Architectures, Protocols and Standards*. CRC Press; 2006.

[2]   ITU-T. *Quality of Telecommunication Services: Concepts, Models, Objectives and Dependability Planning - Terms and Definitions Related to the Quality of Telecommunication Services*. Technical Report; 2008.

[3]   Aurrecoechea C., Campbell A.T., Hauw L. 'A survey of QoS architectures'. *Multimedia Systems*. 1998;6(3):138–51.

[4]   ITU Y.1541. *Network performance objectives for IP-based services [online]*. 2006. Available from http://www.itu.int/rec/T-REC-Y.1541/en [Accessed 22 Feb 2006].

[5]   Bi T., Yuan Z., Muntean G.M. 'Network reputation-based stereoscopic 3D video delivery in heterogeneous networks'. 2014 IEEE International Symposium on Broadband Multimedia Systems and Broadcasting; Beijing, China; 2014. pp. 1–7.

[6]   BEREC. *A framework for Quality of Service in the scope of net neutrality [online]*. 2011. Available from http://berec.europa.eu/eng/document_register/subject_matter/berec/download/0/117-a-framework-for-quality-of-service-in-th_0.pdf [Accessed 13 Nov 2016].

[7]   Carnec M., Le Callet P., Barba D. 'Full reference and reduced reference metrics for image quality assessment'. Seventh International Symposium on Signal Processing and Its Applications; Paris, France; 2003. pp. 477–80.

[8]   ITU-T RECOMMENDATION, P.910. 'Subjective video quality assessment methods for multimedia applications'. *Series P: Telephone Transmission Quality, Telephone Installations, Local Line Networks*. 1999.

[9]   Lee S.B., Muntean G.M., Smeaton A.F. 'Performance-aware replication of distributed pre-recorded IPTV content'. *IEEE Transactions on Broadcasting*. 2009;55(2):516–26.

[10] Ke C.H., Shieh C.K., Hwang W.S., Ziviani A. 'An evaluation framework for more realistic simulations of mPEG video transmission'. *Journal of Information Science and Engineering*. 2008;24(2):425–40.

[11] Wang Z., Lu L., Bovik A.C. 'Video quality assessment using structural distortion measurement'. International Conference on Image Processing; Rochester, NY, USA; 2002. pp. 65–8.

[12] Pinson M.H., Wolf S. 'A new standardized method for objectively measuring video quality'. *IEEE Transactions on Broadcasting*. 2004;50(3):312–22.

[13] McClaning K., Vito T. *Radio Receiver Design*. Noble Publishing; 2000.

[14] Li G., Xu Z., Xiong C., *et al.* 'Energy-efficient wireless communications: tutorial, survey, and open issues'. *IEEE Wireless Communications*. 2011;18(6):28–35.

[15] Rice A., Hay S. 'Measuring mobile phone energy consumption for 802.11 wireless networking'. *Pervasive and Mobile Computing*. 2010;6(6):593–606.

[16] Floyd S., Handley M., Kohler E. Datagram congestion control protocol (DCCP); 2006. Available from https://datatracker.ietf.org/doc/html/rfc4340 [Accessed Mar 2006].

[17] Floyd S., Kohler E. Profile for datagram congestion control protocol (DCCP) congestion control ID 2: TCP-like congestion control; 2006. Available from https://datatracker.ietf.org/doc/html/rfc4341 [Accessed Mar 2006].

[18] Floyd S., Kohler E., Padhye J. Profile for datagram congestion control protocol (DCCP) congestion control ID 3: TCP-friendly rate control (TFRC). IETF RFC 4342; 2006. Available from https://datatracker.ietf.org/doc/html/rfc4342 [Accessed Mar 2006].

[19] Ford A., Raiciu C., Handley M., Barre S., Iyengar J. Architectural guidelines for multipath TCP development (No. RFC 6182); 2011. Available from https://datatracker.ietf.org/doc/html/rfc6182 [Accessed Mar 2011].

[20] Ford A., Raiciu C., Handley M., Bonaventure O. TCP extensions for multipath operation with multiple addresses (No. RFC 6824); 2013. Available from https://datatracker.ietf.org/doc/html/rfc6824 [Accessed Jan 2013].

[21] Bonaventure O., Handley M., Raiciu C. 'An overview of multipath TCP'. *USENIX login.* 2012;37(5):17–23.

[22] Dugue G., Diop C., Chassot C., Exposito E. 'Towards autonomic multipath transport for infotainment-like systems'. 2012 IEEE International Conference on Communications (ICC); 2012. pp. 6453–7.

[23] Benveniste M. 'QoS for wireless mesh: MAC layer enhancements'. *International Journal on Advances in Internet Technology.* 2009;2(1).

[24] ANSI/IEEE. *Part 11: wireless LAN medium access control (MAC) and physical layer (PHY) Specifications.* 1999 edition (R2003). STD 802.11; 2003.

[25] 'IEEE 802.11e-2005, IEEE standard for information technology — telecommunications and information exchange between systems — local and metropolitan area networks — specific requirements part 11: wireless LAN medium access control (MAC) and physical layer (PHY) specifications Amendment 8: medium access control (MAC) quality of service enhancements'. *IEEE Std 802.11e-2005 (Amendment to IEEE Std 802.11, 1999 Edition.* 2005:1–212.

[26] Raman B., Chebrolu K. 'Design and Evaluation of a New MAC Protocol for Long-Distance 802.11 Mesh Networks'. MOBICOM; Cologne, Germany; Aug–Sept 2005. pp. 156–69.

[27] Haas Z.J., Jing Deng, Deng J. 'Dual busy tone multiple access (DBTMA)-a multiple access control scheme for ad hoc networks'. *IEEE Transactions on Communications.* 2002;50(6):975–85.

[28] Hasan M.I., Rana M.M., Ahmed K.E.U., Karim R., Tareque H.M. 'An Adaptive and Efficient MAC Protocol for Wireless Mesh Networks: An Extension of Maca-P'. IEEE International Conference on Computer Engineering and Technology (ICCET); Singapore; 2009.

[29] Acharya A., Misra A., Bansal S. 'Design and analysis of a cooperative medium access scheme for wireless mesh networks'. *BROADNETs.* 2004;4:621–31.

[30] Shankar N. S., van der Schaar M, Schaar Mvander. 'Performance analysis of video transmission over IEEE 802.11a/e WLANs'. *IEEE Transactions on Vehicular Technology*. 2007;56(4):2346–62.

[31] Jiao H., Li F.Y. 'A Mini-Slot-based Cooperative MAC Protocol for Wireless Mesh Networks'. IEEE Globecom Workshops; Miami, USA; 2010. pp. 89–93.

[32] Santamaria R., Bourdeau O., Anjali T. 'MAC-ASA: A New MAC Protocol for WMNs'. ICCCN; Hawaii, USA; 2007. pp. 973–8.

[33] Huang F., Yang Y. 'Energy efficient collision avoidance MAC protocol in wireless mesh access networks'. IWCMC; Hawaii, USA; Aug 2007. pp. 272–7.

[34] Hassanzadeh A., Stoleru R., Shihada B. 'Energy efficient monitoring for intrusion detection in battery-powered wireless mesh networks'. Proceedings of the 10th International Conference on Ad-hoc Mobile, and Wireless Networks; 2011. pp. 44–57.

[35] McQuillan J., Richer I., Rosen E. 'The new routing algorithm for the ARPANET'. *IEEE Transactions on Communications*. 1980;28(5):711–9.

[36] Korf R.E. 'Linear-space best-first search: summary of results'. Proceedings of the Tenth National Conference on Artificial Intelligence; 1992.

[37] Clausen T., Jacquet P. 'Optimized link state routing protocol'. Hipercom, RFC-3626; 2003.

[38] Perkins C.E., Royer E.B., Das S.R. 'Ad hoc on-demand distance vector routing'. IETF RFC 3561; 2003.

[39] Sampaio S., Souto P., Vasques F. 'DCRP: a scalable path selection and forwarding scheme for IEEE 802.11s wireless mesh networks'. *EURASIP Journal on Wireless Communications and Networking*. 2015;2015(1):211.

[40] Schrijver A. 'On the history of combinatorial optimization (till 1960)'. *Handbooks in Operations Research and Management Science*. 12. Elsevier; 1960. pp. 1–68.

[41] Kubale M. *Graph Colorings*. American Mathematical Society; 2004.

[42] Mathews G.B. 'On the partition of numbers'. Proceedings of the London Mathematical Society; 1896. pp. 486–90.

[43] Kirkpatrick S., Gelatt C.D., Vecchi M.P. 'Optimization by simulated annealing'. *Science*. 1983;220(4598):671–80.

[44] Norris J.R. *Markov Chains*. Cambridge University Press; 1998.

[45] Rozner E., Seshadri J., Mehta Y., Qiu L. 'Simple opportunistic routing protocol for wireless mesh networks'. 2nd IEEE Workshop on Wireless Mesh Networks; Reston, VA, USA; 2006. pp. 1024–30.

[46] Krishnaswamy D., Shiang H., Vicente J., *et al.* 'A cross-layer cross-overlay architecture for proactive adaptive processing in mesh networks'. IEEE Workshop on Wireless Mesh Networks; 2006. pp. 1050–8.

[47] Liu C., Shu Y., Zhang L. 'Backup routing for multimedia transmissions over mesh networks'. IEEE International Conference on Communications (ICC); Glasgow, Scotland; 2007. pp. 3829–34.

[48] *ISO/IEC 23009-1:2014 Information technology - Dynamic adaptive streaming over HTTP (DASH) - Part 1: Media presentation description and segment formats*

[online]. 2014. Available from http://www.iso.org/iso/home/store/catalogue_ics/catalogue_detail_ics.htm?csnumber=65274 [Accessed 13 Oct 2014].

[49]　3GPP. *Transparent End-to-End Packet-Switched Streaming Service (PSS) (release 12)*. Technical Report TS-26.234; 2014.

[50]　Open IPTV Forum. *OIPF Release 2 Specification Volume 2a - http Adaptive Streaming [v2.3]*. Technical Report; 2014.

[51]　Stockhammer T. 'Dynamic adaptive streaming over HTTP-standards and design principles'. ACM Multimedia Systems; San Jose, CA, USA; 2011. pp. 133–44.

[52]　Wu F., Li S., Sun X., Zhang Y.Q., Microsoft Corporation. Seamless Switching of Scalable Video Bitstreams. U.S. Patent 6,996,173; 2006.

[53]　Hassoun D. *Dynamic streaming in flash media server 3.5 part 1: overview of the new capabilities [online]*. 2009. Available from http://www.adobe.com/devnet/adobe-media-server/articles/dynstream advanced pt1.html [Accessed 13 Dec 2016].

[54]　Varsa V., Curcio I. Transparent end-to-end packet switched streaming service (PSS); RTP usage model release 5. 3GPP TR26, 937, p.V1; 2003.

[55]　Tu W., Sreenan C.J., Jha S., Zhang Q. 'Multi-source video multicast in internet-connected wireless mesh networks'. *IEEE transactions on Mobile Computing*. 2017;16(12):3431–44.

[56]　Pan C., Liu B., Zhou H., Gui L. 'Multi-path routing for video streaming in multi-radio multi-channel wireless mesh networks'. 2016 IEEE International Conference on Communications (ICC); 2016. pp. 1–6.

[57]　Lee C.-N., Kao Y.-C., Tsai M.-T. 'A vEB-tree-based architecture for interactive video on demand services in peer-to-peer networks'. *Journal of Network and Computer Applications*. 2010;33(4):353–62.

[58]　Borcoci E., Negru D., Timmerer C. 'A novel architecture for multimedia distribution based on content-aware networking'. 3rd International Conference on Communication Theory, Reliability, and Quality of Service; 2010. pp. 162–8.

[59]　ALICANTE. *MediA Ecosystem Deployment Through Ubiquitous Content-Aware Network Environments [online]*. 2010. Available from http://www.ict-alicante.eu/ [Accessed Mar 2011].

[60]　Boskovic D., Vakil F., Dautovic S., Tosic M. 'Greening of video streaming to mobile devices by pervasive wireless CDN'. *Journal of Green Engineering*. 2011;2:1–27.

[61]　Lertsinsrubtavee A., Mekbungwan, N P. 'Weshsuwannarugs: comparing NDN and CDN performance for content distribution service in community wireless mesh network'. AINTEC'14 Proceedings of the AINTEC 2014 on Asian Internet Engineering Conference; 2014. pp. 43–50.

[62]　Mandal U., Habib M.F., Zhang S., Lange C., Gladisch A., Mukherjee B. 'Adopting hybrid CDN–P2P in IP-over-WDM networks: an energy-efficiency perspective'. *Journal of Optical Communications and Networking*. 2014;6(3):303–14.

[63] Ghaffari Sheshjavani A., Akbari B., Ghaeini H.R. 'CMPVoD: a cluster mesh-based architecture for VoD streaming over hybrid CDN-P2P networks'. 8th International Symposium on Telecommunications (IST); 2016. pp. 783–8.

[64] Saengarunwong A., Sanguankotchakorn T. 'A two-step server selection in hybrid CDN-P2P mesh-based for video-on-demand streaming'. 2018 International Conference on Information and Communication Technology Convergence (ICTC); 2018. pp. 499–504.

[65] Yang H., Liu M., Li B., Dong Z. 'A P2P network framework for interactive streaming media'. 2019 11th International Conference on Intelligent Human-Machine Systems and Cybernetics (IHMSC); 2019. pp. 288–92.

[66] Asheralieva A., Niyato D. 'Game theory and Lyapunov optimization for cloud-based content delivery networks with Device-to-Device and UAV-Enabled Caching'. *IEEE Transactions on Vehicular Technology.* 2019;68(10):10094–110.

[67] Al-Abbasi A., Aggarwal V., Lan T., Xiang Y., Ra M.-R., Chen Y.-F. 'FastTrack: minimizing stalls for CDN-based over-the-top video streaming systems'. *IEEE Transactions on Cloud Computing.* 2019;2019:1–5.

[68] Alghamdi F., Mahfoudh S., Barnawi A. 'A novel FOG computing based architecture to improve the performance in content delivery networks'. *Wireless Communications and Mobile Computing.* 2019;2019(5):1–13.

[69] *Luman CDN Mesh Delivery Data Sheet [online].* 2020. Available from https://assets.lumen.com/is/content/Lumen/cdn-mesh-delivery?Creativeid=28ef947a-fc9d-41ef-a57e-211cd47117eb [Accessed Aug 2020].

[70] *SmartMesh: Wireless Mesh for Tough Industrial IoT Applications [online].* 2017. Available from https://www.analog.com/media/en/technical-documentation/product-selector-card/smartmesh_rev_d.pdf [Accessed 17 Mar 2017].

[71] Google. *Rendering Omni-directional Stereo Content [online].* 2016. Available from https://developers.google.com/vr/jump/rendering-ods-content.pdf [Accessed 30 Nov 2016].

[72] YouTube. *Upload virtual reality videos [online].* 2016. Available from https://support.google.com/youtube/answer/6316263?hl=en&ref_topic=2888648 [Accessed 30 Nov 2016].

[73] Yuan Z., Bi T., Muntean G.-M., Ghinea G. 'Perceived synchronization of Mulsemedia services'. *IEEE Transactions on Multimedia.* 2015;17(7):957–66.

Chapter 9

Toward intelligent extraction of relevant information by adaptive fuzzy agents in big data and multi-sensor environments

Zakarya Elaggoune[1], Ramdane Maamri[1], Allel Hadjali[2], and Imane Boussebough[1]

The speed at which data is flooded from big multi-sensor networks and the noise generated with such a large volume of data pose various new challenges in the field of big data and, especially, those relating to data quality.

The aim of this research is to combine two different artificial intelligence techniques, represented by multi-agent technologies and fuzzy logic, and to provide a generic approach for noise elimination and relevant data extraction in big data environments.

The specificity of the multi-agent system in this approach is that agents are powered by machine learning algorithms so it can learn to adjust the parameters of the fuzzy logic inference system to produce the best quality of data by eliminating only the noisy data and avoiding the loss of information. Therefore, agents will learn to strike a balance between extracting relevant information and losing information, which means that they will learn how to produce optimal performances in terms of data quality and energy consumption. This will extend the lifetime of the sensor network also.

9.1 Introduction

The current world is one of the data. We have data all around us; this data is huge in volume and is generated exponentially from multiple sources, such as social networks, forums, messaging systems, research articles, online transactions, and so on. Data in its native form also has multiple formats. In addition, this data is no longer static; on the contrary, it evolves rapidly over time. These characteristics belonging

[1]LIRE Labs, University of Constantine 2, Constantine, Algeria
[2]LIAS Labs, ENSMA, Chasseneuil Cedex, France

to the mass of current data are the source of many challenges on the storage and calculation of this data. As a result, conventional data storage and management techniques as well as IT tools and algorithms have become incapable of handling it, and this is where the era of big data began.

Michael Cox and David Ellsworth [1] were among the first to literally use the term big data, referring to the use of very large volumes of scientific data for visualization and decision support. Currently, there are several numbers of big data definitions, and the most well-known definition is that of IBM [2], which suggested that big data could be characterized by one or all of the three "V" words used to investigate situations, events, or other information: "Volume," "Variety," and "Velocity."

Big data is not limited to volume, variety, and velocity, but it is volume, variety, and velocity in constant evolution. Consequently, big data has received considerable attention since distributed and parallel computing has made it possible to process a large volume of data, notably via the Hadoop ecosystem.

Once familiar with the Hadoop ecosystem, the data scientists managed to develop new techniques that started to give good results, but quickly, the flood of data overwhelmed them. Indeed, as the volume of data increased and the sources multiplied, the raw data became increasingly poor and the relevant information increasingly scarce (thirst despite the floods).

The usefulness and reliability of the data and its sources have been increasingly questioned. Hence the appearance of two new challenges bringing the 3V from big data to 5V [3] defines these new "Vs" by "Value," which emphasizes the usefulness of the data or more precisely the quantity of relevant information in the data stream; and "Veracity," which represents the reliability and confidence attributed to the data and its sources. In addition, with the recent increase in the number of smart and portable devices, as well as other measuring instruments in ambient applications, we are just beginning to face all the "Vs" of this new era of data. As a result, the importance of the current challenges is renewed and new complementary challenges in terms of data processing and management appear. As mentioned in [4], we take the next step in the analysis of big data, which is essentially represented by three new challenges:

- Genericity: as we can see in [5], most of the analysis tools depend on the domain treated and require specific expertise for the creation of these tools. To adapt these tools to other fields of application, the designer must therefore reconstruct his/her processing technique. Therefore, the design of a generic analysis tool for big data should constitute a new challenge.
- The identification of relevant information in a noisy data flow: as already mentioned, the increase in the volume of data and sources of information led to the appearance of two new "Vs." This leads to an increase in the noise rate, the impoverishment of the data and a decrease in the relevant information, that is to say thirst despite the deluge.
- Decentralization of the data analysis process: the Hadoop ecosystem and the MapReduce model use a distributed logic but not a decentralized control. The MapReduce architecture essentially comprises two types of nodes; the master nodes playing the role of coordinators and the distributed slave nodes which

execute the same algorithms in parallel mode by following the instructions of the master node. This non-autonomy of the slave nodes causes a static and rigid dimension to the systems, which leads to failures in the management of dynamic data whose content and structure are constantly changing. From there, we consider that the management of data dynamics with intelligent and decentralized processing capacities comprising autonomous computation entities must be recognized as a new crucial challenge.

The aspect of genericity means that users have the possibility of adjusting the different parameters of fuzzy logic to produce the best quality of data according to the field of application; therefore, they must configure the various parameters of the fuzzy inference system such that: fuzzy concepts and inputs, fuzzy rules, aggregation and implication methods, and so on. In addition to this, users must carefully define the optimal relevance rate to produce the relevant information.

When we implemented the approach and after doing some experiments, we found that we deviated from our main goal which is the proposal of a decision-support system. Because the task of finding the optimal relevance rate is too slow and very difficult for the users, and if users cannot find the optimal parameters, the data produced will be of poor quality. Therefore, in this chapter, we intend to power the agents with machine learning techniques so that they can learn to find the optimal relevance rate, and this task will no longer be the task of users.

Big data analysis mainly aims to convert raw data into knowledge to help end users in decision-making tasks. However, the speed at which new noisy data is disseminated from various sources creates difficulties in terms of management and storage of this data. Therefore, transforming big data into smart data by reducing the amount of redundancy and eliminating irrelevant data becomes a necessary task. Therefore, the smart data extraction process improves the quality of the data and will help make better decisions.

The necessary steps involved in extracting relevant information can be summarized in two steps [6]: the first step is to transform data into information by eliminating redundancy, and the second step is the procedure that transforms this information into relevant information which must fulfill three conditions to be valid as knowledge: (1) the information must be agile, (2) the information must be precise, and (3) the information must be actionable. If these steps of extracting relevant information are summarized in an equation, it will probably be:

$$relevant - information = (data - redundancy)$$
$$+(agile, accurate, actionable) \tag{9.1}$$

Current research on noisy big data issues is focused on redundancy reduction methods, such as data compression, redundancy detection, and data filtering.

The noise generated from the collected data can appear clearly in video or image formats, because it contains considerable redundancy [7], including sensing redundancy, temporal redundancy, statistical redundancy, and spatial redundancy. In [8], it was shown that the video compression method based on the contextual redundancy related to background and foreground in a scene can solve the problem of data redundancy in the video surveillance sensor networks. However, data compression and decompression cause an extra computational burden. Therefore, the benefits of data compression and the negative effects, such as the additional cost should be carefully balanced.

A more general method of redundancy reduction is repeated data deletion, which aims to filter out the repeated data copies [9]. This method uses hash algorithms to identify data segments or data blocks before storing them, and if a new data block arrives with an identifier equal to one of the blocks already stored, it will be considered redundant and replaced by the corresponding stored data block. Repeated data deletion can contribute greatly to the reduction of storage requirement, but the problem with this method is that it takes too much time to browse the list of identifiers, especially in large-scale systems that contain large amounts of data.

Usually, after the redundancy reduction process comes the feature extraction process, where high-dimensional feature vectors will be transformed into relevant feature vectors with a smaller dimension. Such operations play an important role in eliminating noisy data.

On the other hand, much research on the optimization of wireless sensor networks (WSNs) has been done, including the maximization of their lifetime. The traditional client/server (C/S) model goes through direct transmission to communicate data from sensor nodes to the base station, resulting in higher power consumption; what is more, the data is sent directly to the base station without doing any cleaning treatment, which causes additional power consumption. Much research has been done to optimize the performance of this model, some works are listed below:

- Incremental data fusion of a maximum number of sensors [10]: This model tries to concentrate the data of the sensors that follow the same itinerary to the base station and send them all together in the same package at the same time. This solution is not scalable, and it is suitable only for networks that do not contain a large number of nodes. Furthermore, the sensor nodes do not have always relevant information to send, and they do not filter out redundant and irrelevant information.

- The ant agent: In [11], the authors present a data aggregation based on ant colony algorithms to find the shortest possible path between the emitter sensor and the base station. However, the construction of such a model depends on the deployment of the nodes, which is generally random. Furthermore, ant colony algorithms do not perform well in large-scale WSN; they take a lot of time to find the shortest path.

- Mobile agent-based directed diffusion [12]: The role of the mobile agent in this kind of model is the collection of sensor readings throughout the network and transmitting these readings to the base station. The only disadvantage of such type of model is the overload of mobile agents, especially in large-scale WSN, where thousands or millions of sensor readings are collected every second. As a result, mobile agents cannot transmit sensor readings in real time.

- There are several works that have proposed a structured strategy like a multi-cast tree [13, 14]. But, because of excessive communication costs and centralized management of the WSN structure, the structured approaches do not adapt well with dynamic scenarios.

However, most of the previous studies do not take into account the large scale of the WSNs and cannot be adapted for big data applications. Moreover, most of the previous studies focus only on the energy consumed and do not take into account the quality of the data transmitted. There are some approaches that deal with data quality [15, 16], but these approaches only eliminate redundancy, and they do not produce smart data.

Another key limitation of these approaches is that they use specific metrics, which means that they are applicable only to specific domains.

In our previous work, we have proposed a generic fuzzy agent approach to handle those three challenges. We have chosen the agent technology for its distributed control and their autonomous nature. Apart from the characteristics of the agents (autonomy and distributed control), their role in our previous approach was limited in the execution of the algorithms of fuzzy logic. On the other hand, the fuzzy logic technique has taken the main role in our previous approach, because its role was to estimate the relevance of the data, and compared to this estimation and to the expertise of users, the data will be kept or eliminated.

9.2   Our previous work

9.2.1   The fuzzy agent approach

In our previous work, we proposed a fuzzy agent approach to eliminate the noise accumulated with big data, allowing the user to reach the relevant data. We have designed a system that meets the limits of research works in the field of data preprocessing and, more specifically, of cleaning up noisy data and managing their quality. The main objective was to define a generic data quality management system based on multi-agent systems and fuzzy logic.

The role of fuzzy agents* can mainly appear in the preprocessing phase, where they will be integrated into the various big data nodes. As shown in Figure 9.1, these agents apply fuzzy logic reasoning to determine the relevance rate of the data disseminated. Thus, if the result of the defuzzification is greater than or equal to the minimum relevance percentage predetermined by the user, the data will be considered relevant; otherwise, the data will be considered irrelevant and will be neglected.

The major novelties of using multi-fuzzy agent systems are multiple, and we can list as follows:

- *Better data quality:* The ability to estimate the relevance rate and filter out the noise through the use of Multi-Agents Systems (MAS) technology that executes fuzzy algorithms;

---

*Fuzzy agents are agents that apply fuzzy logic algorithms.

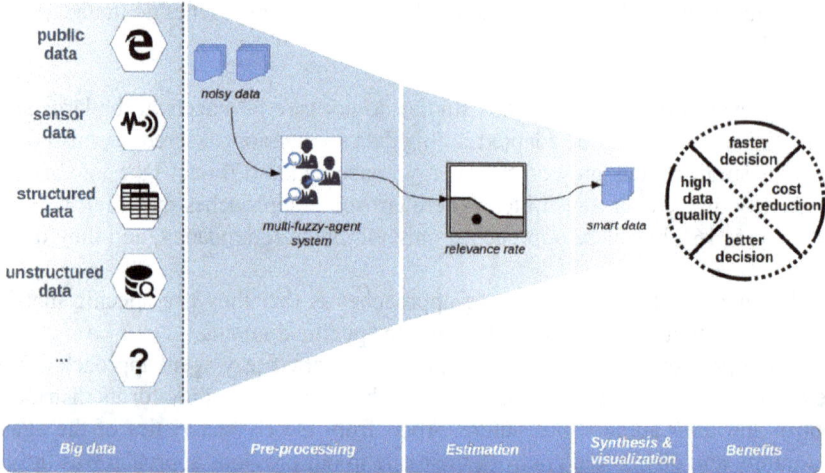

Figure 9.1   An overview of the proposed approach

- *A precise problem-solving methodology:* Fuzzy agents look at the data in imprecise terms and respond with precise actions, which can relieve the problem of incompleteness and uncertainty data;
- *Lower cost:* Fuzzy logic enables low-cost microcontrollers to perform functions traditionally achieved by more powerful expressive machines enabling lower-cost products to execute advanced features;
- *Generacity:* The relevance rate threshold accepted is not constant, which means that this approach does not depend on a specific application, and users can configure the relevance rate threshold accepted taking into consideration the domain of the application.

The proposed multi-fuzzy agent system represents a network of fuzzy agents distributed on the different computing nodes of the big data system, and the network is decomposed into a cluster of neighboring fuzzy agents, with a coordinator agent in the head of each cluster, as shown in Figure 9.2. The clustering technique is implemented to improve coordination between the agents, to transmit the data easily to the end user and, especially, to eliminate the internodes redundancy.

The neighboring agents act as reactive agents in this case, and their principal role is limited in sharing the captured data to the coordinator agent for processing. But as the big data systems are scalable, the number of data nodes and the structure of the fuzzy agent network will change continuously, which means that the fuzzy agent network will be repartitioned periodically, and the reactive agent can become a cognitive coordinator agent. That is to say that the type of agents is not static and the role of agents differs periodically.

Regarding the behavior of the coordinator agent, after having received the data from its neighboring agents, it implements a fuzzy logic inference system to estimate the percentage of the relevance of the data and to filter out the irrelevant ones.

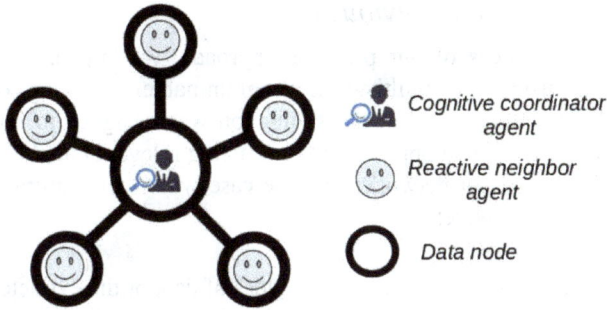

Figure 9.2   Fuzzy agents cluster

Like any fuzzy inference system, the fuzzy logic inference system applied by the coordinator agents following steps: (1) implement fuzzy concepts and fuzzy inputs, which are represented by the parameters that allow the agents to measure the relevance of the data; (2) apply fuzzy operation (AND/OR); (3) apply implication and aggregation methods; (4) deffuzify the aggregate output.

At the end of the execution of fuzzy logic, we come to the last stage, where the fuzzy agent compares the defuzzification output with the user-defined threshold of relevancy to decide whether the information is relevant or not. If the percentage of relevance resulting from the application of this fuzzy logic is equal to or greater than the percentage predefined by the user, then the information will be classified as relevant, otherwise, the information will be classified as irrelevant and it will not be stored (Figure 9.3).

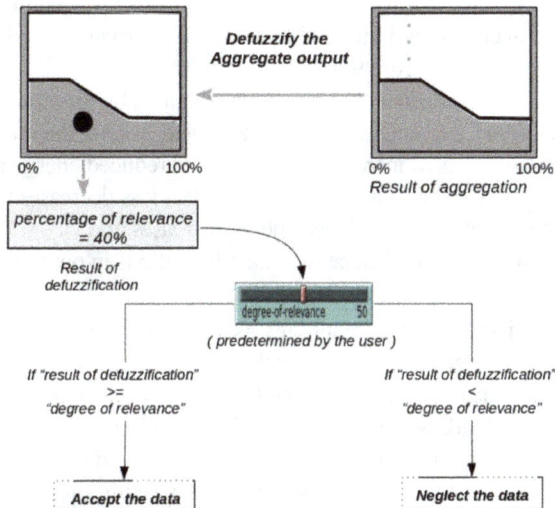

Figure 9.3   Degree of relevance of data

9.2.2   *The limits of the previous work*

To test the performances of our previous approach, we implemented the fuzzy agent approach using the multi-agent programmable modeling environment, *NetLogo* [17]. The model used in the simulation was designed to investigate the efficiency of the fuzzy agent approach for extracting relevant data in decentralized large-scale wireless sensor networks; the use case was "temperature monitoring"; and the main questions were:

- Does the use of this approach cause the loss of data or the extracted smart data will have the same value as that captured with better quality?
- How was the energy efficiency being achieved in networks employing the fuzzy agent approach?
- What parameters gave the minimum percent of the variation between the results obtained with this approach compared to raw data transmission?
- What parameters gave the optimal energy savings compared to raw data transmission?

In this model, we compared the results obtained using the fuzzy agent approach transmission with the raw data transmission. We used the same multi-hop protocol for the fuzzy agent approach transmission and the raw data transmission.

There is a very important factor in this approach, which is "the relevance percentage threshold accepted," this factor is not constant, and only the user has the authority to set the percentage threshold accepted. To verify the effectiveness of our approach, we have chosen to simulate three different scenarios as follows: (1) accept all the data (even the irrelevant data), (2) accept relevant data only, and (3) accept very relevant data only.

The overall measurement results that show the ratio between the relevance percentage threshold accepted and the total average percent variation/the total reduced energy are summarized in Figure 9.4.

On the whole, we can deduce that the results obtained by this approach necessarily depend on the relevance percentage threshold accepted, where, the more the relevance percentage threshold accepted increases, the more the reduced energy and the average percent variation increase, that is to say, the rate of data loss decreases (the average percent variation increase) when the reduced energy increases and vice versa.

From those results, we can deduce two main limitations in our previous work:

- Users are decisive not the machine: as can be seen in the three previous scenarios, users must always monitor the results obtained to correct and reconfigure the relevance threshold accepted to avoid the loss of information or the redundancy overload, which is the role of the machine usually.
- The agents are reactive: apart from their autonomy and their distributed nature, the role of agents in this approach was being limited in the application of fuzzy logic algorithms, we have not really exploited all the capacities of the agents and in particular the capacity of using machine learning techniques.

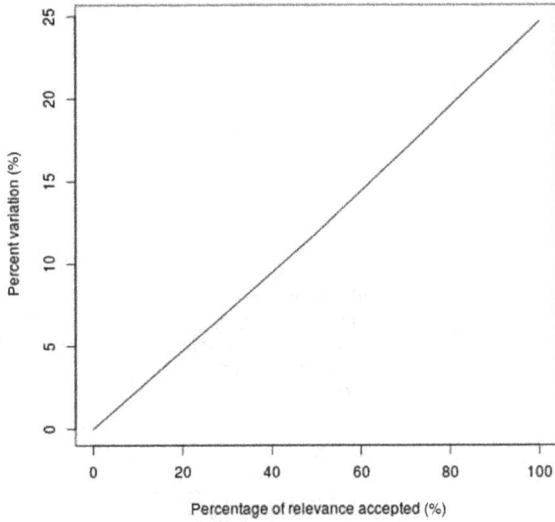

(a) The ratio between the accepted percentage of relevance and the percent variation

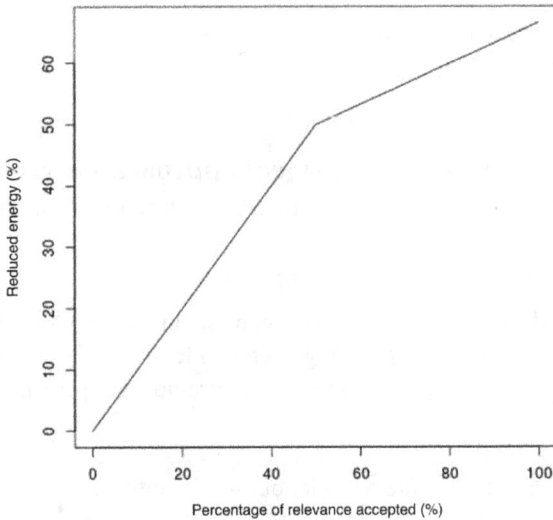

(b) The ratio between the accepted percentage of relevance and the energy consumption reduction

Figure 9.4   *(a) the ratio between relevance percentage and the energy reduced / (b) the ratio between relevance percentage and the percent variation*

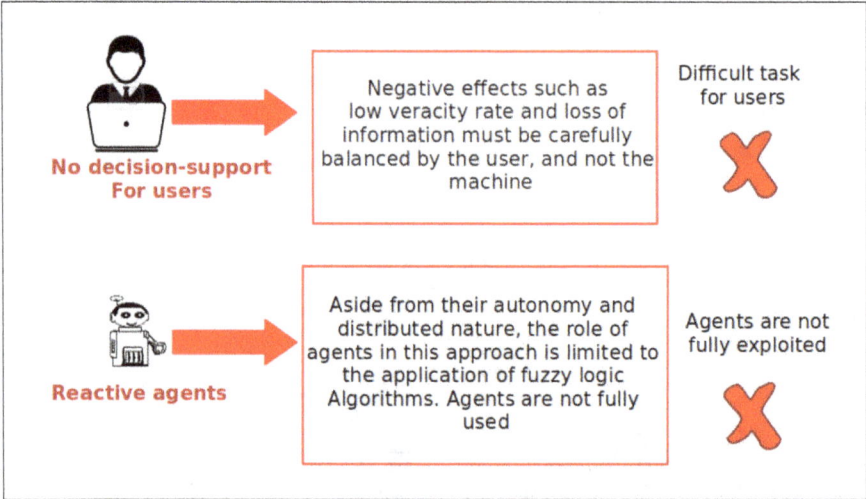

Figure 9.5   The limits of our proposed approach

As shown in Figure 9.5, the proposed multi-agent system offers a structure for extracting relevant information, but it is up to the users to configure the percentage of relevance necessary to produce high data quality without loss of information or lack of veracity. The machine (the multi-agent system) therefore does not fully play its role of decision support.

9.3   Toward a learning fuzzy agent approach for relevant data extraction in big data and multi-sensor environments

9.3.1   An overview of the novel approach

Our proposal to deal with these problems consists in supplying fuzzy agents with a machine learning algorithm, allowing agents to learn to adapt to the application context and the data environment in order to provide the percentage of optimal relevance.[†]

We therefore remain in line with our previous work with the combination of multi-agent systems and fuzzy logic, but the specificity of this new approach lies in the fact that agents use machine learning to learn how to produce the best quality of data to assist in decision-making while avoiding loss of information. Agents will, therefore, learn to find a balance between extracting relevant

---

[†]The optimal relevance percentage produces relevant information without loss of data and with a maximum rate of veracity.

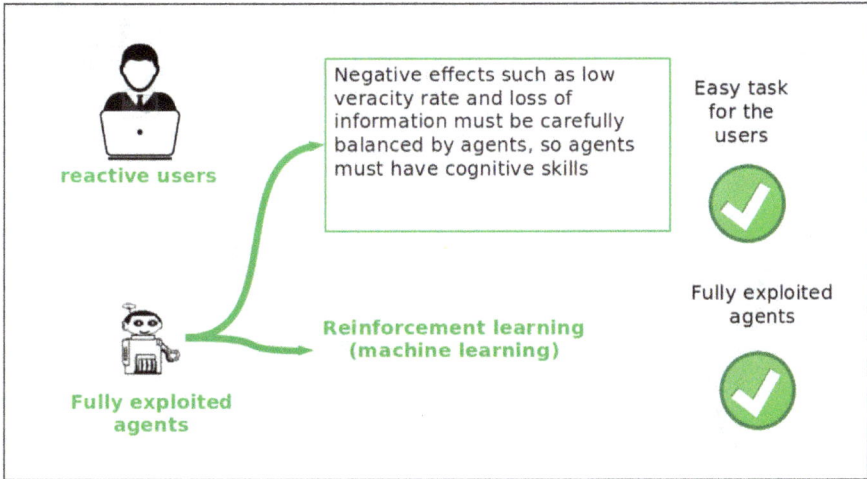

Figure 9.6   The new proposed approach

information and losing information, and this will no longer be the task of the users (Figure 9.6).

The objective is to create a set of fuzzy agents distributed on the different nodes of the big data system whose goal is to reach the optimal relevance percentage. The agents start with random strategies, but the system then uses an evolutionary approach: the agents improve their strategies over time to reach the optimal relevance percentage.

The model initially creates a group of agents and a list of entries, such as the different relevance percentages: 0, 10, 20, etc., and operators, such as +, −, etc. It then defines the strategy of each agent on a random set of combinations of inputs/operators. Each strategy causes the percentage of relevance used by the agent to change. When the model is executed, each agent executes his strategy and reports the best performance produced by this strategy. On the basis of this measure, 75% of the worst-performing agents will be killed and replaced by new agents using other random strategies but will start their strategy with the entries of the 25% of the best agents in the hope to produce better performance. This process will then be repeated several times until the percentage of optimal relevance is found. Therefore, the percentage of optimal relevance will be used by all agents.

As the optimal percentage of relevance is that which produces the highest percentage of veracity and the lowest coefficient of variation between the initial data and the extracted data (minimum loss of information), the agents' performances will be calculated as follows (9.2):

$$Performance = veracity\% - variation\% \tag{9.2}$$

This implies that the best performing agent will obtain the best results in terms of veracity and loss of data.

We can summarize the whole process in the activity diagram (Figure 9.7).

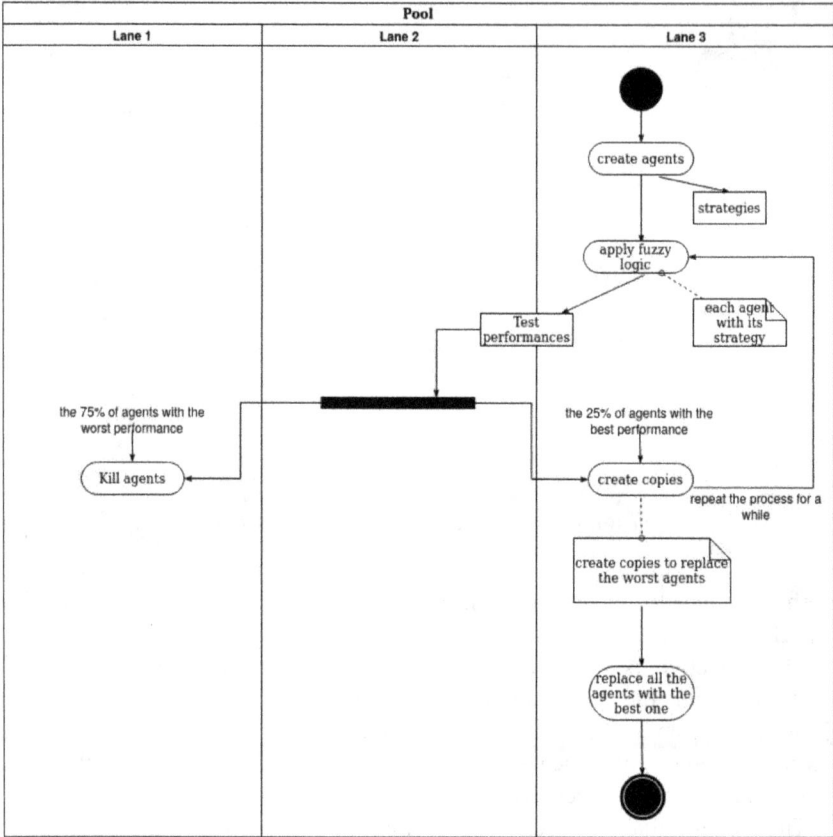

Figure 9.7   Activity diagram

The machine learning Algorithm model is described in Figures 9.8 and 9.9 (NetLogo Model).

9.4   Conclusion

The era of big data has brought new challenges in data processing and management. Existing analytical tools are now close to facing ongoing challenges thus providing satisfactory results at a reasonable cost. However, the velocity at which new data are flooded and the noise generated from such a large volume leads to various new challenges.

In our previous work, we have proposed fuzzy agent approach to extract relevant data from big data, but the only gap was the difficult tasks that users have to perform manually, since the system was only powred by reactive agents and not by intelligent proactive agents.

```
;; agents have a strategy
agents-own [ strategy ]

;; we need to keep track of the goal and what operators and inputs the agents can use
globals [ goal operators inputs ]

to-report performance
  report percent-veracity - percent-information-lost
end

;; create a new strategy of how to set the percent of relevance for each agent
to-report random-strategy
  ;; each strategy starts consists of 5 inputs and 4 operators alternating
  let strat (list one-of inputs)
  repeat 5 [ set strat (sentence strat one-of operators one-of inputs) ]
  report performance
end

to setup
  clear-all
  set operators ["+ " "- "]  ;; set up the usable operators
  set inputs random-seed 100  ;; set up the usable inputs

  create-turtles 1000 [
    set strategy random-strategy
  ]

  ;; set the goal to the max performance
  set goal max performance
  reset-ticks
end
```

Figure 9.8   Algo Part 1

```
to go
  if ticks < 25 [
    ;; kill off 75% of the agents with the worst performances
    ;; (i.e., the ones further from the goal)
    let number-to-replace round (0.75 * count agents)
    ask min-n-of number-to-replace turtles [ performance ] [ die ]
    ;; store agents that survived in an agentset that won't expand when
    ;; we hatch new agents.
    let best-agents agent-set agents
    ;; hatch new mutant turtles
    repeat number-to-replace [
      ask one-of best-agents [

        set strategy random-strategy
        set strat (list one-of agent-set agents)

      ]
    ]
  ]

  set percent-relevance = agent (max-n-of performance) [performance]
]
```

Figure 9.9   Algo Part 2

In the present research, we proposed to feed agents with machine learning so that they become smarter and that they can help users in decision-making tasks.

References

[1]   Cox M., Ellsworth D. 'Managing big data for scientific visualization'. *ACM Siggraph*. 1997;97:21–38.

[2]   Zikopoulos P., Deroos D., Parasuraman K., Deutsch T., Giles J., Corrigan D. *Harness the Power of Big Data the IBM Big Data Platform*. McGraw Hill Professional; 2012.

[3]   Jagadish H.V., Gehrke J., Labrinidis A., *et al.* 'Big data and its technical challenges'. *Communications of the ACM*. 2014;57(7):86–94.

[4]   Fan W., Bifet A. 'Mining big data: current status, and forecast to the future'. *SIGKDD Explorations*. 2012;14(2):1–5.

[5]   Katal A., Wazid M., Goudar R.H. 'Big data: issues, challenges, tools and good practices'. 2013 Sixth International Conference on Contemporary Computing (IC3); 2013. pp. 404–9.

[6]   Garcia-Gil D., Luengo J., Garcia S., Herrera H. 'Enabling smart data: noise filtering in big data classification'. *Computer Science*. 2017. ArXiv:1704.01770.

[7]   Chen M., Mao S., Liu Y. 'Big data: a survey'. *Mobile Networks and Applications*. 2014;19(2):171–209.

[8]   Tsai T.-H., Lin C.-Y. 'Exploring contextual redundancy in improving object-based video coding for video sensor networks surveillance'. *IEEE Transactions on Multimedia*. 2012;14(3-2):669–82.

[9]   Sarawagi S., Bhamidipaty A. 'Interactive deduplication using active learning'. Proceedings of the E ighth ACM SIGKDD International Conference on Knowledge Discovery and Data Mining; ACM; 2002. pp. 269–78.

[10]   Patil S., Das S.R., Nasipuri A. 'Serial data fusion using space-filling curves in wireless sensor networks'. 2004 First Annual IEEE Communications Society Conference on Sensor and Ad Hoc Communications and Networks; IEEE SECON 2004; 2004. pp. 182–90.

[11]   Liao W.-H., Kao Y., Fan C.-M. 'Data aggregation in wireless sensor networks using ant colony algorithm'. *Journal of Network and Computer Applications*. 2008;31(4):387–401.

[12]   Chen M., Kwon T., Yuan Y., Choi Y., Leung V.C.M. 'Mobile agent-based directed diffusion in wireless sensor networks'. *EURASIP Journal on Advances in Signal Processing*. 2006;2007(1):036871.

[13]   Al-Karaki J.N., Ul-Mustafa R., Kamal A.E. 'Data aggregation and routing in wireless sensor networks: optimal and heuristic algorithms'. *Computer Networks*. 2009;53(7):945–60.

[14]   Upadhyayula S., Gupta S.K.S. 'Spanning tree based algorithms for low latency and energy efficient data aggregation enhanced convergecast (DAC) in wireless sensor networks'. *Ad Hoc Networks*. 2007;5(5):626–48.

[15] Sardouk A., Rahim-Amoud R., Merghem-Boulahia L., Gaïti D. 'Information-importance based communication for large-scale WSN data processing'. *Wireless and Mobile Networking*. Springer. Berlin, Heidelberg: Ronan Nugent; 2009. pp. 297–308.

[16] Bendjima M., Feham M. 'Intelligent communication of WSN based on a multi-agent system'. *International Journal of Computational Science*. 2013;2013:10:10:.

[17] Wilensky U. *Netlogo [online]*. Available from http://ccl.northwestern.edu/netlogo/ [Accessed 2021].

Chapter 10

Artificial intelligence-aided resource sharing for wireless mesh networks

Indrakshi Dey[1], Georgios Ropokis[2], and Nicola Marchetti[3]

10.1 Introduction

Recently both academic and industrial communities have turned their attention towards next-generation networking concepts capable of providing extremely high data rates and supporting novel applications. Next-generation networks are expected to automatically take information about users and make the world surrounding us react accordingly. They will learn from the accumulated information to determine optimal system configuration using advanced learning and decision-making techniques. Machine learning (ML) and deep learning (DL), a subset of special ML techniques, have emerged as a promising tool for autonomous decision-making [1, 2].

An intelligent telecommunication system will be capable of accessing the available resources like spectrum, transmission power, computational memory, etc., if assisted by ML and DL algorithms. ML and DL can be applied to a wide range of functionalities like image/audio processing, financial and economical decision-making, social behavior analysis, project management etc. [3], through a combination of processes like data mining, autonomous discovery, database updating, programming by examples, etc [4]. Over 60 different types of ML algorithms exist, which can be categorized based on functionality and structure, including regression, instance-based, regularization, decision tree (DT), Bayesian, clustering, association-rule-based, neural networks, DL, dimension reduction, ensemble learning algorithms, etc.

ML has found a wide range of applications in telecommunication networks due to its appropriateness in modeling various wireless networks like large-scale multiple-input-multiple-output (MIMO), device-to-device (D2D) networks, heterogeneous networks made up of femtocells and small cells, etc [5]. In Sections 2 and 3, we introduce ML and DL algorithms, explain their functionalities and how they can

[1]Maynooth University, Maynooth, Kildare, Ireland
[2]CentraleSupélec, Campus de Rennes, & the Institute of Electronics and Telecommunications of Rennes, Rennes, France
[3]Trinity College Dublin, Dublin, Ireland

be applied in different scenarios for wireless mesh networks (WMNs) for accessing and sharing resources among multiple communicating devices within the network.

Recent studies testify that the energy cost of ML is becoming unsustainable. For instance, the process of training a common ML model can emit more than 626,000 pounds of carbon dioxide equivalent, which is nearly five times the lifetime emissions of an average American car [6]. Hence, besides the immense potential of future generations of networks, there is a huge risk that they might become energetically unsustainable, fail to offer services due to high energy consumption and become a burden to our society due to their large carbon footprint. Certain applications are latency-sensitive with a processing time of only 1 or 2 milliseconds for each action within the network. However, average response time of traditional ML algorithms ranges between 200 and 300 milliseconds. This response time increases exponentially with the increase in the number of communicating devices and decrease in the amount of resources (energy, bandwidth, etc.) available [7].

Game-theoretic modeling and mechanism design combine distribution of intelligence and resources by exploiting energy-efficient high-speed decision-making methods, thereby reducing both latency and energy requirements. Such game-theoretic approaches can be applied to solve several key problems in different kinds of wireless networks, like Internet of Things (IoT), ad-hoc networks, cognitive radio networks, wireless sensor networks, etc, [8].

Modeling these networks as a set of point-to-point communications may be strongly suboptimal and even inappropriate. Game theory becomes relevant for environments where the communicating nodes interact with each other in a structured way, each node with knowledge (complete or incomplete) about the situation, while no single node completely controls the final outcome [9]. Game-theoretic approaches combine means of proposing excellent strategies, designing interaction models, studying the conditions under which some outcomes can be reached. Section 4 introduces and explains the applicability of first distributed learning and then, game-theoretic approaches for sharing resources like energy, power, routes, spectrum and time among communicating nodes within different kinds of wireless networks.

10.2   ML-assisted resource sharing

ML is a group of algorithms that allow communicating devices or nodes to learn from experience $\mathcal{E}$ without being explicitly programmed. With the capacity of finding hidden patterns within complex data, ML algorithms can be used to analyze large amounts of data, variables and parameters in wireless communication networks to execute complex tasks $\mathcal{T}$ without any existing formula. The performance of ML algorithms can be measured in terms of $\mathcal{P}$. The task $\mathcal{T}$ can be *classification* (assigning a class label to an input object or data structure) or *regression* (assigning a real value to an example). The performance measures $\mathcal{P}$ can typically be the accuracy of classification or precision of regressive categorization of examples. Based on how the experience $\mathcal{E}$ is gained by ML algorithms, we classify them into three groups.

- Supervised learning – Data set used for learning contains examples and their associated labels.
- Unsupervised learning – Data set used for learning contains only examples, and the algorithms themselves try to learn the properties of the data set.
- Reinforcement learning (RL) – Interact with the environment to extract the examples and try to learn the properties of the data set through iterative interaction and feedback.

Before applying any specific algorithm to a wireless networking problem, we summarize the generic steps that are followed from problem to solutions.

- *Problem definition* – A wireless networking problem is formulated as a data training task.
- *Data collection and preparation* – Necessary amount of recommended kind of data is collected for solving problem at hand. Collected data goes through preprocessing like cleaning and transformation of data pattern into vectors, $\mathbf{x} \in \mathbb{R}^n$, where $\mathbf{x}$ is the feature vector belonging to the real space $\mathbb{R}$ consisting of n possible features and feature extraction where each pattern becomes a single point in an n-dimensional input space.
- *Model training* – The next step is to train an ML algorithm to develop a model. If N feature vectors are available, a training set $\mathcal{E}$ can be defined such that it contains N training examples, i.e., $(\mathbf{x}_i, y_i), i = 1, \dots, N$,

$$\mathcal{E} = \{(\mathbf{x}_1, y_1), (\mathbf{x}_2, y_2), \dots, (\mathbf{x}_N, y_N)\} \tag{10.1}$$

where $\mathbf{x}_i$ is the th feature vector, $\mathbf{x}_i = [x_{i1}, x_{i2}, \dots, x_{in}]^T, i = 1, \dots, N$, and y_i is the label associated with the feature vector $\mathbf{x}_i$ and $\mathbf{y} = [y_1, y_2, \dots, y_N]^T$, where T denotes transpose of a vector or matrix.

- *Model learning* – Given the set $\mathcal{E}$, the main idea is to obtain a suitable model such that $\mathcal{T} : \mathbf{x} \to y$ and $\mathcal{T}$ offers a good estimate of future unobserved data, $y \approx \hat{y} = \hat{\mathcal{T}}(\mathbf{x}_{\text{new}})$. The choice of $\mathcal{T}$ should be such that it can effectively predict the outcome y_j depending on the input value $\mathbf{x}_j$, where $\mathbf{x}_j$ is estimated from the observation vector $\mathbf{x}_i$. The predictor or rather the task $\mathcal{T}$ can be a *classifier* (if y is discrete $y \in \{0, 1\}$) or *regressor* (if y is continuous $y \in \mathbb{R}$). Since the process involves estimation and prediction, the generic data model can be expressed as, $y = \mathcal{T}(\mathbf{x}) + \epsilon$, where $\mathcal{T}(\mathbf{x})$ is the model for learning and ϵ denotes the cumulative errors and discrepancies associated with the prediction process. The main aim of any ML algorithm is to find $\mathcal{T}$ that offers optimum $\mathbf{x} \to y$ correlation. In order to achieve that we can parameterize $\mathcal{T}$ in terms of $\Theta = [\theta_1, \theta_2, \dots, \theta_n]^T \in \mathbb{R}^n$ and the estimate Θ by solving a convex optimization problem. The objective will be to minimize the cost/loss function $l(\mathbf{x}, y, \theta)$

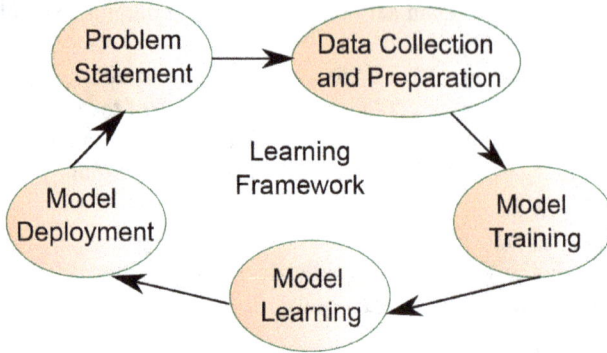

Figure 10.1   Generic steps of ML

representing the error between observation y_i and prediction $\hat{\mathcal{F}}(\mathbf{x}_{\text{new}})$ for each value of θ. Therefore, the optimization problem that needs to be solved is,

$$\arg\min_{\Theta \in \mathbb{R}^n} J(\Theta) = J(\mathcal{E}, \Theta) \frac{1}{k} \sum_{(x_j, y_j) \in \mathcal{E}} l(\mathbf{x}_j, y_j, \Theta) \tag{10.2}$$

where $J(\Theta)$ is the average loss over all training examples and the optimum model can be expressed as $\hat{y} = \Theta^T \mathbf{x}$.

- *Model deployment* – The final step is to deploy the obtained model into a practical wireless network scenario and then make predictions in terms of the solution to the problem at hand.

A diagrammatic representation of the above-mentioned generic steps is presented in Figure 10.1.

10.2.1   ML for resource sharing in WMN

Resource allocation in large WMNs is an age-old problem and is getting more and more complicated with increasing number of communicating entities in the network, increasing quality-of-service (QoS) requirements and scarce radio resources like energy, spectrum, computational memory, etc. In such a scenario, ML algorithms can be invoked to adopt optimal or near-optimal solutions by comparing historical scenarios, optimal solutions in historical scenarios and the current scenario. A conceptual framework for ML-aided resource allocation in WMNs involves ML algorithm being deployed at a centralized control center or base station (BS), which is in charge of allocating resources to the other communicating nodes present in the network. A large amount of historical data is collected on the number of nodes, channel side information at the nodes, energy consumption at each node, etc.

Feature vectors are created from the selected key attributes of the historical data through feature selection. All feature vectors are then randomly split into a training and a test set. Using learning algorithms, similarities in the data are analyzed to build a predictive model by solving the following optimization problem

$$\text{Minimize}_{y \in \mathscr{E}} \; f(\mathbf{y}, \mathbf{x}) \quad \text{subject to} \quad g_i(\mathbf{y}, \mathbf{x}) \leq 0 \quad i = 1, \dots, n,$$
$$h_i(\mathbf{y}, \mathbf{x}) = 0 \quad i = 1, \dots, k \tag{10.3}$$

where $f(\cdot, \cdot)$ is the objective function vector of allocated amount or configuration of radio resources, $\mathbf{x}$ is the vector of system parameters, $\{g_i\}_{i=1}^{n}$ and $\{h_i\}_{i=1}^{k}$ are inequality constraint function of specific scenarios and equality constraint function of the limitations on resource allocation and $\mathcal{E}$, respectively. In this way, solution to the resource allocation optimization problem can be obtained offline for each training feature vector, and all training vectors yielding the same solution are grouped in a single class.

Using several classes obtained from historical data is used to predict the class of a future scenario, where classifier is $Y = \mathscr{T}(F)$, F is the input feature vector, $\mathscr{T}$ is the classification task and Y is the output class to which the scenario belongs to. The solution can be further refined iteratively by evaluating it against the test set. Here, we provide the fundamental description of few ML algorithms and examples of WMN problems that can be solved using those algorithms.

10.2.2 Supervised learning

From definition, we know that in supervised learning we can have a data set $\mathbf{D}$ that contains both $\mathbf{x}$ and y, and we try to predict $\hat{y}$ from $\mathbf{x}$ or equivalently, the conditional distribution $p(y|\mathbf{x})$. Assuming a set of independent and identically distributed data $\mathbf{D} = \{\mathbf{x}_1, \mathbf{x}_2, \dots, \mathbf{x}_n\}$ with distribution $p_d(x)$, we can estimate the model parameter θ using maximum-likelihood (*ML*) estimator (for example) given by

$$\theta_{ML} = \arg \max_{\theta} p_m(\mathbf{D}; \theta) = \arg \max_{\theta} \prod_{i=0}^{n} p_m(x_i; \theta) \tag{10.4}$$

where p_m is the function space of probability distributions over the parameter θ. Taking logarithm on both side, (10.4) can be written as

$$\theta_{ML} = \arg \max_{\theta} \sum_{i=0}^{n} \log \left(p_m(x_i; \theta) \right) \tag{10.5}$$

Taking expectation of the log-probability of the model over the empirical data distribution, we can express (10.4) as

$$\theta_{ML} = \arg \max_{\theta} \mathbb{E}_{x \sim \hat{p}_d} \log(p_m(x_i; \theta)) = \arg \max_{\theta} \sum_{i=0}^{n} \log \left(p_m(y_i|x_i; \theta) \right) \tag{10.6}$$

where $\mathbb{E}$ denotes expectation. The *ML* estimator can be replaced by Maximum *a*-posteriori (*MAP*) estimator if a regularization function or a prior knowledge over the model parameters is introduced such that,

$$\theta_{MAP} = \arg \max_{\theta} p(\theta \mid \mathbf{D}) = \arg \max_{\theta} \left(p(\mid \mathbf{D}; \theta)) + \log \left(p(\theta) \right) \right) \tag{10.7}$$

An interesting observation is that, if we put a Gaussian prior, $p(\theta) \sim \mathcal{N}(0, \frac{1}{\lambda}\mathbf{I}^2)$, then $\log(p(\theta)) = \lambda \theta^T \theta$, where λ is an arbitrary parameter controlling the variance of the model parameter θ and T denotes the transpose of a vector or a matrix. In the following subsections, we discuss the major supervised learning algorithms that are extensively used to solve several challenges in WMNs, such as clustering and data aggregation, resource allocation [10], traffic classification and mobility prediction, power allocation and interference management, resource discovery and channel selection [11].

10.2.2.1   *k*-Nearest neighbors (*k*-NN)

A *k*-NN algorithm memorizes all previous instances, analyzes the closeness (distance) between input instances and predicts the output by searching for *k* closest instances. It accomplishes classification by predicting the majority class of *k* closest instances and regression by predicting the output value by calculating the average of the values obtained from *k* closest neighbors. The main aim is to calculate the distance $d(\mathbf{x}_i, s)$, where *s* is the unknown sample and then collect the indices of *k* smallest distances $d(\mathbf{x}_i, s)$ within a set $\mathbf{I}$.

Let us consider a training set of $(y_i, \mathbf{x}_i)$, where $\mathbf{x}_i$ is the feature vector and y_i is the label associated with $\mathbf{x}_i$. Then, the mean value/majority label for $\{y_i, i \in \mathbf{I}\}$ is calculated to obtain the predicted output $\mathscr{T}(\mathbf{x})$. Next, we will see how *k*-NN can be used to solve a resource allocation problem in a massive WMN.

Example – We tackle the problem of beam allocation in a massive IoT-like network with large number (*K*) of single-antenna sensors communicating with a data fusion center equipped with a linear array of multiple (*N*) identical isotropic antenna elements, such that $N > K$. Each sensor is served by a fixed beam allocated to it with polar coordinates (ρ_k, θ_k) for the *k*th sensor. The problem that we intend to solve here is how to efficiently select and activate *K* out of *N* beams such that the sum rate is maximized. Deciding on which beams to activate can be solved by the *k*-NN framework. We can formulate the following algorithm to arrive at the solution with user layout as the input data with $u = [(\rho_1, \theta_1), (\rho_2, \theta_2), \dots, (\rho_K, \theta_K)]$, where ρ_i denotes the radial distance and θ_i is the phase information for the *i*th sensor. Since beam gains vary with their phase, the feature vector can be formulated as,

$$\mathscr{F} = \left[\cos \phi^{(1)}, \cos \phi^{(2)}, \dots, \cos \phi^{(K)} \right] \tag{10.8}$$

where $\phi^{(1)} \leq \phi^{(2)} \leq \dots \leq \phi^{(K)}$. Similarly, feature vectors are formed for each training sensor layout to obtain,

$$\mathscr{F} = \left[\cos \phi_j^{(1)}, \cos \phi_j^{(2)}, \dots, \cos \phi_j^{(K)} \right] \tag{10.9}$$

for the *j*th sensor. For a new user layout u_i, the feature vector $\mathscr{F}_{u_i}$ is formed and the distance $d_i = \|\mathscr{F}_{u_i} - \mathscr{F}_j\|^2$ is calculated. Out of the distances, *k* smallest ones are chosen. The most common class among the *k* smallest distances is chosen as the predictive class of the sensor layout, and the predictive model outputs the associated active beam solution. Finally, based on the active beam information, each active beam is allocated to the sensor depending on the instantaneous received signal-to-noise-plus-interference ratio (SINR) over the selected beam.

10.2.2.2   Support vector machine

For classification, support vector machine (SVM) maps input data into a higher dimensional feature space, such that data becomes linearly separable by a hyperplane and the separated data stream is used for classification. For regression, continuous value output is predicted by using the above-mentioned hyperplane. The mapping function of the input data to the feature space is generally non-linear in nature and is known as kernel function. A large range of kernel functions are available and are applied depending on the application scenario. A few examples are

linear kernel : $k(x_i, x_j) = x_i^T x_j$ (10.10)

polynomial kernel : $k(x_i, x_j) = (x_i^T x_j + 1)^m$ (10.11)

radial basis function kernel : $k(x_i, x_j) = \exp\left(-\dfrac{x_i^2 x_j^2}{\sigma^2}\right)$ (10.12)

where σ is a user-defined parameter. SVM can be applied to solve different problems in WMNs, like the problem of energy-efficient subcarrier and power allocation in a multi-cast cognitive orthogonal frequency division multiplexing (OFDM) network subjected to probabilistic interference constraints. The above-mentioned problem can be converted from a constrained optimization problem to a deterministic optimization problem, which can be further solved by multi-objective optimization technique [12].

10.2.2.3   Random forest

Before talking about random forest, we need to first of all acquaint ourselves with DT. DT is a supervised learning algorithm in its own right that maps possible outcomes to specific inputs using a tree-like graph. A DT consists of three different kinds of nodes: (a) root node, (b) decision nodes used for testing input against experience or examples and c) leaf nodes used for representing the final class.

If bagging is applied to DT, resultant learning technique is referred to as random forest (RF). Bagging is a method of training multiple classifiers and then calculating the average output of the ensemble. The main aim of RF is to sample k data sets $\mathbf{D}_1, \dots, \mathbf{D}_k$ from the training set $\mathscr{D}$ with replacement. Next a DT classifier $h_i()$ is trained for each $\mathbf{D}_i$, and the tree is split only by considering a subset of the features and the parameters defining the feature subset can be denoted by l, such that, $l \leq d$, where d is the number of features in each training set. In this way, a sequence of classifiers are generated $h_1(), \dots, h_k()$ for each $\mathbf{D}_1, \dots, \mathbf{D}_k$ and the ensemble classifier can be calculated from the mean of all the DTs or the majority vote output decision. A diagrammatic representation of the RF algorithm with DT as a part of RF is provided in Figure 10.2.

Example – In order to elaborate on how RF can be applied to solve computationally complex problems in WMNs, we present the example of indoor positioning system using a network of smart devices. Such a network can be used to locate people or objects where different satellite technologies fail to offer precise location services based on the spatial division technique. Before applying RF, the input data set in the form of radio map needs to be developed. The radio map is a collection of data sets

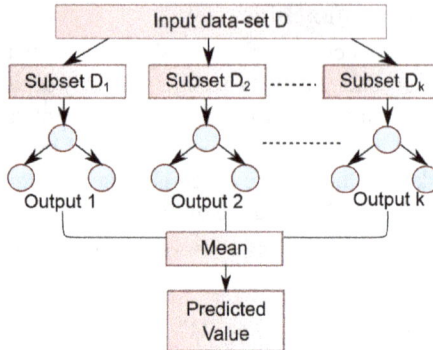

Figure 10.2   Diagrammatic representation of RF-based supervised learning

that matches the received signal strength (RSS) with position coordinates. Using the RSS and location information, the electromagnetic environment around the space of interest can be represented statistically. Therefore, the input database of radio maps is first developed by sampling the RSS values received from the access points (APs) and reference points (RPs) surrounding the object or people of interest.

An RF classifier can be developed by first of all selecting *m* features randomly to develop a decision criteria for each node of a DT and then calculate the best split accordingly. This procedure is repeated for all the nodes in the DT. For each DT, a subset of the radio-map database available, i.e., the number of RPs, is randomly selected along with their associated label of class. It is worth mentioning here that only a subset of the available radio-map database is used, and the rest of the map is used for testing error rates. As the RF classifiers are developed, they can be used to differentiate between the most probable spaces capable of holding the object of interest. In a nutshell, the group of RSS values received by an user or object will serve as the input to the RF classifier, the output of which will statistically indicate the specific space where the user can be located.

10.2.2.4   Neural networks

Neural networks (NNs) mimic the functioning of the human brain being capable of deriving complex non-linear decision boundaries for classification tasks and predicting real-value outputs by training regression models. NNs are also known as artificial neural network (ANN) consisting of traditionally one input and one output layers and several hidden layers (they are called so as their values are not available outside the processing unit) between the input and the output ones. Each layer can be mathematically expressed as,

$$\mathbf{G} = \alpha(\mathbf{W}^T\mathbf{X} + \mathbf{b}) \tag{10.13}$$

where $\mathbf{X}$ is the training or input data that corresponds to the input layer, $\mathbf{G}$ is the predicted output that forms the output layer, $\mathbf{W}$ are the weights associated with the input data, $\mathbf{b}$ is the bias term and α denotes the activation or the transfer function. Each hidden layer is formed of small processing units called neurons that transform

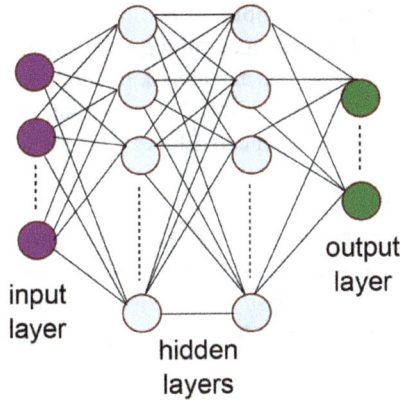

Figure 10.3   *Diagrammatic representation of generic structure for ANN*

the input into the output by using the activation function. An activation function can be any form of mathematical operation (linear or non-linear). A diagrammatic representation of a generic ANN is provided in Figure 10.3. Different kinds of NN have evolved over the years depending on how they use the input data to predict the output data and the processes involved in this transfer process. A few common ones are the dense NN (DNN), convolutional NN (CNN), recurrent NN (RNN), deep belief network (DBN), etc.

1. CNN – CNN involves non-linear transformation from multi-dimensional input data arrays (also known as tensors) to output predicted data vectors. The hidden layers in a CNN extract abstract features through a series of convolution, pooling and fully-connected layers. Each convolution layer uses kernels (finite impulse response filter banks) to convolve with the input for transforming data at the output. Mathematically, we can express the above relation as

$$(x * h)_i = x[i] * h[i] = \sum_n x[n] \cdot h[i-n]$$

(10.14)

where $x[i]$ is the input data, $h[i]$ is the kernel and $(\cdot)$ is the dot product between their weights. After convolution, a non-linear activation function is applied to the input and a bias term is added to create a feature map at the output, such that, $h_i^l = \alpha((W^l * x)_i + b^l)$, where h^l is the lth feature map at a certain convolution layer. Common activation functions include the rectifier $(\alpha(x) = \max(0, x))$, hyperbolic tangent function $(\alpha(x) = \frac{2}{1+exp(-2x)} - 1)$ and the sigmoid function $(\alpha(x) = \frac{1}{1+exp(-x)})$. Kernels using rectifiers as the activation function are often referred to as ReLU (Rectified Linear Unit).

The convolution layer is followed by a pooling layer that is responsible for merging similar features into one, thereby reducing the spatial representation. Common examples of pooling techniques are max pooling and neighbor pooling. In max pooling, the maximum value of a local patch of units in feature map is calculated. In neighbor pooling, only the inputs from patches that are shifted by more than one row or column are considered. The fully connected layer consists of neurons that are all connected to each other and every feature map in the pooling and convolution layers.

The output layer computes the posterior probability of each class label with respect to K classes and measures the difference between the true class labels y_i and the estimated probabilities $\hat{y}_i$, where $\hat{y}_i = \dfrac{\exp(z_i)}{\sum\limits_{i \neq j, j=1}^{K} \exp(z_j)}$, $i = 1, \ldots, K$ and z_i are the z-scores (numerical measurement describing a value's relation to the mean of a group of values) associated with each class label. The final aim of CNN is to minimize the loss function, $\min_\theta \sum_{i \in E} l(\hat{y}_i, y_i)$, where $l(\cdot)$ can be the mean squared error $\|y - \hat{y}\|^2$ or the categorical cross-entropy $\sum_{i=1}^{n} y_i \log(\hat{y}_i)$, θ is the CNN parameter and $\{x_i, y_i\}, i \in E$ is the training set. The output layer of the CNN is therefore trained by solving the loss minimization problem $\hat{\theta} = \arg\min J(E, \theta)$ through traditional optimization algorithms like gradient descent.

Example – In order to exemplify the applicability of CNN, we implement control of transmit power of a device in a device-to-device (D2D) wireless network through a CNN such that spectral efficiency of the network is maximized. The underlying aim in this case is to determine what will be the transmit power from each device in the D2D network depending on the instantaneous channel gain **H** by exploiting the spatial feature of the channel gain matrix. Let us employ a CNN that uses ReLU as the kernel function for the convolutional layer and sigmoid function as the activation function.

First of all, channel gain samples are collected for different transmitter and receiver locations and different channel configurations to form the input data set. After passing the input data through the convolution layer, we can obtain the z-score for all channel samples as

$$h_{i,j'}^k = \frac{h_{i,j}^k - \mathrm{E}(h_{i,j}^k)}{\sqrt{\mathrm{E}\left[(h_{i,j}^k - \mathrm{E}(h_{i,j}^k))^2\right]}} \tag{10.15}$$

where $h_{i,j}^k$ is the channel gain over the kth channel between the ith transmitter and the jth receiver. At the output layer, the loss function can be calculated as $L_P = \sqrt{(P_{\text{norm}}^* - P_N)^2}$, where P^* is the predicted optimal transmit power of the ith transmitter, P_{norm}^* is the standardization value of P^* and P_N is the recorded transmit power. Consequently, the output layer is trained by minimizing L_P. If P^* minimizes L_P to a satisfactory limit, P^* will yield the maximum value for net spectral efficiency $\Sigma_{i,k} C_i^k$

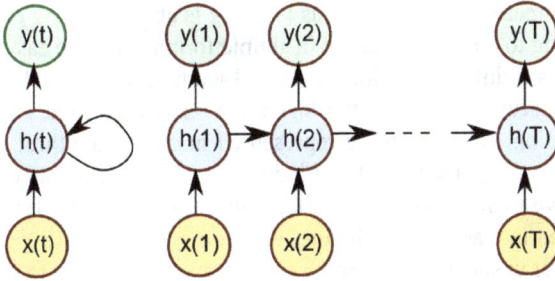

Figure 10.4   Info-graphic representation of RNN

where, $C_i^k = \log_2(1 + \text{SINR}_i^k) = \log_2\left(1 + \frac{h_{i,j}^k P^*}{N_0 B_k + \sigma_m h_{m,i}^k P_m^k}\right)$, where C_i^K is the spectral efficiency of the ith device over the kth channel, B_k is the band-width of the kth channel, N_0 is the channel noise spectral density, $h_{m,i}^k$ is the kth channel between the mth receiver and the ith transmitter and P_m^k is the transmit power from the mth device over the kth channel.

- RNN – This kind of neural networks contain nodes connected in the form of a directed graph along a temporal sequence with recurrent connections exist-ing between the nodes in the hidden layers. Mathematically, we can write, $h^{(t)} = f(h^{(t-1)}, x^{(t)}; \mathbf{W})$, where f is the activation function, $h(t)$ are the temporal states of the hidden nodes at time t, $x^{(t)}$ is the input data sequence at time t, $y(t)$ is the predicted output at time t and $\mathbf{W}$ are the weights associated with the input data. The RNN is graphically represented in Figure 10.4. There are numerous forms of RNNs used in literature to solve different problems in wireless net-works. The most popular among them is the long short-term memory (LSTM). It adds a self-loop on the network state to retain relevant information over a long period of time.

We have so far discussed only a few examples of supervised learning that can be used for resource allocation problems in WMNs. A few other kinds of techniques like Bayesian statistics, DT have also been used for solving different problems. One example application of Bayesian inference technique is assessment of event con-sistency in WSNs. It uses incomplete information on the environment to arrive at decisions on event consistency. On the other hand, using information like packet loss rate, corruption rate, mean time between failures, etc., techniques like DT can be used for judging link reliability in WMNs.

10.2.3   Unsupervised learning

The main aim of unsupervised leaning is to learn specific properties about the exam-ples or observations. This set of learning algorithms does not have any target outputs as opposed to supervised learning. This learning is about creating experiences and, in essence, trying to represent data set in simple formats that are easy to understand. Another important difference with supervised learning is the kind of loss function used. For unsupervised learning, the loss function also considers the error due to data representation. Data has to be represented in one of the three formats: lower dimensional format, sparse format and independent format. If data does not exist in one of the above-mentioned three formats, then it is also considered as erroneous data and is taken into account in the loss function. In the following subsections, we discuss the major unsupervised learning algorithms that are used to solve challenges like node clustering and data aggregation in WMNs.

10.2.3.1   *K*-means clustering

The main aim is to group input data or input points into K different clusters in a way that a high intra- and low inter-cluster similarity is ensured. The similarity is measured in terms of the average of the input data points belonging to a single cluster. In order to achieve the above-mentioned goal, the first step is to randomly select K nodes that can behave as the centroid of each cluster. The next step is to associate labels with all the nodes in the network with the closest possible centroid by calculating a distance function. Next, the centroids are recalculated using the current scenario and continue until convergence is achieved. To evaluate convergence, a predefined threshold is chosen for the sum of all the distances between the nodes and their possible cluster centroids or cluster heads. Once the convergence is reached, the cluster mean is updated for all the K-clusters. Data points before and after clustering using K-means are presented in Figure 10.5.

Example – K-means clustering can be a good solution for the interesting problem of how hotspot-slave configuration can improve overall performance in a dense WMN. The hotspot-slave refers to the condition where the network has to decide whether a node can be a slave or a hotspot, which hotspot a slave should link to, which channel to fire on after the link and cluster is formed and with how much

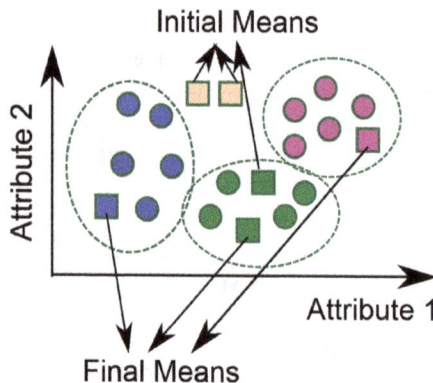

Figure 10.5   Graphical representation of K-means-clustering-based learning

power the slave should fire with. The above-mentioned challenges can be solved through three steps under certain relaxing assumptions.

- The first step is to cluster the nodes using K-means.
- The second step is to assign available spectral slots or frequency sub-bands to each cluster using different collaborative spectrum-sensing approaches.
- The final step is to allocate resources to individual nodes based on instantaneous channel state information.

The main aim of the clustering step is to group neighboring nodes such that the nodes are at shortest distance from the prospective hotspot and experience favorable channel conditions. We can start by defining a threshold γ on the number of nodes that can belong to each cluster and if there are K clusters and N number of nodes $K = N/\gamma$. The clustering optimization problem can be formulated as,

$$
\begin{aligned}
&\text{Minimize}_{y,C} \ \sum_{n=1}^{N} \sum_{k=1}^{K} \left(y_{nk} \| l_n - C_k \|_2^2 \right) \\
&\text{subject to} \ \sum_{k=1}^{K} y_{nk} = 1, 2 \le \sum_n y_{nk} \le \gamma
\end{aligned}
\tag{10.16}
$$

where $y_{nk} \in \{0, 1\}$ is a binary variable such that $n \in \{1, 2, \ldots N\}$, $k \in \{1, 2, \ldots K\}$, C_k is the location of the cluster centroid, l_n is the location of the nth node. The constraints are set to reflect the conditions that any node can belong to only one cluster, and the cluster size should be at least two with one node and one hotspot. If $l_n \in [0, L]$, cluster centroids can be randomly selected from the set, $C_k \in [0, L], k \in \{1, 2, \ldots K\}$ where L is the total number of clusters. By solving (10.16), we can assign node n to cluster k given $C_k^{(t)}$ at time t is known. After the clusters are formed, the centroid $C_k^{(t+1)}$ is updated by calculating the mean location of all nodes in the kth cluster. These steps are repeated iteratively until $C_k^{(t+1)} = C_k^{(t)}$. In this way, neighboring nodes are clustered within a WMN and then allocate resources depending on the instantaneous channel conditions.

10.2.3.2 Gaussian mixture model

Gaussian mixture model (GMM) is also a clustering technique like K-means, but based only on the probability distribution of the input data set. Each cluster is modeled as a Gaussian distribution with mean and co-variance and data distribution is calculated as a weighted linear combination of Gaussian distributions. For a given input or training data set $\mathbf{x}$ that is to be grouped into K clusters, each of the input data point is distributed as,

$$
P_M(\mathbf{x}) = \sum_{n=1}^{N} \alpha_n P(\mathbf{x}|\mu_n, \in_n)
\tag{10.17}
$$

where μ_n and $\in_n$ are the mean and co-variance of the nth Gaussian mixture component, respectively, and $\sum_n \alpha_n$ is the probability of selecting the nth Gaussian mixture component.

Example – Let us consider a WMN with N nodes, each of whose two-dimensional location information is available as the input data set, $\mathbf{X} = \{\mathbf{x}_j | \mathbf{x}_j \in \mathbb{R}^2, j = 1, 2, \ldots, N\}$, where $\mathbf{x}_j$ denotes the location coordinates of the jth node. Let $\mathbf{x}_j$ is represented by its Gaussian mixture component, $Z_j \in \{1, 2, \ldots, n\}$. If the prior probability of Z_j is given by $P(Z_j = n) = \alpha_n$, the posterior distribution $\mathbf{x}_j$ can be given by,

$$P_M(Z_j = n | \mathbf{x}_j) = \frac{\alpha_n P(\mathbf{x}_j | \mu_n, \epsilon_n)}{\sum_{j=1}^{K} \alpha_j P(\mathbf{x}_j | \mu_j, \epsilon_j)} \tag{10.18}$$

The main objective of the GMM clustering algorithm is to divide the $\mathbf{X}$ location data set into K clusters, $C = \{C_1, C_2, \ldots, C_K\}$ and the cluster tag λ_j can be calculated as,

$$\lambda_j = \arg \max_{n \in \{1, \ldots, N\}} P_M(Z_j = n | \mathbf{x}_j) \tag{10.19}$$

We can use the maximum likelihood approach here to estimate the GMM parameters

$$L(X) = \sum_{j=1}^{K} \ln \left(\sum_{n=1}^{N} \alpha_n P(X_j | \mu_n, \varepsilon_n) \right) \tag{10.20}$$

Using expectation maximization (EM) algorithm, we can calculate the mixture coefficient, $\alpha_n = \frac{1}{K} \sum_{j=1}^{K} P(Z_j = n | \mathbf{x}_j)$, mean, $\mu_n = \frac{\sum_{j=1}^{K} P(Z_j = n | \mathbf{x}_j) \mathbf{x}_j}{\sum_{j=1}^{K} P(Z_j = n | \mathbf{x}_j)}$ and co-variance $\epsilon_n = \frac{\sum_{j=1}^{K} P(Z_j = n | \mathbf{x}_j)(\mathbf{x}_j - \mu_n)(\mathbf{x}_j - \mu_n)^T}{\sum_{j=1}^{K} P(Z_j = n | \mathbf{x}_j)}$. Using the GMM for clustering in the aforementioned way, we can cluster nodes within a WMN, and it will be possible to allocate resource to each of the newly formed node clusters depending on the instantaneous propagation environment.

There are other unsupervised learning techniques like principal component analysis (PCA), density-based clustering, auto-encoders and self-organizing maps (SOM), which can also be used for solving specific challenges in WMNs. Density-based clustering can be used in WMNs to assign clusters to regions where the nodes are particularly dense. If a node belongs to a low density area, it can be considered as an outlier of the present network, and the network will decide not to allocate any resource to that particular node. The PCA method can be employed for data compression and dimensionality reduction in presence of a huge amount of transmit data from sensors in a WSN by retaining only a small set of uncorrelated linear combinations of the actual sensor data.

The SOM algorithm can also be applied for data clustering and dimensionality reduction in WSNs. The SOM achieves that by representing the sensor data in a two-dimensional grid, where sensor nodes close to each other are clustered. Once clusters are formed, only a small amount of uncorrelated sensor data is retained from all the transmit data within each cluster, i.e., dimensionality reduction is applied on cluster-by-cluster basis. Auto-encoders are very similar in structure to the supervised neural networks where the input data is mapped to a unique data representation that will be useful to extract important information for delivering a particular task or objectives.

10.2.4 Reinforcement learning

This is a group of learning algorithms that help an agent or the object to react according to its observation of the environment surrounding it. A policy defines how the agent will act at a given point in time. A reward signal is sent back by the environment to the agent indicating how well the agent is reacting accordingly. On the whole, the agent aims to maximize the cumulative reward through the observation of the environment and then takes further actions based on those inputs. The maximized cumulative reward is measured in terms of a value function. The main difference between the reward and the value function is that the reward reflects prospective solution in the current situation, and the value function reflects how much reward can be gained in future based on the previous and current states of the agent. Essentially, RL obtains a model of the environment which is estimated by the agent to characterize the dynamics of the environment and devise its plan to react accordingly.

10.2.4.1 Markov decision process

RL problems can be mathematically formulated as finite Markov decision process(MDP) whose dynamics can be defined by the tuple $(S, A, P_a(\cdot, \cdot), R_a(\cdot, \cdot))$, where S is the state space, A is the action space, $P_a(\cdot, \cdot)$ is the transition probability between the current and the future state s' and $R_a(\cdot, \cdot)$ is the reward received from the environment when moving to a new state. Both $P_a(\cdot, \cdot)$ and $R_a(\cdot, \cdot)$ are conditioned on the action a and can be mathematically expressed as,

$$
\begin{aligned}
P_a(s, s') &= Pr\big(S_{t+1} = s'|S_t = s, A_t = a\big) \\
R_a(s, s') &= E\big[R_{t+1}|S_t = s, A_t = a\big]
\end{aligned}
\tag{10.21}
$$

where E[] denotes expectation, Pr denotes the probability of, R_t is the reward function at time t, S_t is the state space at time t and A_t is the action taken at time t.

The agent devices a policy $\pi(a/s)$ for reacting to the environment and constructs a value function and an action-value function $q_\pi(s, a)$ depending on the reward it observes from the environment. The action-value function can then be written as,

$$
q_\pi(s, a) = E_\pi\left[\sum_{k=0}^{\infty} \gamma^k R_{t+k+1}|S_t = s, A_t = a\right]
\tag{10.22}
$$

where γ is a scaling parameter. For the optimum value of q_π, the solution to the MDP will be the action with the highest action value.

Example – MDP can be employed to model traffic variation within a radio access network and an RL framework can be applied to dynamically turn BSs on/off based on the network conditions, thereby making the network highly energy-efficient. Solution to the MDP can be achieved in numerous ways. One of the best ways to solve the MDP problem mentioned above is to employ the *actor-critic* method which can generate actions directly from the stored policy and can learn stochastic policies. In this case, we can define the traffic load state as s_t at the beginning of time t, and the action a_t can be selected according to a stochastic policy with

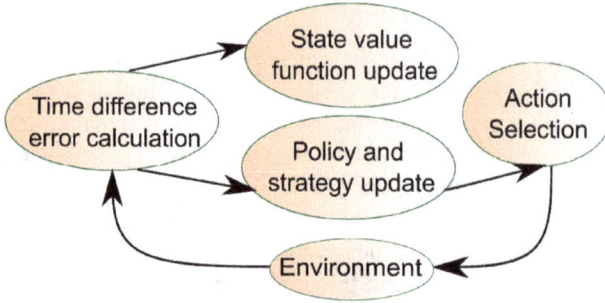

Figure 10.6   *Representation of actor-critic-based learning*

a probability $\pi_t(s_t, a_t) = \frac{e^{p(s_t, a_t)}}{\sum_{a_t \in A} e^{p(s_t, a_t)}}$, where $p(s_t, a_t)$ is the probability of selecting an action a_t at state s_t. Based on π_t, a_t and s_t, the network can select to switch off some of the BSs, and the node at a certain location will connect to one of the switched on BSs and start communicating.

Once transmission at time t is completed, the traffic load state at time $t + 1$ is collected as s_{t+1}. Let us define the total cost of this transmission as $C_t(s, a)$ and the state-value function as $V(s_t)$ and γ as the scaling parameter. Next the difference between the state-value functions at the current state and the estimated future state is calculated as,

$$\delta(s_t) = C_t(s, a) + \gamma V(s_{t+1}) - V(s_t) \tag{10.23}$$

and state-value function can be updated as $V(s_t) \leftarrow V(s_t) + \alpha \delta(s_t)$, where α is the update step size. Similarly, the policy is also updated at the end of time t as $p(s_t, a_t) \leftarrow p(s_t, a_t) - \beta \delta(s_t)$, where β is the update step size for the policy. A pictorial demonstration of the actor-critic learning framework is provided in Figure 10.6.

10.2.4.2 *Q-Learning*

The most common RL algorithm is the Q-learning. In Q-learning, the Q-value measures the total reward achievable by an agent when performing an action a_t in response to the state of the environment s_t, and the Q-value can be computed as,

$$Q(s_{t+1}, a_{t+1}) = Q(s_t, a_t) + \gamma (r(s_t, a_t) - Q(s_t, a_t)) \tag{10.24}$$

where $r(s_t, a_t)$ is the reward function and γ is the learning rate of an agent, a network or any other entity. A demonstration of the Q-learning method is presented in Figure 10.7.

Let us apply Q-learning algorithm to a multi-agent system. If $\mathscr{S}$ is the set of possible states, $\mathscr{S} = \{s(1), s(2), \ldots, s(n)\}$, and $\mathscr{A}$ is the set of possible actions, $\mathscr{A} = \{a(1), a(2), \ldots, a(m)\}$, for each agent, the ith agent's state can be defined as, $s_t^i = s \in \mathscr{S}$. Based on s, the ith agent selects the action $a_t^i = a \in \mathscr{A}$. In the next time instant $(t + 1)$, the ith agent generates the cost function $c_t^i = c \in \mathbb{R}$ for the state transition, which is passed back to the agent and the process is repeated. The ith agent

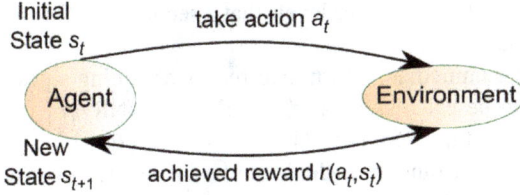

Figure 10.7   *Generic representation of Q-learning*

finds the optimal policy $\pi^*(s) \in \mathscr{A}$ for each s that minimizes the cumulative cost $c_t^i = c(s, a)$ at time t. Finally, the Q-function is calculated as,

$$Q(s, a) = \mathrm{E}\left\{\sum_{t=0}^{\infty} \gamma^t c(s_t, \pi(s)) \mid s_0 = s\right\} \quad 0 \leq \gamma < 1 \text{ is the discount factor} \tag{10.25}$$

Q-function will be minimized based on the current state if the action a_t following the policy $\pi(s)$ is equal to the optimal policy $\pi^*(s)$. Minimization of the Q-function can be mathematically expressed as

$$Q^*(s, a^*) = \min_{a \in \mathcal{A}} [Q^*(s, a)] \tag{10.26}$$

The main aim of the Q-learning process is to calculate $Q^*(s, a)$ in an iterative manner using information on the states, action at time t and $t + 1$ and the cost function at time t. It updates the Q-value through,

$$Q(s_t, a_t) \longleftarrow Q(s_t, a_t) + \alpha \left[c_t + \gamma \left(\min_{a_t} Q(s_{t+1}, a_t) - Q(s_t, a_t) \right) \right] \tag{10.27}$$

where α is the learning rate.

Example – Let us apply the concept of Q-learning discussed so far to the problem of power allocation in a secondary cognitive radio system for wireless regional area networks. Let us model this scenario as a multi-agent system where the secondary BSs are multiple agents and are in charge of allocating power to the users belonging to the secondary cells. The objective of these secondary BSs is to allocate power to the active users in such a way that interference among the secondary users is minimized. In this scenario, the system state for agent i can be defined as $s_t^i = \{I_t^i, d_t^i, P_t^i\}$, where i is the secondary cell index. $I_t^i \in \mathscr{I}$ is the binary indicator for whether the aggregate interference among the secondary active users is above or below the threshold set by the primary users. The threshold measure is calculated based on the instantaneous SINR at the BS of the ith secondary cell. Therefore,

$I_t^i = \begin{cases} 1 & \text{if SINR} < \text{SINR}_{Th} \\ 0 & \text{otherwise} \end{cases}$, where SINR$_{Th}$ is the threshold measure set up by the

primary users. The variables $d_t \in \mathscr{D} = \{1, 2, \dots, d\}$ represent the distance between the secondary user and the primary user cell boundary and $P_t^i \in \mathscr{P} = \{p_1, p_2, \dots, p_l\}$ is the power at which the secondary user currently transmitting its information. Since the state space is defined, we can now define the set of possible actions, which is equivalent to the set $\mathscr{P}$ of power levels that a secondary BS can allocate to the ith interfering secondary user.

The cost of the immediate return incurred by assigning a power level P_t^i to the interfering SU can be calculated as, $c_t^i = (\text{SINR}_t^i - \text{SINR}_{Th})^2$, where SINR_t^i is the instantaneous SINR at the BS of the ith cell. The objective of Q-learning algorithm is to minimize this cost function, which can be formulated as a quadratic function. SINR can be calculated as a function of the transmit signal power, noise power, interfering power and signal and noise bandwidth. All of these parameters can be used to represent the cost function and minimizing the cost function can be solved by any traditional optimization technique.

10.2.4.3   Multi-armed bandits

It is a kind of RL where the actions are called arms and an agent takes an action at successive trials to receive a certain predefined unknown reward. The agent considers only the reward from the actual pulled arm or token action and calculates the difference between the maximum achievable reward with the optimal arm and the actual reward gained. This difference is referred to as regret. The main objective of this learning process is to pull arms in consecutive trails in a way that the average regret is minimized and the average reward is maximized, while maintaining a balance between scoring immediate reward over the present trial and large rewards over all the trials. A large range of bandit games are formulated based on the application at hand and can be grouped as,

* games with random reward processes,
* games with different density and type of agents,
* games with available side information,
* games with random available actions.

A variety of algorithmic solutions have also been developed for multi-armed bandits (MABs), including the well-known upper confidence bound (UCB) strategy and ϵ-exploration algorithms.

Example – Next, we employ MAB to solve the challenge of how an unmanned aerial vehicle (UAV) can be deployed or rather which will be the most efficient patch for UAV deployment such that it can act as a wireless BS for providing post-disaster communication services to the maximum possible number of users while satisfying the battery constraint. The optimization task can be formulated as,

$$\max_{s_p \in S} \sum_{t=1}^{l_p} r_{n_t} \text{ subject to } t_p e_h T_h + \sum_{t=1}^{l_p} e_f \frac{d_{n_t, n_{t+1}}}{V_f} \leq E \tag{10.28}$$

where E is the battery capacity, e_h is the average engine power for hovering, e_f is the average engine power for flying, V_f is the flying velocity, T_h is the hovering interval, $s_p = \{n_1, n_2, \ldots, n_{l_p-1}, n_{l_p}\}$ is the UAV path, where each element $n_i \in \mathcal{N}$ represents a grid in the area, r_{n_t} is the traffic demand of the users within the grid n_t

and d_{n_t}, n_{t+1} is the distance between the centers of grid n_t and n_{t+1}. The UAV path starts at grid n_1, serves t_p grids and returns to grid n_1 to recharge, and S is the set of all possible paths that start and end at the center grid.

Let us now model the above-mentioned scenario as an MAB problem where multiple arms of an agent are denoted by $\mathscr{J} = \{1, 2, \ldots, J\}$. At round t, the agent pulls the ith arm to get a reward, $x_i(t)$. The goal of the agent is to maximize the total reward and let us employ UCB to solve this MAB problem. Therefore, at round t, the agent pulls the arm $j^* \in \mathscr{J}$ that satisfies,

$$j^*_{UCB} = \arg\max_{j \in \mathscr{J}} \left\{ \bar{x}_j(t) + \sqrt{2\ln(t)/T_{j,t}} \right\} \tag{10.29}$$

where $\bar{x}_j(t)$ is the average reward over the jth arm and tth round, and $T_{j,t}$ is the number of times the jth arm can be pulled. For solving the UAV path-planning problem, let us introduce the terms $\beta d_{n_{t-1}}$, n as the flight cost and γd_{n_1}, n/B_r as the remaining battery power for the UAV to visit the next grid.

Here, $d_{n_{t-1},n}$ is the distance between the next grid and the current one, $d_{n_1,n}$ is the distance between the next grid and the final recharging grid and B_r is the remaining battery power. The algorithm follows like this:

1. The first K^2 trials are dropped to set an initial reward for each grid where the set of all grids are denoted by $\mathscr{N} = \{1, 2, \ldots, K^2\}$ considering that the entire area of interest is divided into $K \times K$ equal grids.
2. Next the remaining battery power at each round t is calculated.
3. If the remaining battery power is sufficient for another trip, the UAV selects the next grid, n^*_{D-UCB} from $\mathscr{N} = \{1, 2, \ldots, K^2\}$ such that it satisfies,

$$n^*_{D-UCB} = \arg\max_{n \in \mathscr{N}} \left\{ \bar{r}'_{n,t} + \alpha\sqrt{\frac{2\ln t}{T_{n,t}}} - \beta d_{n_{t-1}} - \gamma\frac{d_{n_1,n}}{B_r} \right\} \tag{10.30}$$

where $\bar{r}'_{n,t}$ is the average reward of grid n.

Other RL algorithms like Joint Utility and Strategy based Learning or deep RL have also found their applications in wireless network design. In Joint Utility and Strategy based Learning, each agent estimates utility and updates it based on the intermediate reward received. The probability of selecting a particular action is also calculated and is termed as strategy. The strategy is also updated in every iteration based on the utility estimation, and the reward is calculated locally. Based on the immediate rewards and utility estimation, it is possible to estimate regret and then update strategy based on estimated regret. On the other hand, deep RL is a combination of neural networks and K-means. State transition samples are generated through interaction with the environment and are stored in the memory for training the neural network, and modified target neural network can be adopted to generate target values for stabilizing the training procedure.

10.3   DL-assisted resource sharing

A group of ML algorithms based on ANN with multiple hidden layers between the input and output layers are referred to as *deep learning* algorithms and the associated NN is referred to as deep NN (DNN). A DNN is able to formulate a functional relationship between input data sets in a way that this relationship will be applicable to data sets even lying outside the input training data set. Features in the data set are learned at multiple abstraction levels and complex functions are developed at the system level such that the input data can be directly mapped to the output data using those functions. The main difference between traditional ML and DL techniques is that ML considers feature extraction and classification as separate processes, while DL performs both the functions at different layers and the results from the previous layer percolates to the next one. The process continues in successive layers extracting and classifying more and more complex features.

The major advantage of DL over ML is that it is capable of handling huge amount of data and use them efficiently for improving model training and accuracy. DL simplifies and automates the entire process of extracting complex features from heterogeneous sources at multiple layers and maps out the correlation between them to develop an optimum and accurate model of the scenario. DL also can use unsupervised learning techniques like restricted Boltzmann machine (RBM) and generative adversarial network to train models from unbalanced data. The possibility to use a single training model to do multiple tasks, without actually doing re-training or re-modeling, also reduces the time and computational complexity of the entire procedure. Therefore, DL makes the modeling process easier and faster.

RBM is a two-layered ANN with the capability to learn from a probability distribution over its set of inputs. The two layers are, one with visible units and one with hidden units, which are connected by a fully bipartite graph (every unit in the visible layer is connected to every unit in the hidden layer with no connection between the units in the same layer). Also it is worth mentioning that RBMs are stochastic neural networks with each unit exhibiting random behavior when activated.

Generative adversarial networks refer to a group of ANN that can generate new outputs or examples automatically based on the regularities and patterns in the input data. The problem at hand is framed as a supervised learning problem divided into groups of tasks. The first group trains to generate new examples, and the second group classifies the examples as either true or false. The two groups form a zero-sum game (adversarial) and are trained simultaneously until the probability of false outcome from the classifier falls below 0.5.

10.3.1   Deep learning

In order to exemplify the working of DL and its applicability to WMNs, let us consider the problem of channel estimation and signal detection in OFDM systems. OFDM systems form the core of many large-scale heterogeneous WMNs involving cellular and Wi-Fi networks. The core of any DL operation is the DNN, the diagrammatic representation of which is presented in Figure 10.8. DNN consists of one

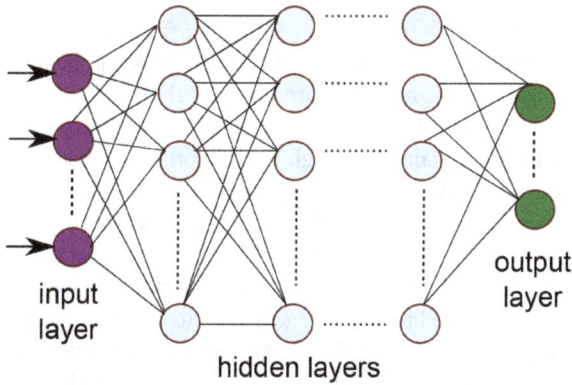

Figure 10.8   Diagrammatic representation of generic structure for DNN

input, one output and multiple hidden layers, each of which consists of units (neurons). Neurons in each layer outputs a nonlinear function formed by the weighted sum of the sum of the previous layer's neurons. Therefore the output z of the DNN can be mathematically expressed as,

$$\mathbf{Z} = f(\mathbf{I}, \theta) = f^{(L-1)}(f^{(L-2)}(\ldots f^{1}(\mathbf{I}))) \tag{10.31}$$

where $\mathbf{I}$ is the input data, L is the number of layers, θ are the weights of the NN and $f(\ldots)$ is the nonlinear function (can be sigmoid, $f(a) = \frac{1}{1+e^{-a}}$ or Relu, $f(a) = \max(0, a)$). The main aim is to optimize the weights for online deployment using the input training data.

Let us employ this DNN structure to an OFDM system. It involves inverse discrete Fourier transform for converting parallel data streams containing transmit symbols and pilots in frequency domain to time domain. Cyclic prefixes are inserted to mitigate intersymbol interference on the transmit side. On the receive side, cyclic prefix is removed followed by DFT to reconstruct back the original frequency domain signal. The DNN model takes one pilot block and one data block initially as the input and recovers the transmit data in an end-to-end manner. Once the initial model is set up, a full-fledged DNN model is deployed for joint channel estimation and symbol detection. The model processes the input-received data in two deployment stages. For the offline training stage, the model is trained with received OFDM samples that are generated with various information sequences and under diverse channel conditions with certain statistical properties, such as typical urban or hilly terrain delay profile. For the online deployment stage, the DNN model generates the output that recovers the transmitted data without explicitly estimating the wireless channel.

DL is recommended for solving a huge range of problems in WMNs like prediction, classification, resource allocation, spectral state estimation, object detection, speech recognition and language translation. The major drawback of DL is that a large data set is required for training and testing learning models and, therefore, may fail in situations of data insufficiency and where data are hard to generate or come by.

10.3.2   Deep RL

In order to avoid training with enormous data sets, DNN can be as an agent for RL in order to formulate a new group of learning algorithms called deep reinforcement learning (DRL). Then, in this case, DNN guess the optimal policy function after being trained using data obtained through interaction with the environment without requiring any prior data. Therefore, the training data are basically obtained through RL procedure. Therefore, combining DNN and RL, the agents can self-learn to maximize their long-term reward, or, in a nutshell, RL obtains experiences that can assist the DNN model to take the best action at any given state.

DRL techniques can be broadly categorized into two groups : Deep Q-network (DQN) and policy gradient. For DQN, a Q value is calculated for every plausible action, and the action with the highest Q value is selected. For policy gradient technique, probability distribution of every action is calculated, and the action with highest probability is chosen without considering the actual Q-value of the action. Q-learning is used for updating the experience, which can help the DNN model to select the best action in DQN. Gradient descent-based expected reward and policy optimization for updating experience are combined with DNN in policy gradient technique.

In order to elucidate the applicability of DRL to WMNs, let us select the problem of decentralized resource allocation in vehicle-to-vehicle (V2V) communication network. Let us assume a vehicular communication network that consists of $\mathcal{M} = \{1, 2, \dots, M\}$ users demanding vehicle-to-infrastructure (V2I) links (orthogonally allocated spectrum bands) and $\mathcal{K} = \{1, 2, \dots, K\}$ pairs of D2D links (links for sharing traffic safety information). In this case, the SINR of the mth user is given by

$$\gamma_m^C = \frac{P_m^C h_m}{\sigma^2 + \sum_{k \in \mathcal{K}} \rho_{m,k} P_k^d \bar{h}_k} \tag{10.32}$$

where SINR originates from background noise and interference due to signal flowing over the V2V links, P_m^C and P_k^d are the transmit powers from the mth user and the kth device respectively, σ^2 is the noise variance, h_m is the channel gain over the mth user link, $\bar{h}_k$ is the interfering power gain from the kth device and $\rho_{m,k}$ is the spectral allocation link. In this case, $\rho_{m,k} = 1$ for kth device reusing the spectrum of the mth user and $\rho_{m,k} = 0$ if the spectrum is not reused.

The capacity for the mth user can be given by $C_m^C = W \log(1 + \gamma_m^C)$ with W as the bandwidth. On the other hand, the SINR of the kth device can be expressed as $\gamma_k^d = \frac{P_k^d g_k}{\sigma^2 + G_C + G_d}$, where g_k is the power gain of the kth device, G_C is the matrix of the interfering power gain from the mth user and G_d is the matrix of the interfering power gain seen at the k'th device from the kth device. The capacity for the kth device can be given by, $C_K^d = W \log(1 + \gamma_k^d)$.

The main aim of the V2V resource allocation problem is that the interference of the V2I links on the V2V links is minimized while satisfying the latency constraints. The interference to the V2I and other V2V links are measured through sum capacities over V2I and V2V links, while latency is formulated as penalty, which increases linearly as the remaining time decreases. The reward function can, in turn, be expressed as,

$$r_t = \lambda_C \sum_{m \in \mathcal{M}} C_m^C + \lambda_d \sum_{k \in \mathcal{K}} C_k^d - \lambda_p(T_0 - U_t) \tag{10.33}$$

where T_0 is the constraint time, U_t is the remaining time for the device to fire its signal and λ_C, λ_d and λ_p are the weights associated. In this scenario, the goal of RL is to maximize the return, mathematically expressed as $G_t = \mathrm{E}|\sum_{n=0}^{\infty} \beta^n r_{t+n}|$, where G_t accounts for the expected cumulative discounted rewards and β is the discount factor. Here we will use the Q-learning technique for RL and therefore the DRL technique is used in the DQN method. In this case, the optimal policy with Q-values Q^* can be calculated without information on the system dynamics using the following equation:

$$Q_{\mathrm{new}}(s_t, a_t) = Q_{\mathrm{old}}(s_t, a_t) + \alpha[r_{t+1} + \beta \max_{s \in S} Q_{\mathrm{old}}(s, a_t) - Q_{\mathrm{old}}(s_t, a_t)] \tag{10.34}$$

where α is the update step size, s_t is the state at time t, a_t is the action at time t, (s_t, a_t) is the state-action pair, $Q(s_t, a_t)$ is the Q-value for a given state-action pair and $Q(s, a_t)$ is the Q-value over all the states and the action only at time t.

A DNN function approximator with weights $\{\theta\}$ is used as a Q-network. The basic idea is to determine $\{\theta\}$ in order to calculate the output Q-values, $Q(s_t, a_t)$ of the DNN. The weights are updated in each iteration to minimize the loss function resulting from the application of old weights on the data set D

$$\mathrm{Loss}(\theta) = \sum_{s_t, a_t \in D} (y - Q(s_t, a_t, \theta))^2 \tag{10.35}$$

where $y = r_t + \max Q_{\mathrm{old}}(s_t, a, \theta)$ and r_t is the corresponding reward. DRL is a promising technique for solving a wide range of problems in WMNs like multi-variable optimization problems in IoT networks [13], predictive analysis and automation [14] and in networks with large state and action spaces [15].

10.3.3 Graph neural network

Another kind of DL is the graph neural network (GNN) capable of generalizing graph convolutional operation in each layer. For a given graph, $G = \{V, \epsilon\}$ with m vertices or nodes $V = \{1, 2, \ldots, m\}$ and the set of weighted edges $\epsilon = \{e_{i,j}|i,j \text{ connected}\}$ which connects the nodes. The structure of a GNN can be represented by a graph shift operator matrix $\mathbf{S} \in R_+^{m \times m}$ whose (i,j)th component $s_{ij} = e_{ij}$ for all edges belonging to the set ϵ and 0 otherwise. A graph constructed in this way depicts the relationship between the elements of a graph signal $\mathbf{z} \in R^m$ with component z_i denoting the value of the signal at the ith node. The graph convolution operation for any input $\mathbf{z} \in R^m$ and graph filter $\alpha \in R^K$ with respect to the graph shift operator, $\hat{\mathbf{w}} \in R^m$ with the component,

$$\hat{w}_j := [\alpha_{*\mathbf{S}}\mathbf{z}]_j := \sum_{k=0}^{K} \alpha_k[\mathbf{S}^k\mathbf{z}]_j \tag{10.36}$$

where the term $\mathbf{S}^R$ shifts the elements of $\mathbf{z}$ in k turns according to the structure defined in S.

With graph convolution formulated, a GNN is constructed from a sequence of L hidden layers, each with a set of F_l graph filters of size K_l. The output of layer l is set of F_l filter outputs, which is then fed to the input to the $(l+1)$th layer. The input of the lth layer $\mathbf{Z}_l := [\mathbf{Z}_l^1, \mathbf{Z}_l^2, \dots, \mathbf{Z}_l^{F_l}] \in R^{q_l}$, where the dimension $q_l = mF_l$. Each feature in $\mathbf{Z}_l$ is passed through a graph filter, $\alpha^{ij} \in R^{K_l}$ and an intermediate Vector $\mathbf{u}_{l+1}^{ij}$ is generated from the feature $\mathbf{Z}_l^i$, which is then transported to the next layer feature $\mathbf{z}_{l+1}^j$, which can be mathematically expressed as,

$$z_{l+1}^j := \sigma_l \left(\sum_{i=1}^{F_l} \alpha_l^{ij} *_{\mathbf{S}} Z_j^i \right) \tag{10.37}$$

where $\alpha_l(\cdot)$ is the non-linear activation function. The output from the lth layer is then constructed as, $\mathbf{z}_{l+1} := [\mathbf{z}_{l+1}^1; \dots; \mathbf{z}_{l+1}^{F_l}] \in R^{q_{l+1}}$ with dimension $q_{l+1} = mF_{l+1}$. The filter operations for each layer $l = 1, \dots, L$ compose the full GNN operation to obtain the output $\mathbf{z}^l$. At the lth layer, the filter coefficients $\alpha_l^{ij} \in R^{K_l}$ are learned for all the feature combinations (i, j) for all $K_l \times F_l \times F_{l+1}$ coefficients.

10.4   Distributed intelligence-assisted resource sharing

Managing large-scale data to enhance the efficiency and scalability of ML algorithm has been a major challenge in wireless networks. Wireless networks often involve large amount of data produced and distributed over billions of devices. The ML techniques, described in Sections 10.1 and 10.2, cannot handle distributed data sets as everything has to be processed at a central entity. This creates the requirement for distributed techniques that imparts intelligence to all the communication entities so that the data privacy is preserved and the learning models are trained locally, thereby reducing the bandwidth and energy requirement and the latency in information transfer.

One approach to incorporate distributed intelligence is to resort to distributed learning techniques. In such schemes, an aggregator arranges locally generated data in a format such that a more accurate estimation and consensus can be reached. There are different variations of distributed learning. One option is parallel learning, where the training data set at the central gateway is divided into subsets of data and each data set is assigned to a group of local processing units, such that each of the data sets follows an equivalent distribution. The local processing is executed in parallel and is then fed back to the central gateway.

Another option is to combine multiple entities acting as learners (both classifiers and regressors) for performance improvement; such a learning approach is regarded as distributed ensemble learning. Again, the entire data set is partitioned into smaller portions, each of which is used to train different models that are used to improve the overall performance of the training procedure. In this case, a mixture of different models are used for learning instead of a global model. In the following subsections, we will discuss different kinds of special techniques that are used to solve decentralized intelligence-based problems in WMNs. These techniques are quite different from each other; however, they all offer considerable advantage over traditional ML techniques especially in terms of energy consumption, bandwidth requirement, latency in response, computational complexity and data privacy. The first group of techniques that we are going to look into operates within the paradigm of learning.

10.4.1 Federated learning

Federated learning (FL) is one of the most popular and efficient ways of learning from fragmented sensitive data. It does not rely on data aggregation and replication design but involves training a shared global model at a central location without sharing the local data among the users or the central platform. The learning model is shared without sharing the actual data.

Let us consider a WMN with K activated communication nodes where the kth node is associated with data D_k consisting of n_k number of samples. Therefore, the total sample size is, $N = \sum_{k=1}^{K} n_k$. The FL problem can be mathematically expressed as a loss function's minimization problem as,

$$\min_{W \subset \mathbb{R}^d} f(W) := \sum_{k=1}^{K} \frac{n_k}{N} f_k(W) \quad \text{where } f_k(W) = \frac{1}{n_k} \sum_{x_i \in D_k} f_i(W) \tag{10.38}$$

where $f_i(W)$ is the loss function dependent on the input-output pair $\{\bar{x}_i^*, y_i\}$ with $\mathbf{x}_i \in \mathbb{R}^d$ and $y_i \in \{-1, 1\}$ and W is the learning model. The loss function can be defined using common supervised learning models like $f_i(W) = -\log(1 + \exp(-y_i \mathbf{x}_i, W))$ used in logistic regression or $f_i(W) = \max\{0, 1 - y_i \mathbf{x}_i W\}$ used in SVMs. Once the loss function has been defined, we have to look for techniques that can be used to solve the minimization problem. These techniques can be broadly classified into two groups – (a) consensus solution and (b) pluralistic solution.

In cases of consensus solution, the global model is trained to minimize the loss with respect to the uniform distribution, $\tilde{D} = \sum_{k=1}^{K} \frac{n_k}{N} D_k$, where $\tilde{D}$ is the target data distribution of the learning model. If the data is drawn from non-independently and identically distributed (iid) distribution or any other distributions other than uniform distribution, pluralistic solution is the best option. Multi-task learning is a natural form of pluralistic learning that leverages the relation between non-iid distributions and the data drawn from them.

Example – Let us consider an Internet-of-Medical-Things (IoMT) architecture based on collecting inputs from wearable and ambient sensors and taking appropriate predefined actions, as shown in Figure 10.9. The ambient sensors such as camera, room temperature, humidity and air-quality sensors will inform the system about environmental conditions of the users. If the need be, the user will be sent alert to adjust these parameters, or these can also be integrated with actuators or smart home automation systems. The weight and blood pressure sensors shall be used to collect and report the body mass index (BMI) information, which is the major contribution of the proposed architecture. The patient may choose to measure their weight once a day so the patterns in BMI data may be identified and the alerts may

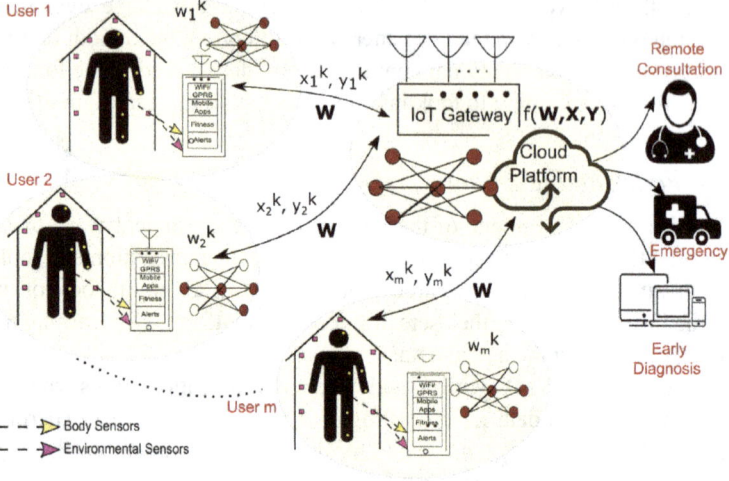

Figure 10.9   Proposed IoMT architecture for remote fitness monitoring

be generated indicating the obesity risk. Moreover, the wearable sensors shall be used to monitor the health status of the user regularly.

The proposed IoMT-based architecture involves a cloud computing platform and a set $\mathcal{M}$ of M-monitored individuals that jointly execute an FL algorithm for collecting body parameters using wearable and environmental sensors, calculating the BMI at regular intervals and formulating and delivering fitness strategies on the smart phones or any other portable user devices. The mth user device collects N_m samples of body parameters from portable sensors for training its local FL model w_m. Let us consider the kth input data sample of the mth user is denoted by $\mathbf{x}_m^k \in \mathbb{R}_{N_i \times 1}$ and the corresponding output by $\mathbf{y}_m^k \in \mathbb{R}_{N_o \times 1}$, where $\mathbf{y}_m^k$ reflects the corresponding fitness strategy. It is worth mentioning here that the input data samples follow the same distribution for all the M individuals.

The main aim of the FL training process is to solve the following minimization problem

$$\min_{w_1, w_2, \ldots, w_M} \frac{1}{\mathcal{N}} \sum_{m=1}^{M} \sum_{k=1}^{N_m} f(\mathbf{w}_m, \mathbf{x}_m^k, \mathbf{y}_m^k) \tag{10.39}$$

subject to $w_1, w_2, \ldots, w_M = \mathcal{W}$

where $\mathcal{N} = \sum_{m=1}^{M} N_m$ is the total number of training data accumulated from all the individuals and

$$f(\mathcal{W}, \mathbf{x}_m^k, \mathbf{y}_m^k) = -\mathbf{y}_m^k \log(\mathbf{x}_m^{kT} \mathcal{W}) + (1 - \mathbf{y}_m^k) \log(1 - \mathbf{x}_m^{kT} \mathcal{W}) \tag{10.40}$$

is the loss function and $\mathcal{W}$ is the global FL model. The final objective is to optimize the local FL model minimizing the loss function in (10.40), and it can be achieved through the following steps :

- Step 1 : The learning task information is broadcasted from the cloud platform (or the IoT gateway) for generating local FL models $\mathbf{w}_m$ at the mth user.
- Step 2 : The mth user trains its own local model $\mathbf{w}_m$ using its own collected data and the trained $\mathbf{w}_m$ is transmitted back to the gateway.
- Step 3 : The gateway modifies its own FL model based on the trained models from the users and broadcasts back the modified model to the user devices.
- Step 4 : The steps 2 and 3 are repeated until the optimum FL model is arrived at.

It is to be noted here that we assume that each FL iteration is completed within a single time slot, and the FL model parameters are transmitted over the wireless networks.

At each FL iteration, each user can exhibit his/her current BMI range or can exhibit BMIs in one of the four ranges: *Under*, *Normal*, *Over* and *Obese*. The probability of each user's BMI falling in one of the four ranges over a time slot of duration t can be expressed as a vector

$$\mathbf{p}_{m,t} = [p_{m,t,0}, p_{m,t,1}, p_{m,t,2}, p_{m,t,3}, p_{m,t,4}] \tag{10.41}$$

where $p_{m,t,0}$ is the probability that a user experiences the current BMI while $p_{m,t,1}, p_{m,t,2}, p_{m,t,3}$ and $p_{m,t,4}$ are the probabilities that the user BMI changes to under, normal, over and obese ranges. If the BMI of the mth user at time t can be represented by $\alpha_{m,t}$, then the BMI of that user at time $t + 1$ is,

$$\alpha_{m,t+1} = \begin{cases} \alpha_{m,t}, \text{ with probability } p_{m,t,0} \\ \alpha_{m,t} + \Delta\alpha < 18.5, \text{ with probability } p_{m,t,1} \\ 18.5 \leq \alpha_{m,t} \pm \Delta\alpha < 25, \text{ with probability } p_{m,t,2} \\ 25 \leq \alpha_{m,t} \pm \Delta\alpha < 30, \text{ with probability } p_{m,t,3} \\ \alpha_{m,t} \pm \Delta\alpha > 30 \text{ with probability } p_{m,t,4} \end{cases} \tag{10.42}$$

where $\Delta\alpha$ is the increment or decrement in the current BMI.

In order to mathematically formulate the FL training procedure, we consider the fact that the time required by each user device to transmit its local FL model to the gateway depends on the distance between the gateway and the individual. It is also to be noted that the gateway has to wait for receiving the local FL models before formulating its global model in each iteration and the threshold for the waiting time is denoted by γ. The updated global FL model at time t can be given by,

$$\mathscr{W} = \frac{\sum_{m=1}^{M} N_m w_{m,t} 1_{\{\omega_m \leq \gamma\}}}{\sum_{m=1}^{M} N_m 1_{\{\omega_m \leq \gamma\}}} \tag{10.43}$$

where ω_m is the waiting time after which the IoT gateway receives the local FL model from the mth user and $1_{\{\omega_m \leq \gamma\}}$ is the indicator function and is equal to 1 for the time taken by the local FL model from the mth user to reach the gateway, i.e., ω_m is less than the threshold time γ and is equal to 0 for all the transport times that exceed γ. Using (10.41) and (10.43), we can calculate the probability that mth user's local FL model is used to generate the global FL model at time t as,

$$\mathscr{P}_{m,t} = \sum_{k=1}^{5} p_{m,t}^k 1_{\{\omega_m \le \gamma\}} \tag{10.44}$$

Following (10.43) and (10.44), the local FL models are updated using the full gradient descent [16] to obtain,

$$w_{m,t+1} = \mathscr{W} - \frac{\lambda}{N_m} \sum_{k=1}^{N_m} \nabla f(\mathscr{W}, \mathbf{x}_m^k, \mathbf{y}_m^k) \tag{10.45}$$

where λ is the learning rate, $\nabla f(\mathscr{W}, \mathbf{x}_m^k, \mathbf{y}_m^k)$ is the gradient of $f(\mathscr{W}, \mathbf{x}_m^k, \mathbf{y}_m^k)$ with respect to $\mathscr{W}$. The uploaded local FL model is sent back to the gateway to update the global FL model to obtain,

$$\mathscr{W}' = \frac{\sum_{m=1}^{M} N_m w_{m,t+1} 1_{\{\omega_m \le \gamma\}}}{\sum_{m=1}^{M} N_m 1_{\{\omega_m \le \gamma\}}} \tag{10.46}$$

which is distributed back to the user devices.

Similarly, FL can be used in a variety of IoT applications like smart healthcare, smart transportation, smart cities and industries. Similar to the example provided in this subsection in case of smart health, FL can be used in smart transportation system by coordinating vehicles, traffic signals and other roadside units to execute autonomous driving, road safety prediction, nearby vehicle detection, etc.

10.4.2   Collective awareness

WMNs are rapidly evolving from communication hubs to advanced ecosystems with computation capabilities. These systems will have profound impact on our life through a number of real-time applications. This digital revolution is expected to be particularly relevant due to the increasing number of IoT devices that are being deployed in cities and industry environments. The sensor/field data that IoT sensors generate is currently processed by cloud servers; but in near future, WMNs will be likely to offer the needed computation and intelligence right at the communicating nodes. Artificial self-awareness (SA) will play a crucial role in the such networks. SA imparts every node with the capability to model its own state with respect to the environment.

Collective awareness is a technique of estimating all SA models and then learn and react accordingly within a WMN. Collective awareness involves combining outputs $\mathbf{y}_1, \ldots, \mathbf{y}_L$ corresponding to an input $\mathbf{x}$ for L-trained models either in parallel or in sequence. The parallel combination can be achieved either globally (all nodes generate an output and all the outputs are combined) or locally (a gating technique selects one node that is responsible for generating the final output). In the sequential combination, the nodes are sorted in a sequence based on their response to the input until one of them is confident enough.

A common method for achieving parallel combination is voting and averaging. Voting can be (a) majority vote: the class with most votes wins, (b) weighted vote: through posterior probability, $p_l(C_1|\mathbf{x}), \ldots, p_l(C_K|\mathbf{x})$ for each training model,

$l = 1, \ldots, L$, (c) average vote: calculate the output using $\mathbf{y} = \frac{1}{L} \sum_{l=1} \mathbf{y}_l$ and (d) median vote: the class that is most robust to outlying outputs. After voting, Bayesian model-based combination can be employed where the posterior probability is marginalized over all the models through $p(C_k|\mathbf{x}) = \sum_{\text{all models} \mathscr{M}_i} p(C_k|\mathbf{x}, \alpha \mathscr{M}_i) p(\mathscr{M}_i)$. This process can be viewed as weighted averaging using the model prior probabilities as weights. Combination is followed by ensemble averaging. If iid random variable $\mathbf{y}_1, \ldots, \mathbf{y}_L$ has an expected value of $E\{\mathbf{y}_l\} = \mu$ and variance $\text{Var}\{\mathbf{y}_l\} = \sigma^k$. If the $y_1, \ldots, y_L$ are identically but not independently distributed, then the variance can be calculated as $\text{Var}\{\mathbf{y}\} = \frac{1}{L}\sigma^2 + \frac{2}{L^2} \sum_{i,j=1, i \neq j}^{L} \sigma_{ij}$. Generally, the main target of good ensemble combination is to reduce or possibly eliminate positive correlation between the nodes with self-awareness.

A common method for sequential combination is bagging. Bagging starts with bootstrapping followed by bootstrap aggregation. If a training set $\mathscr{X} = \{\mathbf{x}_1, \ldots, \mathbf{x}_N\}$ contains N samples, bootstrapping technique generates L subsets $\mathscr{X}_1, \ldots, \mathscr{X}_L$ of $\mathscr{X}$, each of size N, where subset $\mathscr{X}_l$ is random using N replacement points from $\mathscr{X}$. Using bootstrapping, L partly different subsets of the training data are generated. Then self-aware nodes are trained on different subsets using an unsupervised learning algorithm. The nodes' outputs are averaged out or voted out to generate the ensemble output, a process called bootstrap aggregation.

Another technique for achieving sequential combination is boosting. Boosting actively generates complementary self-aware nodes by training the next self-aware nodes based on the errors of the previous aware node. A self-aware node can be a strong learner that exhibits arbitrarily small probability of error. On the other hand, a self-aware node can be called a weak learner if it exhibits a probability of error less than 0.5. At the beginning of the boosting stage, each node $l = 1, \ldots, L$ is trained on the entire training set $\mathcal{X} = \{\mathbf{x}_1, \ldots, \mathbf{x}_N\}$, where each point $\mathbf{x}_n$ has an associated probability $p_n^{(l)}$ indicating how "important" the data point is for the node.

Another technique for sequential combination is cascading. For an instance $\mathbf{x}$, each self-aware node gives an output and an associated confidence level, $p(C_k|\mathbf{x})$. Node L is trained on instances for which previous nodes are not confident enough. When applying the ensemble to a text instance, the nodes learn in sequence until one is confident enough, and node is used for learning, only if the previous nodes $1, \ldots, l-1$ are not confident enough about the outputs.

10.4.3 Game-theoretic approach

Resource allocation in a large-scale WMN can be optimized if the central computing platform has complete and perfect knowledge of the wireless network environment. However, the optimization techniques aided by traditional ML techniques get more and more complicated and the signaling load increases exponentially with the increase in the number of communicating nodes. This condition has become the largest bottleneck for the development of future WMNs. Moreover, ML techniques are slow as they operate offline and incur latency in decision-making in response to the rapidly varying network environments. Game theory can act as a powerful tool in this respect for developing self-organizing WMNs, especially with respect to learning and imparting intelligence at each and every participating nodes and executing decisions online.

Game theory is a crucial candidate for resource allocation in such WMNs as wireless networks are highly structured and the nodes naturally interact among themselves. With a game-theoretical setup, nodes not only can act with incomplete information about the environment, but they can make decisions in a distributed manner. Each node maximizes its own interest and can act selfishly at a certain point in time in the context of network operation. Using some basic assumptions, tractable models for relevant communication scenarios can be constructed.

Assumption 1 : Each participating node has a definite ordering of preferences in the form of utility function or payoff for the overall outcomes in a given scenario.

Assumption 2 : The nodes are considered to be rational, capable of thinking through all possible outcomes and selecting the best possible action that maximizes their payoffs. In this subsection, we start with basic definitions and concepts of game theory that need explaining before applying it to WMNs for resource allocation.

10.4.3.1   Static games with complete information

A game is referred to as static, if all players make decisions simultaneously without any knowledge of the strategies of the other players. A model of interacting decision-makers is referred to as a strategic game, and the decision-makers are referred to as players. Each such player will have a possible set of strategies and a preference on the action profile referred to as a payoff function.

Let us define a set of players, $\mathcal{N} = \{1, \ldots, N\}$ with a set of strategies $\mathcal{S} = S_1 \times S_2 \times \ldots S_N$ and a set of payoffs $\mathcal{U} = \{u_1, \ldots, u_N\}$, where S_n is the strategy set and u_n is the payoff or utility function of the nth player. If $\mathbf{s} = \{s_1, \ldots, s_N\}$ is the strategy profile, in which strategy of each player is $s_n \in S_n$, then $\{s'_n, \mathbf{s}_n\}$ is the strategy profile in which every (jth) player other than the nth player selects his/her action s_j with the nth player choosing s'_n. In that case, the payoff of the nth player will be $u_n(s'_n, \mathbf{s}_n)$, where nth player goes for s'_n and all the other players except n go for s'_n.

The set of strategies for each player where no player has incentive to unilaterally change his/her strategy is referred to as Nash Equilibrium (NE). This can be mathematically expressed as,

$$u_n(s_n^*, \mathbf{s}_n^*) \geq u_n(s_n, \mathbf{s}_n^*)$$

(10.47)

for all $\mathcal{S}_n \in \mathbf{S}_n$ and $n \in \mathcal{N}$, where s^* is the pure strategy Nash equilibrium. Therefore, NE represents the steady state of any system or a WMN as long as the environment and other conditions do not change. In order to analyze NE, three conditions are considered. The first condition is whether an equilibrium exists. This can be checked through the Debreu's sufficient condition [17]. It states that if the strategy sets S_n are non-empty, compact and convex subsets of an Euclidean space and if the payoff functions u_n are continuous in $\mathbf{S}$ and quasi-concave in s_n, then a pure strategy NE will exist. The second condition is whether one unique or multiple equilibria exist. There exists no generic results that can test this condition; however, different results have been derived over literature depending on the problem at hand [18]. If multiple equilibria exist, then the final condition is which equilibrium to select out of the multiple choices. Again there is no generic rule for this. There are different options including max-min criteria, proportional fairness, global optimization, etc. [19].

10.4.3.2 Static games with incomplete information

NE considers the fact that within a static game, all players know about the structure of the game. However, if some players do not have complete knowledge on the structure of the game, it is referred to as static game with incomplete information. Such a game can be modeled as a Bayesian game $\mathscr{G}$, which can be mathematically expressed as

$$\mathscr{G} = \langle \mathscr{N}, \{T_n, A_n, \rho_n, u_n\}_{n \in \mathscr{N}} \rangle \tag{10.48}$$

where $\mathscr{N} = \{1, \ldots, N\}$ is a set of players, $T_n(\mathscr{T} = T_1 \times T_2 \times \ldots T_N)$ is a set of type, $A_n(\mathscr{A} = A_1 \times A_2 \times \ldots A_N)$ is a set of action, ρ_n is a probability function, $\rho_n : T_n \to f(T_{-n})$ and $u_n : \mathscr{A} \times \mathscr{T} \to \mathbb{R}$ is a set of payoff functions. The variable "type," $\mathscr{T}_n \in T_n$ represents any kind of private information belonging to the nth player and is unknown to other players though can be relevant to their decision-making. Now the strategy S_n of the nth player refers to the function that maps type set T_n of the nth player to its action set A_n. The probability function ρ_n measures what the nth player believes about the types of other players based on its own type, i.e., $\rho_n(T_{-n}|T_n)$. Finally, the payoff function u_n of the nth player can be expressed as,

$$u_n(S(\mathscr{T}), \mathscr{T}) = u_n(S_1(T_1), \ldots, S_N(T_N), T_1, \ldots T_N) \tag{10.49}$$

If $\{(s'_n, \mathbf{s}_{-n})\}$ denotes the strategy profile for all players $\mathbf{s}(\ldots)$, except the nth player that plays $s'_n(\ldots)$, then the nth player's payoff can be defined as $u_n(s'_n(T_n), s_{-n}(T_{-n}), \mathscr{T}) = u_n(s_1(T_1), \ldots, s'_n(T_n), \ldots, s_N(T_N), \mathscr{T})$. In case of Bayesian game model, instead of NE, we can define Bayesian equilibrium, where a strategy profile $s^*(\cdot) = \{s_n^*(\cdot)\}_{n \in \mathscr{N}}$ is a pure strategy Bayesian equilibrium if for all $n \in \mathscr{N}$, $s_n \in S_n$ and $s_{-n} \in S_{-n}$,

$$E_{\tau_n}[u_n(s_n^*(T_n)), s_{-n}^*(T_{-n}), \mathscr{T})] \geq E_{\tau_{-n}}[u_n(s_n^*(T_n)), s_{-n}^*(T_{-n}), \mathscr{T})] \tag{10.50}$$

where, $E_{\tau_{-n}}[u_n(s_n^*(T_n)), s_{-n}^*(T_{-n}), \mathscr{T})] \underline{\Delta} \sum_{T_n \in \mathscr{T}} \rho_n(T_{-n}|T_n)u_n(s_n^*(T_n)), s_{-n}^*(T_{-n}), \mathscr{T})$ is defined as the expected payoff of the nth player. Intuitively, a Bayesian equilibrium for $\mathscr{G}$ can be considered as a mixed strategy NE of the transformed game $\hat{\mathscr{G}} = < \mathscr{N}, \hat{\mathscr{A}} = S_1 \times S_2 \times \ldots S_N, \hat{\mathbf{u}} = u >$.

10.4.3.3 Potential games

Potential games are a subgroup of strategic games that involve a potential function. A potential function quantifies the difference in the payoffs due to the unilateral deviation of each player either exactly, in sign or deviation to the best response. A potential game can be mathematically expressed as $\mathscr{G}^p = < \mathscr{N}, \mathscr{S}, v >$, where v is the potential function satisfying the following:

- Exact response $-v(s_n, \mathbf{s}_{-n}) - v(s'_n, \mathbf{s}_{-n}) = u_n(s_n, \mathbf{s}_{-n}) - u_n(s'_n, \mathbf{s}_{-n})$
- Ordinal response $-v(s_n, \mathbf{s}_{-n}) - v(s'_n, \mathbf{s}_{-n}) > 0 \;\leftrightarrow\; u_n(s_n, \mathbf{s}_{-n}) - u_n(s'_n, \mathbf{s}_{-n}) > 0$
- Best response $-\arg\max_{s_n \in \mathscr{S}_n} v(s_n, \mathbf{s}_{-n}) = \arg\max_{s_n \in \mathscr{S}_n} u_n(s_n, \mathbf{s}_{-n})$

for all $n \in \mathscr{N}$ and $(s_n, \mathbf{s}_{-n}), (s'_n, \mathbf{s}_{-n}) \in \mathscr{S}$, where $s_n = s'_n$. Potential games exhibit some special properties that are useful under certain scenarios. For example, if the strategy set $\mathscr{S}$ is convex and V is continuously differentiable on $\mathscr{S}$, then every NE of V. If V is concave, then every NE of $\mathscr{G}$ is a maximum point of V and if V is strictly concave, then the NE is unique.

10.4.3.4 Nash bargaining games

If in a game-theoretic model, where a group of two or more players choose their payoffs from a set of feasible alternatives, any one of the outcomes can be agreed upon by the bargainers, then the corresponding game model is referred to as a Nash Bargaining Game. Such games can be mathematically described by a set $\mathscr{N} = \{1, \ldots, N\}$ of players and a pair $(S, \mathbf{d})$, where $\mathscr{S} \in \mathbb{R}^N$ is a compact convex set of feasible payoffs and $\mathbf{d} = \{d_1, \ldots, d_N\}$ is the threat point. A bargaining solution on a class B of bargaining problems is defined as a function $f : B \to \mathbb{R}^N$ such that it assigns an allocation $f(S, \mathbf{d}) = \{f_1, \ldots, f_N\} \in \mathscr{S}$ to each bargaining problem $(S, \mathbf{d})$ in B. In the following subsections, we will apply different game-theoretic approaches to resource allocation problems within a WMN.

10.4.3.5 Non-cooperative games with complete information

Let us consider a scenario where N access points (AP) are simultaneously communicating with M mobile terminals over M independent frequency sub-bands over an OFDM network. The main aim of each AP (player) is to choose its power with which it will fire information (i.e., $\mathbf{p}_n = [p_{n,1}, \ldots, p_{n,M}]^T$, transmit power vector of the nth AP) subject to total power constraint $\sum_{m=1}^{M} p_{n,m} \leq P_n^{max}$, such that it maximizes its sum rate r_n. Let us assume that the outcome of the game is the power vector $\mathbf{p} = [\mathbf{p}_1^T, \ldots, \mathbf{p}_N^T]^T$ for all N APs over M sub-bands. In this case, the non-cooperative game can be formulated as $\mathscr{G} \triangleq \langle \mathscr{N}, \mathscr{P}, \mathscr{U} \rangle$, where $\mathscr{N} = \{1, \ldots, N\}$ is the set of APs, $\mathscr{P} = \mathscr{P}_1 \times \ldots \times \mathscr{P}_N$ is the set of strategies and the strategy of the nth player can be given by

$$\mathscr{P}_n = \left\{ \mathbf{P}_n : P_{n,m} \geq 0, \sum_{m=1}^{M} P_{n,m} = P_n^{max} \right\} \tag{10.51}$$

Here, $\mathscr{U} = \{u_1, \ldots, u_N\}$ is a set of utility facility or payoff and the payoff for the nth player can be given by,

$$u_n(\mathbf{p}_n, \mathbf{p}_{-n}) = \sum_{m=1}^{M} \log\left(1 + \frac{g_{n,m}p_{n,m}}{\sigma^2 + \sum_{j \neq n} g_{j,m}p_{j,m}}\right) = r_n \tag{10.52}$$

where $\mathscr{G} \in \mathbb{R}^{N \times M}$ is the channel gain matrix with $g_{n,m}$ as the channel power gain of the link between the nth AP and the mth mobile terminal on the mth subchannel, σ^2 is the variance of the white Gaussian noise, $\mathbf{p}_{-n}$ is the power vector of length $(N-1)M$ consisting of elements of **p** other than the nth element, $\mathbf{p}_{-n} = [\mathbf{p}_1^T, \ldots, \mathbf{p}_{n-1}^T, \mathbf{p}_{n+1}^T, \ldots \mathbf{p}_N^T]^T$. Once $\mathbf{p}_{-n}$ is known, the best power strategy $\mathbf{p}_n$ can be found by solving the following maximization problem:

$$\max_{\mathbf{p}_n} u_n(\mathbf{p}_n, \mathbf{p}_{-n}) \text{ subject to } \sum_{m=1}^{N} P_{n,m} \leq P_n^{max}, p_{n,m} \geq 0 \tag{10.53}$$

$$\begin{aligned} \mathscr{L}_m(\mathbf{p}, \lambda, \nu) &= \sum_{m=1}^{M} \log\left(1 + \frac{g_{n,m}p_{n,m}}{\sigma^2 + \sum_{j \neq n} g_{j,m}p_{j,m}}\right) - \lambda_n\left(\sum_{m=1}^{M} P_{n,m} - P_n^{max}\right) \\ &+ \sum_{m=1}^{M} \nu_{n,m}p_{n,m} \end{aligned} \tag{10.54}$$

with Karush–Kuhn–Tucker (KKT) conditions $\frac{g_{n,m}p_{n,m}}{\sigma^2 + \sum_{j \neq n} g_{j,m}p_{j,m}} - \lambda_n + \nu_{n,m} = 0$, $\lambda_n\left(\sum_{m=1}^{M} P_{n,m} - P_n^{max}\right) = 0$, $\nu_{n,m}p_{n,m} = 0$, where $\lambda_n \geq 0$ and $\nu_{n,m} \geq 0$ are dual variables. The above-mentioned problem can be solved in the form of $p_{n,m} = \left(\frac{1}{\lambda_n} - \frac{\sigma^2 + \sum_{j \neq n} g_{j,m}p_{j,m}}{g_{n,m}}\right)^+$, where $(x)^+ = \max\{0, x\}$ and λ_n satisfies $\sum_{m=1}^{M}\left(\frac{1}{\lambda_n} - \frac{\sigma^2 + \sum_{j \neq n} g_{j,m}p_{j,m}}{g_{n,m}}\right) = P_n^{max}$. In order to analyze the Nash equilibrium in the above-mentioned scenario, we can resort to the following condition [20].

A power strategy set $\{\mathbf{p}_1^*, \ldots, \mathbf{p}_N^*\}$ is a Nash equilibrium of game $\mathscr{G}$ if and only if each player's power strategy $\mathbf{p}_n^*$ is the single-player water-filling result of (10.54) while treating other players' signals as noise, and the corresponding necessary and sufficient conditions are

$$\frac{g_{n,m}p_{n,m}}{\sigma^2 + \sum_{j \neq n} g_{j,m}p_{j,m}} - \lambda_n + \nu_{n,m} = 0; \lambda_n\left(\sum_{m=1}^{M} P_{n,m} - P_n^{max}\right) = 0, \nu_{n,m}p_{n,m} = 0 \tag{10.55}$$

From the condition in (10.55), if $\lambda_n > 0$, $\sum_{m=1}^{M} p_{n,m} = P_n^{max}$, and it is intuitive that every AP (player) in Nash equilibrium must distribute its power over all the sub-channels based on their own observation.

10.4.3.6 Cooperative games

Let us apply the cooperative game theoretic approach for resolving the two-tier downlink interference problem for distributed resource allocation in macrocell- and femtocell-based cellular network. Let the formulated game be expressed as V such that,

$$V^*(S) = V(N) - V(N \setminus S) \text{ subject } \forall S \subset V \tag{10.56}$$

where S is the cooperation among N femtocells (players), and $V(N)$ and $V(N\setminus S)$ are the total number of sub-bands that can be distributed and the demands of the users attached to the femtocells of other players, respectively. Therefore, the main objective of this kind of WMN can be mathematically formulated as,

$$V^*(S) = \text{Max}[0, V(N) - V(N \setminus S)] \tag{10.57}$$

where V is the set of resource blocks that are to be allocated to N femtocells. Let us first divide the players (femtocells) into different groups, and the players in each group cooperate among themselves. If N^* players belong to the group G, $G(N^*, V)$, we can calculate the game core as $C(V) = \sum_{iinN} x_i = V(N)$ and $\sum_{iinS} x_i \geq v(S)$. Calculating the Shapley value is the best way to distribute the total advantage among all the players. In our example, the ith femtocell should have

$$\phi_i(V) = \sum_{S \subset M\{i\}} \frac{|S|!(n-|S|-1)!}{n!}(V(S \cup \{i\}) - V(S)) \qquad (10.58)$$

where the summation extends over all subsets S of N not containing the ith player. Based on the Shapley value, resource blocks can be applied to each player in group G. Once group G is looked after, the network works on the next group, i.e., $G := G + 1$.

10.5   Outlook

In this chapter, we have discussed different techniques for incorporating intelligence and self-organization within a large-scale WMN. The intelligence and the decision-making process are employed either in a centralized way (ML and DL) or a distributed way (collaborative learning and game theory). Learning algorithms use non-linear processing units for feature extraction and transformation. One of the key signature of learning is a multiple level representation of features or data. This hierarchical structure can be represented as a dynamic interactive environment that can be mapped as a strategic game problem. The neuron units of a neural network can be assumed as players, weight functions as actions, loss function as objective function and solution concepts of games like NE as learning tasks.

The major advantage of game theory is that the decision-making process is distributed where both local and global optimal solutions are analyzed and exchanged to reach a consensus. Learning structures can be easily mapped into a game, but the question is "can game theory be used for learning?" This is possible as the problem at hand can be considered as learning and then finding a game-theoretic solution with concepts like NE, Stackelberg solution, Pareto optima, correlated equilibrium, etc. From a game-theoretic perspective, it means that the objective functions of the players have to be replaced by interactive neural networks that learn and approximate them. This refers to the reality that, we need to formulate approximate games (new games) instead of original strategic games. In this case, first of all, in model-based strategic learning algorithm, the developed algorithm must use the model by adjusting the parameters to learn the solution concept.

If we go for model-free strategic learning, since the model is unknown, a procedure for estimating and adjusting the parameters has to be constructed. In certain cases, a strategic hybrid learning has to be adopted. However, how accurate such a game solution is by means of strategic learning for arbitrary games is an open area and needs detailed scrutiny, investigation and analysis depending on the network and communication scenario at hand.

References

[1] Jiang C., Zhang H., Ren Y., Han Z., Chen K.-C., Hanzo L. 'Machine learning paradigms for next-generation wireless networks'. *IEEE Wireless Communications*. 2017;24(2):98–105.

[2] LeCun Y., Bengio Y., Hinton G. 'Deep learning'. *Nature*. 2015;521(7553):436–44.

[3] Qian L., Zhu J., Zhang S. 'Survey of wireless big data'. *Journal of Communications and Information Networks*. 2017;2(1):1–182–18.

[4] Rathore M.M., Ahmad A., Paul A., ICA-ACCA. 'Iot-based smart city development using big data analytical approach'. Proceedings of the 2016 IEEE International Conference on Automatica; Chile, Curico; 2016. pp. 19–21.

[5] Alsheikh M.A., Lin S., Niyato D., Tan H.-P. 'Machine learning in wireless sensor networks: algorithms, strategies, and applications'. *IEEE Communications Surveys & Tutorials*. 2014;16(4):1996–2018.

[6] MIT Technology Review. *Training a single AI model can emit as much carbon as five cars in their lifetimes* [online]. Available from https://www.technologyreview.com/2019/06/06/239031/training-a-single-ai-model-can-emit-as-much-carbon-as-five-cars-in-their-lifetimes/. [Accessed 24 Jul 2020.].

[7] Machine Learning Machinery. *Comparing classical and machine learning algorithms for time series forecasting* [online]. Available from https://machinelearningmastery.com/findings-comparing-classical-and-machine-learning-methods-for-time-series-forecasting/. [Accessed 24 Apr 2021].

[8] Bkassiny M., Li Y., Jayaweera S.K. 'A survey on machine-learning techniques in cognitive radios'. *IEEE Communications Surveys & Tutorials*. 2012;15(3):1136–59.

[9] Marden J.R., Arslan G., Shamma J.S. 'Cooperative control and potential games'. *IEEE Transactions on Systems, Man, and Cybernetics, Part B*. 2009;39(6):1393–407.

[10] Wang X., Li X., Leung V.C.M. 'Artificial intelligence-based techniques for emerging heterogeneous network: state of the arts, opportunities, and challenges'. *IEEE Access*. 2015;3:1379–91.

[11] Park T., Abuzainab N., Saad W. 'Learning how to communicate in the Internet of things: finite resources and heterogeneity'. *IEEE Access*. 2016;4:7063–73.

[12] Vapnik V.N., Vapnik V. *Statistical Learning Theory*. New York, NY, USA: Wiley; 1998. p. 1.

[13] Mnih V., Kavukcuoglu K., Silver D., *et al.* 'Human-level control through deep reinforcement learning'. *Nature*. 2015;518(7540):529–33.

[14] Li Y. 'Deep reinforcement learning: an overview'. *arXiv:170107274v5*. 2017.

[15] Ruder S. 'An overview of gradient descent optimization algorithms'. *arXiv:160904747*. 2017:v2.

[16] Liu W., Chen L., Chen Y., Zhang W. 'Accelerating federated learning via momentum gradient descent'. *IEEE Transactions on Parallel and Distributed Systems*. 2020;31(8):1754–66.

[17] Debreu G. 'Continuity properties of Paretian utility'. *International Economic Review*. 1964;5(3):285–93.

[18] Kalai E., Lehrer E. 'Rational learning leads to NASH equilibrium'. *Econometrica*. 1993;61(5):1019–45.

[19] Foster D.P., Young H.P. 'Learning, hypothesis testing, and NASH equilibrium'. *Games and Economic Behavior*. 2003;45(1):73–96.

[20] Ui T. 'Bayesian potentials and information structures: team decision problems revisited'. *International Journal of Economic Theory*. 2009;5(3):271–91.

Chapter 11

Boosting machine learning mechanisms in wireless mesh networks through quantum computing

Francesco Vista[1,2], Vittoria Musa[1,2], Giuseppe Piro[1,2], Luigi Alfredo Grieco[1,2], and Gennaro Boggia[1,2]

11.1 Introduction

Nowadays, the ever-increasing demand of data rate and node density, along with low latency and reliability features, makes the introduction of wireless mesh networks (WMNs) a key solution for future wireless communication networks [1, 2]. WMNs, in fact, are self-organized and self-configured networks, where every node is able to autonomously establish and manage its connection to the network in real time. In detail, a WMN consists of two different types of nodes, named mesh routers (MRs) and mesh clients (MCs). MRs, as in traditional wireless communication systems, are usually equipped with multiple interfaces to integrate the WMN with Internet and various existing wireless networks (e.g., wireless sensor networks, wireless-fidelity (Wi-Fi), and mobile networks). MCs, instead, correspond to typical wireless devices which, differently from MRs, can be mobile and cannot be used as gateways (e.g., laptops, mobile phones, and tablets) [3]. Based on node functionalities, WMNs can be deployed by following three different network architectures. In backbone WMNs, only MRs build the mesh network by creating an infrastructure for clients and providing access to the backbone by leveraging existing wireless interfaces. In client WMNs, instead, also end users act as relay nodes forwarding incoming packets through the network. To reduce the overall network cost and complexity of the previous architectures, the hybrid WMN considers that each MC can directly communicate with neighboring MCs or access the mesh network exploiting MRs.

Despite the manifold advantages introduced by the adoption of WMNs in terms of reliability, network installation costs, long-range communications, and large-coverage connectivity, several critical factors negatively affect WMN

[1]Department of Electrical and Information Engineering (DEI), Politecnico di Bari, Italy
[2]Consorzio Nazionale Interuniversitario per le Telecomunicazioni (CNIT), Research Unit Politecnico di Bari, Italy

performance, including network capacity and management issues, scalability, and mobility [3, 4]. Some of these drawbacks can be partially solved through the introduction of novel enabling technologies already investigated by the scientific community. For instance, the network flexibility and capacity can be strongly enhanced by the introduction of single-user or multi-user multiple-input/multiple-output systems [5, 6]. Moreover, several works exploit unique capabilities of software-defined networking (SDN) paradigm, such as global visibility, real-time programming, and agility, to guarantee optimal network management and further improve the system performance [7, 8]. The quality of service (QoS) of the communication system can be also enhanced by the adoption of intelligent reflective surfaces, which improve the signal-to-noise ratio both in line-of-sight (LoS) or non-line-of-sight (NLoS) scenarios by exploiting the environment as a controllable signal reflector [9]. A further improvement in terms of capacity and QoS, while guaranteeing secure and fault-tolerant communications, is provided by the application of machine learning (ML) algorithms to solve design and management tasks in WMNs [10]. In the last years, in fact, the scientific community is promoting the adoption of ML techniques to strongly enhance the network adaptability according to the real-time conditions also in highly variable scenarios. However, given the continuous growth of involved devices and, in turn, the amount of data to be exchanged and processed in future WMN applications (e.g., wearable devices [11], vehicular ad hoc networks [12], and smart cities [13]), the computational time required by traditional computers to solve ML algorithms is expected to proportionally increase, demanding much efforts for training and inference procedures [14].

Under these premises, the innovation progress triggered by the emerging of quantum computing (QC) can be considered as a turning point to counteract this issue. Indeed, QC investigates quantum mechanics principles to develop new types of algorithms able to solve complex problems faster than classical approaches. Accordingly, the adoption of QC may speed up ML techniques, making them suitable also for extreme scenarios (i.e., computationally heavy and real-time applications). The scientific community already proposed the usage of QC-aided ML (i.e., quantum machine learning (QML)) for several challenging applications, such as 6G networks [14], chemistry, and physics [15]. However, at the time of this writing and to the best of authors' knowledge, the application of QML algorithms for WMNs remains an unexplored research topic. To bridge this gap, this work provides a two-fold contribution. First, it defines design strategies and new logical entities useful to exploit the potential of QC for WMN intelligence. Specifically, it proposes a centralized and a distributed network architectures according to the quantum computers location. The centralized architecture is supposed to perform QML tasks by exploiting quantum computers already deployed by Tech Giants in their cloud. Traditional computers, instead, are supposed to be spread at the edge of the network to solve simpler ML problems. In this case, a new node is added to the network in order to efficiently allocate computing resources (i.e., traditional and quantum computers) based on the task complexity and the status of computing resources. The distributed architecture, instead, is expected to be feasible only in very far future where quantum computers, equipped with a lower number of quantum bit (qubit), will be placed at the edge of the network and communicate by exploiting the quantum Internet. In this case, a creation of a quantum network allows to distribute computing resources more efficiently and also to solve more complex ML problems. Anyway, this approach requires the introduction of new nodes that setup the quantum network by distributing entangled particles between involved quantum computers. Second, this work presents pros and cons of the proposed WMN architectures, pointing out the main issues to be solved and paving the way for future research activities in this promising topic.

The rest of this chapter is organized as follows. Section 11.2 presents the role of ML for WMNs, highlighting the usage of well-known algorithms to solve typical issues in this context and discussing their main limitations. Section 11.3 describes QC principles and introduces QML algorithms as a possible solution to overcome ML limits. Section 11.4 proposes two network architectures to integrate quantum computers in WMNs and, finally, Section 11.5 draws the conclusions of this work and faces possible future research directions.

11.2 The role of ML in WMNs

ML algorithms are a subset of artificial intelligence (AI) techniques aiming at improving the performance of a system starting from data and information collected during previous tasks. Unlike optimization schemes, they are strongly adaptable to environmental conditions, resulting particularly suitable for time-variable use cases, such as WMN-based applications. Typically, ML techniques are classified as supervised, unsupervised, reinforcement, and deep learning. Given the features of these algorithms, they can be used in WMNs to solve different design and management tasks [16]. Figure 11.1 summarizes the main ML algorithms exploited in WMN scenarios.

11.2.1 Supervised learning

Supervised learning employs a training dataset, composed of input-output pairs, to develop a model that learns over time and computes the corrected output corresponding to new input data. To this end, supervised learning algorithms aim at minimizing a loss function in an iterative manner by modifying the hyperparameters of the model. In this context, the most commonly used algorithms for WMNs are decision tree (DT), support vector machines (SVMs), and K-nearest neighbors (KNN), typically performing classification or regression tasks.

Specifically, DT algorithms exploit a tree-like structure to solve both classification and regression problems. The attributes of the input data are compared with features labeling internal nodes of the tree. Starting from the root node and performing these comparisons, the algorithm traverses the tree until it reaches the leaf nodes, which represent the class or the relationship between dependent and independent variables. SVM, instead, is an ML technique commonly used for classification tasks. In this case, the algorithm constructs hyperplanes aiming at maximizing the width of the gap between points belonging to different classes in order to increase

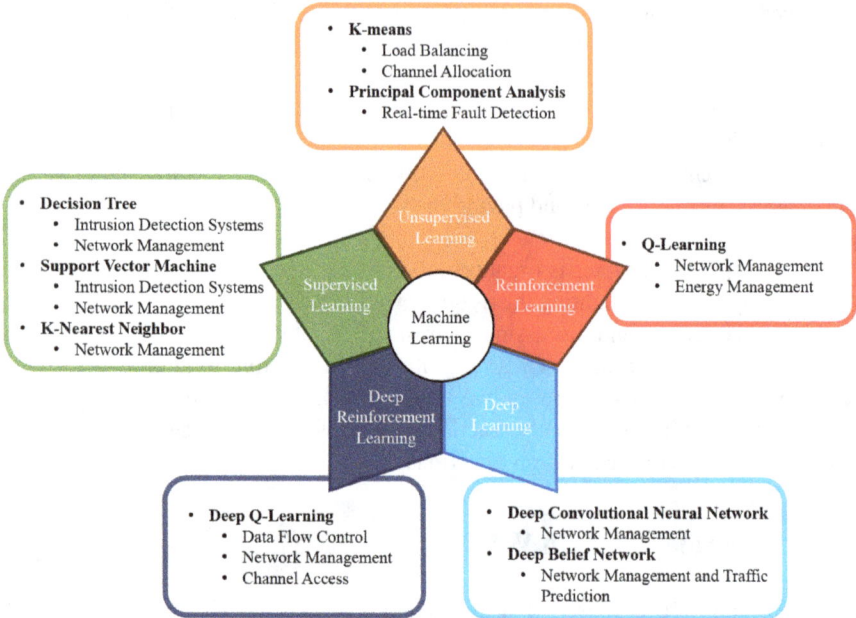

Figure 11.1   Summary of ML techniques and corresponding role in WMNs

the classification precision of successive input data. SVM and DT algorithms are employed in WMNs to build efficient cross-layer-based [17] and network-layer-based [18] intrusion detection systems. In this case, the model is trained starting from packet delivery ratio, packet arrival interval, and end-to-end delay statistics in order to easily detect anomalous behavior and remove malicious nodes. These algorithms are also integrated with a threshold that avoids false decisions. An easier algorithm used to perform both classification and regression tasks is the KNN. Here, the input data are classified by considering a similarity concept (i.e., every data point falling near others belongs to the same class). Considering an internet of things (IoT) network supported by a software-defined WMN, the presented supervised learning approaches (i.e., DT, SVM, and KNN) are used for optimizing the management of the network and perform time granular analysis of the network traffic. The comparison among these learning strategies demonstrated that the KNN algorithm provides the best performance in terms of accuracy [7].

11.2.2   Unsupervised learning

Unsupervised learning algorithms are used to find unknown patterns from a training dataset containing unlabeled data. In particular, this kind of algorithm analyzes the internal structure of the training dataset, thus discovering patterns between data without any external information. The main unsupervised algorithms used in WMNs are K-means and principal component analysis (PCA).

K-means is a clustering algorithm that groups the input unlabeled data in k clusters with an iterative procedure. Specifically, at each iteration, the n observations are grouped in order to minimize the variance intra-cluster and maximize the distance inter-cluster. Here, the distance, usually measured through a Euclidean metric, is computed considering the cluster centers, named centroids. The iterative procedure ends when the algorithm converges. In WMNs, K-means can be used for the load balancing of the network [19] or the channel allocation [10]. In detail, load balancing is performed in order to optimize the resource allocation, increase the overall load of the network, and reduce the congestion at the gateways [19]. Moreover, the K-means clustering can be used to group the MRs efficiently, choosing the cluster head according to the computed centroid [10]. The PCA algorithms, instead, aim at reducing the data dimensionality by describing each data point only with several uncorrelated principal components, while maintaining the highest training-data variance in the first component. Given that the PCA algorithm allows handling high-dimensionality application scenarios, it is particularly suitable for real-time fault detection in high-interference environments, such as WMNs [20].

11.2.3   Reinforcement learning

The training of reinforcement learning (RL) models is an iterative process aiming at choosing the optimal decisions among a set of available actions to maximize cumulative feedback received from the environment, named cumulative reward.

The most known RL algorithm is the Q-learning. Its main feature is the capability to train a model without the knowledge of the environment. Q-learning algorithms, in fact, are based on a Q-value that is updated at each iteration: the optimal action corresponds to the largest cumulative Q-value. In WMNs, RL strategies can be exploited for routing purposes in order to decide the optimal route, among many possible paths, to take from source to destination node. RL fits very well with this kind of problem: the next MR could be chosen, at each iteration, from a set of possible actions in that state. Moreover, Q-learning can be used to avoid critical problems, such as the congestion at the gateway, by dynamically learning an optimal routing scheme that considers several metrics (e.g., loss-ratio, interference, and load at the gateways) [21]. Since classical routing protocols may suffer from excessive energy consumption and do not consider past experience, Q-learning can also optimally enhance the energy balance of the network [22].

11.2.4   Deep learning

Deep learning (DL) is a subfield of ML, which involves multiple layers for the processing of input raw data in order to progressively extract higher-level features. It commonly uses an artificial neural network composed of many perceptrons organized in multiple dense hidden layers. To properly train a model, it also needs an initial step useful to tune the hyperparameters starting from a huge amount of data. Specifically, DL algorithms train the model by minimizing the loss function over the training dataset and extracting the weights of the final model. The main DL architectures used for WMNs are deep convolutional neural network (DCNN) and deep belief network (DBN).

DCNN is an example of DL architectures, mostly used in computer vision. In this case, the classification task is performed by filtering the input data using convolution layers in order to extract low-level information. Then, the size of the extracted features is reduced by pooling layers, thus obtaining the output of the fully connected layer (i.e., a vector which contains the result of the classification process). In WMNs, gateways receive traffic information from both MRs and MCs, leading to a higher probability that several nodes become congested. To overcome this issue, DCNN can be used to periodically train a model in order to make optimal routing decisions based on past events [23]. On the other hand, DBN is a class of deep neural network defined as a stack of restricted Boltzmann machines (RBMs), which is a two-layer undirected graphical model. Each RBM layer is connected with both the previous and next layers and the nodes alongside any layer are not connected with each other. Since RBMs training process is unsupervised, a DBN ending with a Softmax layer can be used both for classification and clustering of unlabeled data. This makes the DBNs algorithms particularly suitable in WMNs to improve the network management operations in terms of network traffic prediction [24].

11.2.5   Deep RL

Deep RL (DRL) combines two subfields of ML: RL and DL. To efficiently use RL, in fact, agents must infer a good representation of the environment, thus choosing the action, which maximizes the reward by following a trial-and-error strategy. However, if the state spaces or action spaces are too large, this decision can be a complex task that requires more computational time. In this context, DL can help agents to make decisions by learning policies directly from high-dimensional and unstructured input data [25]. The most promising example of DRL for WMNs is the deep Q-learning network (DQN).

DQN combines deep neural networks and Q-learning in order to estimate and maximize the Q-values by considering both states and rewards. It can be employed in WMNs to control the data flow and enhance the throughput. In fact, classical control flow methods suffer from the continuous growth of the number of mesh nodes and the complexity of data applications, which make these kinds of scenarios strongly dynamic. DQN, instead, intrinsically has the capability to manage and optimize complex traffic communication flows [26]. DRL algorithms can also be used for optimally planning the network in real time, thus optimally deploying gateways in the WMN and choosing the network topology [27]. Moreover, DRL can manage the channel access in dynamic spectrum scenarios, where multiple discrete channels are shared by different types of nodes without any a priori knowledge [28].

11.2.6   Open issues in the application of ML for WMNs

The network management improvement reached with the introduction of aforementioned ML techniques in WMNs is motivating the scientific community to implement these intelligence strategies in real-time applications. However, features of

WMNs make them particularly suitable for the management of high-dynamic scenarios. In this context, the training dataset must be continuously updated in order to accurately describe the behavior of considered networks, thus requiring a periodical retraining of ML models [7]. Moreover, the number and heterogeneity of devices involved in future networks are expected to grow exponentially, thus increasing the number and dimensionality of data to be managed. At the same time, the training time of traditional ML methods strongly depends on the data-space dimension [14]. Accordingly, the benefits obtained by the application of ML algorithms in WMNs may be invalidated by delays related to real-time training procedures, thus requiring the introduction of novel intelligence strategies.

11.3 Quantum computing: background and QML

QC, a subfield of Quantum Information Science, is a well-known paradigm that is gaining momentum in the last years for its particular features. In fact, it harnesses quantum mechanics principles, completely transforming traditional computing approaches and providing performance enhancements in terms of tasks execution time, accuracy, and computational complexity [29]. For instance, superposition and entanglement principles can be exploited to perform complex tasks that in classic realm would be very challenging. Accordingly, QC emerges as a promising solution to overcome the aforementioned issues for the application of ML in WMNs.

11.3.1 Superposition principle

The quantum computation is based on the concept of qubit. A qubit is a mathematical representation of a discrete two-level quantum system, where the two computational basis states are commonly denoted as 0 and 1. Physically, for instance, a qubit can be described as the polarization of a photon, where the two orthogonal basis states are the horizontal and the vertical polarization of the photon. Differently from classical computers, where bits can assume exactly one binary value at any time (i.e., either 0 or 1), in quantum computers qubits can be in a superposition of two simultaneous values until it is observed. According to the superposition principle, hence, n qubits can encode all the 2^n possible states at once. As a consequence, the power of QC, as well as the information intrinsically kept, grows exponentially with the number of involved qubits [30].

11.3.2 Quantum measurement

In quantum mechanics, it is not possible to establish the state of a qubit by directly observing the quantum state. However, it is allowed to observe the results of the measurements. According to the quantum measurement postulate, after the measurement of the original quantum state, a qubit collapses in either the zero state or the one state (or in one state of the basis states in case of multi-qubit systems). The result of the quantum measurement depends on the amplitude probability associated with each state, and any further measurement will give the same result. This deeply impacts the design of the quantum network: a quantum state cannot be transmitted by simply measuring the qubit and sending the result [30].

11.3.3   No-cloning theorem

The no-cloning theorem states that it is not allowed to make a copy of an unknown quantum state. This is a fundamental concept of the quantum key distribution (QKD), since if an eavesdropper tries to "read" the state of a photon, which travels along the path from a sender to a receiver, it will destroy the state of the photon. Furthermore, since it is not possible to store redundant copies of the qubit, new strategies are needed to send a qubit among remote quantum devices, such as quantum entanglement and teleportation [30].

11.3.4   Entanglement

Entanglement is a quantum phenomenon with no equivalent example in the classic world, where two (or more) distant particles share a quantum state. In this case, a measurement performed on one particle affects the outcome of the entangled one. The maximally entangled quantum states are the well-known Bell states [30]:

$$\left|\Phi^{\pm}\right\rangle = \tfrac{1}{\sqrt{2}}(\left|00\right\rangle \pm \left|11\right\rangle) \tag{11.1}$$

$$\left|\Psi^{\pm}\right\rangle = \tfrac{1}{\sqrt{2}}(\left|01\right\rangle \pm \left|10\right\rangle) \tag{11.2}$$

The entanglement is a key resource to enable teleportation in quantum networks. Accordingly, two remote quantum computers that want to communicate must share entangled particles. In this context, three different methods can be employed to generate and distribute entanglements [31].

The first method, known as spontaneous parametric down-conversion (SPDC), generates a pair of entangled photons using their polarization. A non-linear crystal is hit by a laser beam generating two photons with vertical and horizontal polarization. These photons, called flying qubit, reach the interested nodes through a quantum channel. Then, at each side, flying qubits are transferred to a computation qubit by using a transducer device to execute quantum operations.

The second method uses optical fiber to connect optical cavities to each side. In particular, entanglement is generated at the sender by exciting the atom with a laser beam, which causes the emission of a photon entangled with the atom. This photon reaches the other node passing through the optical fiber, and it is absorbed by the optical cavity. At this point, the atom-photon entanglement is mapped to an atom-atom entanglement.

The latter method also uses optical cavities, but it generates the entanglement at both remote quantum computers by exciting both the atoms at the same time. The emitted photons reach a particular device, which is in charge of performing a Bell state measurement in order to map the atom-photon entanglement on both sides into an atom-atom entanglement between the remote quantum computers.

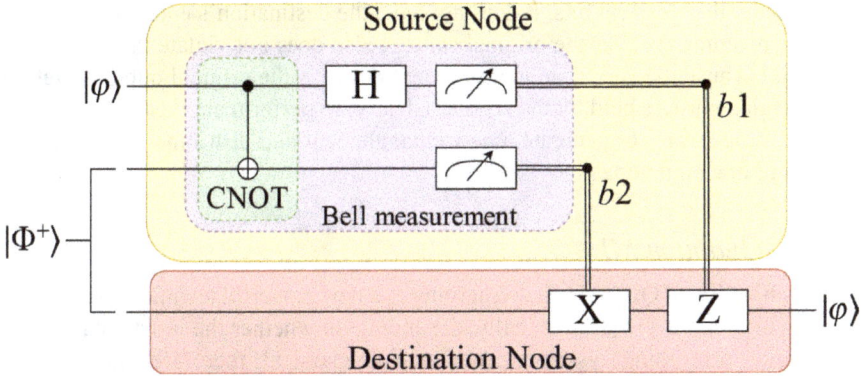

Figure 11.2   Quantum teleportation circuit

11.3.5   Teleportation

Quantum teleportation allows sending qubits without the transmission of the physical particle that stores the quantum state. In fact, although a qubit can be encoded by the photon polarization, if it is lost due to attenuation or altered by the environment while it is transferred to a remote quantum computer, the original quantum state cannot be recovered. Hence, quantum teleportation is a workaround procedure for transferring quantum information leveraging classical communication media.

At the basis of quantum teleportation, it is assumed that a specific node in a quantum network can generate and distribute entanglement between the source and destination node. Given the quantum measurement postulate and the no-cloning theorem, the source node has to send the quantum state φ to a destination node without any a priori knowledge about its state. To this end, the source node must perform some operations, depicted in Figure 11.2, between the qubit to be sent and its owned part of the entangled particle, i.e., Φ^+. In particular, the sender performs a Bell state measurement which consists of a controlled NOT (CNOT) operation followed by a Hadamard operation and two measurements. The CNOT operation acts on two qubits and performs a bit-flip (i.e., a NOT operation) on the target qubit when the control qubit is 1. Otherwise, the target qubit remains unchanged. The Hadamard operation is applied on the first qubit and creates a superposition of the two basis states. Both the CNOT and Hadamard operations are used to rotate the Bell basis into the computational basis of the two qubits. Finally, the outcomes of the measurement on the two qubits, i.e., $b1$ and $b2$, are sent to the destination node through a classical channel.

At this point, based on the measurement outcomes, the receiver can recover the original quantum state from its entangled particle by either applying X or Z operations, both or none. The X and Z operations correspond to a bit-flip or phase-flip on the qubit, respectively.

It should be noted that the teleportation of a quantum state does not violate the relativity principle, since it requires classical communication. Furthermore, quantum teleportation guarantees a safe state transferring, since even though an attacker can intercept the classical bits, it does not have the destination's entangled particle, and thus it cannot recover the original state. It also does not violate the no-cloning theorem, because the Bell state measurement destroys the original qubit as well as the entangled particle held by the sender. Clearly, to perform another qubit teleportation, it is necessary to generate a new entanglement and distribute it between the two communicating nodes.

11.3.6   Quantum ML

The combination of QC and ML is emerging as a new powerful technique to improve learning algorithms [15]. Specifically, depending on whether the input data and the information processing system are quantum or classical, there are four different approaches to merge QC and ML [32, 33]:

- *Classical-classical approach.* It implements quantum-inspired classical algorithms on classical computers. Here, classical data are processed by classical computers, by employing traditional ML algorithms based on quantum principles theory.
- *Quantum-classical approach.* It consists in employing ML techniques in a QC system. In particular, ML can help quantum computers to learn from data. For instance, ML can be used to analyze measurement data, thus reducing the number of measurements of a quantum state.
- *Classical-quantum approach.* It is commonly known as QML. This approach aims at translating classical ML algorithms into a quantum-compliant language to take advantage of quantum mechanics by running it on quantum computers. The adoption of this approach requires a pre-processing step to convert the classical input data into suitable data for quantum computers. Nowadays, the research community proposes several encoding methods, such as basis encoding and amplitude encoding [32, 34].
- *Quantum-quantum approach.* It aims to develop quantum algorithms to manipulate quantum data. In this approach, it is not required to encode data, as the input is directly the quantum state of the system.

In particular, this work considers the third approach, as in the real-world scenario, most of the input data are classical. Moreover, since quantum-inspired algorithms are executed on classical computers, the achievable speed-up is not comparable with running it on quantum computers [15].

Nevertheless, QC and, in turn, the application of QML have to face some hardware problems. Quantum states, in fact, are very fragile (i.e., decoherence principle) and suffer from every gate operation, thus inevitably altering computation tasks and limiting quantum computers capabilities. To avoid these problems, the scientific literature proposes two main strategies: first, quantum circuits must be embedded in a specialized large infrastructure with cooling systems able to maintain a near absolute zero temperature [35]; second, the state of qubits must be preserved through the introduction of quantum error correction schemes which spread the information originally belonging to one logical qubit into several physical qubits [29].

However, due to the continuous growth of the number of devices involved in the WMN and, consequently, the amount of exchanged information, QML can help to speed up algorithms used in WMN. In fact, it can improve the computational time, thus getting results faster and also in real time, as well as increasing the learning capacity and efficiency by discovering more intricate patterns from the input data [36, 37]. In detail, preliminary studies on the performance comparison between QML and ML algorithms demonstrated that the QML is convenient in the case of high-dimensionality input data [38]. Hence, future wireless networks must take into account the possibility to jointly use traditional ML and QML capabilities by supporting the integration of quantum computers.

11.4 Introduction of QML in WMNs: design principles and research challenges

The application of QML methodologies in WMNs can be achieved only with the definition of novel network architectures. In fact, the integration of quantum and traditional computers performing QML and ML algorithms, respectively, requires the introduction of new logical entities embedded with new functionalities. To this end, this section presents design principles for the realization of two innovative network architectures, denoted by centralized and distributed approaches, able to combine the benefits provided by traditional and quantum computers deployed either in the cloud or at the edge of the network.

11.4.1 Centralized architecture

The integration of QML functionalities in future WMNs first requires the introduction of quantum computers in their architectures. Nowadays, some Tech Giants, such as IBM, Google, and Microsoft, have already developed quantum computers with up to a hundred qubits, also envisioning strong improvements in this direction for the next years [35]. Accordingly, a first suitable approach for integrating quantum computers in WMNs is the centralized architecture depicted in Figure 11.3, where quantum computers deployed by Tech Giants in their cloud are supposed to perform QML algorithms.

The proposed architecture is composed of the access network, the wireless mesh backbone, and the remote cloud:

- The access network includes all the application scenarios sustained by the WMN (e.g., mobile networks, wireless sensor networks, and vehicular networks) and the related network attachment points, which provide the connection to the mesh network (e.g., base station (BS) and sink node).
- The wireless mesh backbone hosts MRs (with or without gateway capabilities), a data aggregator node, which stores and transmits dataset for intelligence operations, and traditional computers solving simple and low-dimensionality

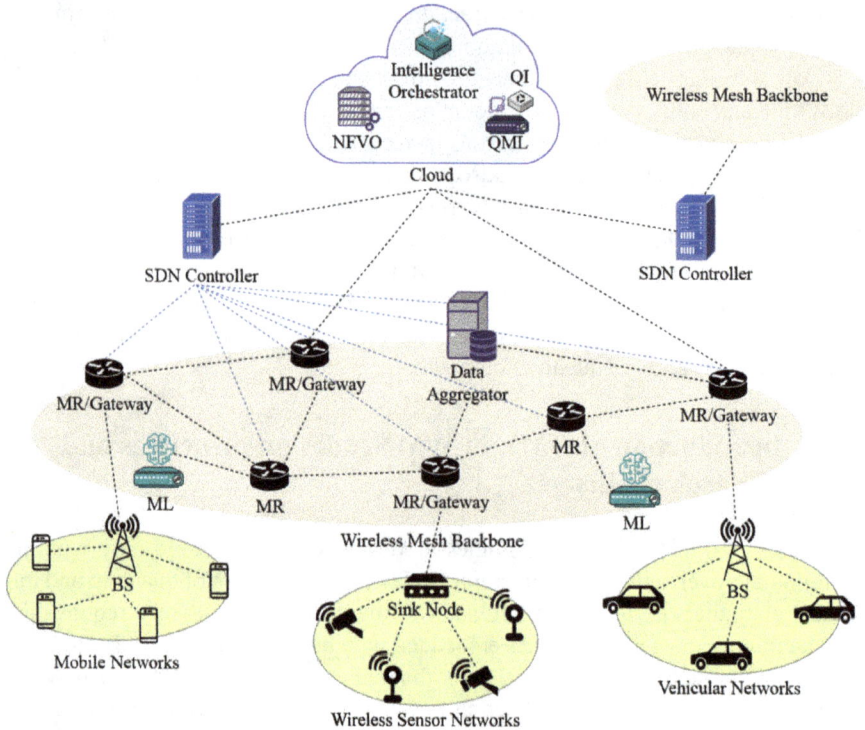

Figure 11.3   QC-guided network intelligence: centralized deployment

ML problems. Given the heterogeneity and complexity of the wireless mesh backbone, the traffic flow is managed by an SDN controller, thus avoiding network congestion issues.

- The remote cloud provides orchestration and high-dimensionality computational capabilities to the overall network. In detail, the intelligence orchestrator performs the allocation of computing resources among ML and QML tasks according to their data-dimensionality, while the network function virtualization orchestrator (NFVO) provides service management functionalities. It is worthwhile to note that quantum computers are equipped with a quantum interface (QI) useful for data pre-processing.

11.4.1.1   The information exchange in the centralized architecture

As illustrated in Figure 11.4, the information exchange in the centralized architecture can be summarized as in what follows.

- *Phase 1: Dataset creation.* Each node belonging to the network generates information data to be processed by traditional or quantum computers for the purposes listed in Section 11.2. The collected data strongly depend on the

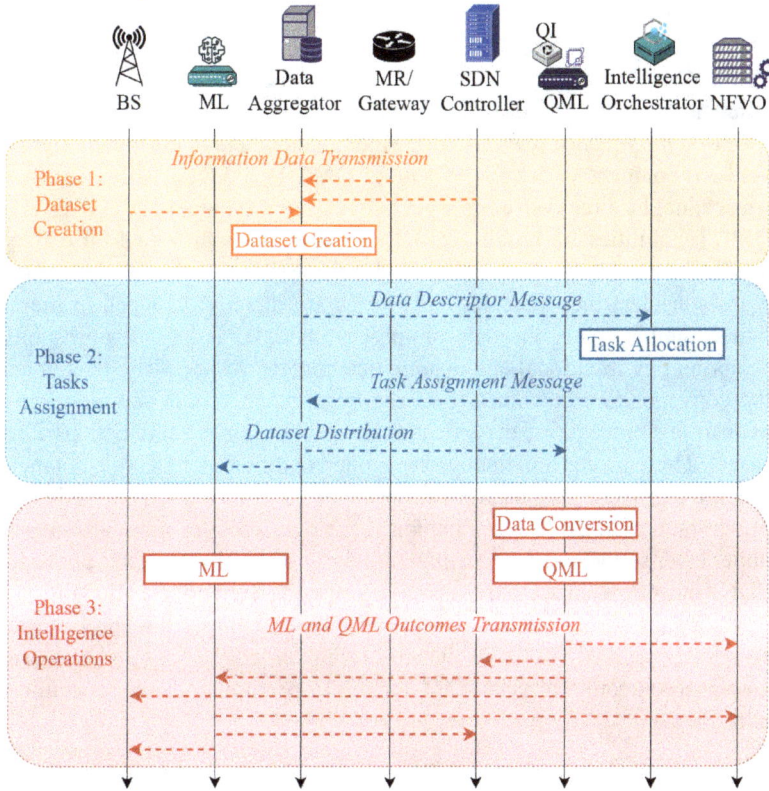

Figure 11.4   Message sequence chart of the centralized architecture

considered node. While end users (e.g., mobile phones, sensors, and vehicles) acquire data from the surrounding environment, such as channel quality indicators and performance levels of high-level applications, network equipment (e.g., MRs, BSs, sink nodes, SDN controllers, and NFVO) provide information related to the network functionalities, such as bandwidth and energy consumption. All the collected data are transmitted to data aggregators by means of Representational State Transfer (REST) or RESTful communication protocols in order to increase the performance, scalability, simplicity, and reliability of the network. Then, data aggregators pre-process incoming data and compare them with existing network information in order to create and/or update datasets useful for intelligence operations.

- *Phase 2: Tasks assignment.* The intelligence orchestrator must assign the generated datasets to computing resources (e.g., traditional or quantum computers). To this end, given the huge amount of data to be exchanged, the data aggregators periodically create and transmit to the intelligence orchestrator a data descriptor message containing high-level information about the available datasets, such as data format, data size, and statistical variability with respect to previous updates. Starting from this information and considering the status of computing resources, the intelligence orchestrator performs the task allocation and sends a task assignment message to the data aggregators in order to efficiently transmit the datasets to designed traditional or quantum computers.

- *Phase 3: Intelligence operations.* Quantum computers in the cloud offer a suitable environment where implementing QML techniques. However, classical data cannot be directly used as input of quantum computers. Accordingly, when QML capabilities are required, a QI logical entity is first used for converting classical data into quantum data and vice versa. Without loss of generality, this work considers that these logical entities are directly equipped in the quantum computer. After the data pre-processing, ML and QML operations are performed by traditional and quantum computers, respectively, thus obtaining the corresponding outcomes (e.g., the hyperparameters of the model in case of learning procedures; classification, prediction, or specific actions in other cases). These results are, finally, transmitted to different network equipment for specific purposes, ranging from service management to network optimization. For instance, the NFVO can exploit these outcomes for optimally managing upper layer services and allocating virtual resources among active applications. SDN controllers, instead, dynamically configure network functionalities (e.g., flow forwarding and load balancing) and solve complex routing problems based on users' mobility and traffic dynamics. Finally, edge nodes and BSs use ML and QML outcomes to update ML models or perform resource scheduling and allocation.

11.4.1.2 Benefits and research challenges

The implementation of the proposed centralized architecture in WMNs presents both advantages and disadvantages, thus providing several research directions to the scientific community. On the one hand, at the time of this writing, the number of available quantum computers and qubits is strongly limited by physical and economical constraints, making the centralized approach the first feasible solution to integrate QML in WMNs. Moreover, the centralized architecture presents only one point of failure (i.e., the central nodes in the cloud). As a consequence, it is the easiest architecture to maintain and control. The simplicity of this approach, in fact, allows to efficiently control the computing resources by directly managing the few quantum computers placed in the cloud. On the other hand, the centralized approach is very unstable, since any issue affecting the central nodes in the cloud inevitably causes impairment throughout the network. Furthermore, when QML operations are required, the network must sustain the transmission of a huge amount of data from the data aggregator to the remote cloud, thus producing very high bandwidth and energy consumption, along with possible congestion episodes. In addition, the centralized approach presents high communication latency, significantly impairing the benefits provided by the introduction of quantum computers. These open issues introduced several possible research challenges. The transmission of data from data

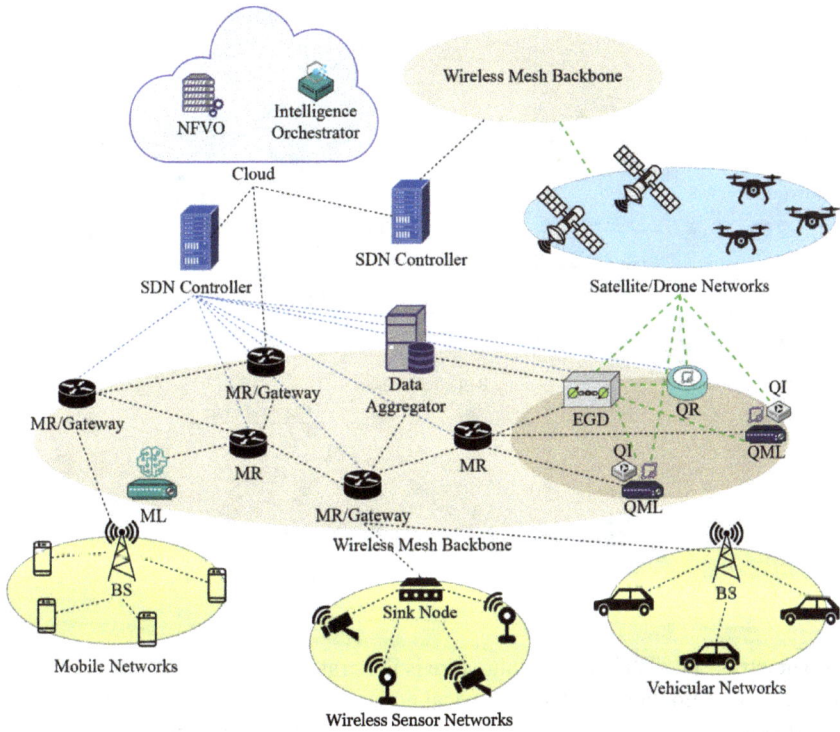

Figure 11.5   QC-aided network intelligence: distributed deployment

aggregators and QIs, in fact, requires the introduction of optimized and energy-aware routing strategies. At the same time, the network scalability can be improved through the optimization of the number and the position of data aggregators based on service details and users' statistics. These problems can be solved by the introduction of the same QC in WMNs. In fact, QML can be used for complex optimization tasks, like routing problems (i.e., select the optimal path of data packets) or location strategies.

11.4.2   Distributed architecture

Despite the current limited number of quantum computers and qubits, the growing interest of the scientific community in QC supports the idea of developing also a distributed architecture in a very far future. The decentralization of QC capabilities entails a higher number of quantum computers, geographically distributed at the edge of the network and equipped with few qubits to mitigate the deployment cost. However, to scale up the number of qubits, it is possible to consider a network of quantum nodes that communicate through the quantum Internet paradigm in order to solve also complex QML tasks [39].

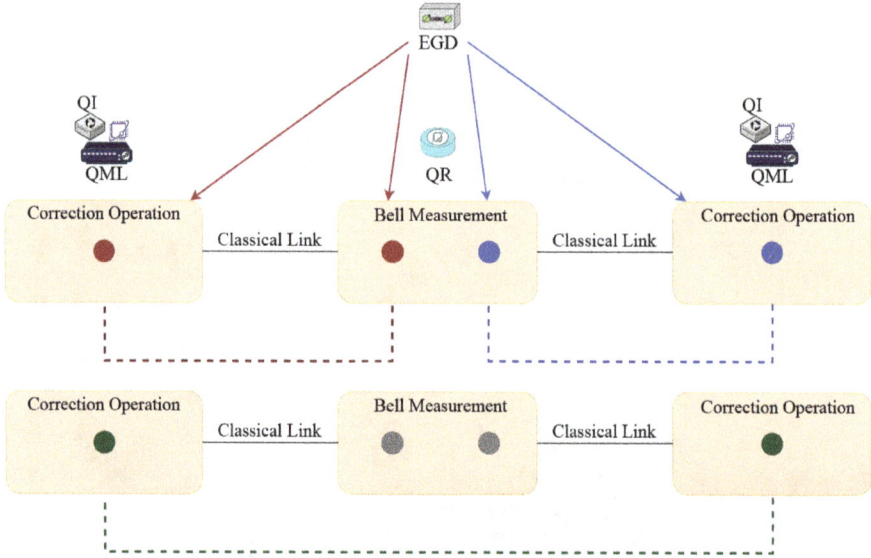

Figure 11.6　　*Representation of the entanglement swapping procedure*

The proposed distributed architecture is illustrated in Figure 11.5. It requires the introduction of novel logical and physical nodes, along with those presented for the centralized architecture, in order to efficiently sustain the deployment of distributed quantum computers:

- The wireless mesh backbone contains two new nodes, named entanglement generator and distributor (EGD) and quantum repeater (QR). The former represents a third party used to generate and distribute entangled particles following, for example, the SPDC method. The latter, instead, allows to distribute entangled particles between remote quantum computers. In fact, the reliability of the entanglement distribution is strongly affected by attenuation effects due to the distance between quantum computers [39]. However, since the no-cloning theorem does not allow to simply read and copy qubits, traditional repeaters must be replaced by QRs, which perform the entanglement swapping to establish longer-distance end-to-end entanglements. Figure 11.6 depicts an example of this procedure involving two distant quantum computers and a QR. In particular, the EGD first creates and delivers a pair of entanglement particles to the involved quantum computers and the QR. Then, the QR performs a Bell state measurement on its particles, thus causing their collapse. The resulting measurement outcomes are transferred to both quantum computers by means of classical channels. Finally, each quantum computer carries out correction operations in order to reconstruct the end-to-end entanglement. Without loss of generality, this procedure can be extended for multiple QRs scenarios where each QR must be able to receive, process, and transmit both classical and quantum data.

- The entangled particle distribution can be also supported by a satellite or a drone network. Satellites and drones, in fact, can act as QRs for the entanglement distribution of two distant quantum remote computers, completely substituting ground QRs or simply supporting them. On the one hand, the main benefit of using satellite is that photons loss takes place at low levels of the troposphere and the transmission path has no photon absorption [40]. On the other hand, since low-orbital satellites serve specific ground quantum computers only for a limited time, drones can be used as QR, receiving a photon and retransmitting it to the involved quantum remote computer, the next drones, or the next ground QR [41].

It is important to note that the EGD can be either a separated physical node of the network or simply a logical entity equipped by involved QRs (e.g., ground QRs, satellites, or drones). Without loss of generality, this work considers the EGD as a physical ground entity placed in the wireless mesh backbone.

11.4.2.1 The information exchange in the distributed architecture

The distributed architecture, shown in Figure 11.7, works as in what follows.

- *Phase 1: Dataset creation.* As for the centralized architecture, in the first phase of the information exchange procedure, end nodes belonging to different use cases and network equipment deployed in the wireless mesh backbone and in the cloud generate a huge amount of data. This information is, then, transmitted to the data aggregator, which creates new datasets or updates existing ones.
- *Phase 2: Task assignment.* Also the second phase, aiming at allocating tasks among computing resources (i.e., traditional and quantum computers), is equivalent to the corresponding phase in the centralized architecture. Here, data aggregators transmit metadata of the generated datasets (such as the format or size of data) to the intelligence orchestrator, thus avoiding the exchange of an excessive amount of information and, in turn, the congestion of the network. The intelligence orchestrator, starting from the aforementioned metadata and from the status of the intelligence network, assigns specific tasks to computing resources and sends a task assignment message to the data aggregator. Involved datasets are, finally, delivered to traditional or quantum computers.
- *Phase 3: Network setup.* Differently from the centralized architecture, in the distributed approach, quantum nodes are deployed at the network edge and dynamically grouped in clusters in order to scale up the number of qubits and efficiently solve more complex QML problems. In this case, the intelligence orchestrator creates quantum computer networks aiming at grouping computing resources based on the number of available qubits and the distance between them in order to reduce attenuation effects. Since quantum computers belonging to the same cluster share quantum states through the aforementioned teleportation protocol, the third phase of the information exchange envisages setting up the QML network by generating and transmitting entangled particles among

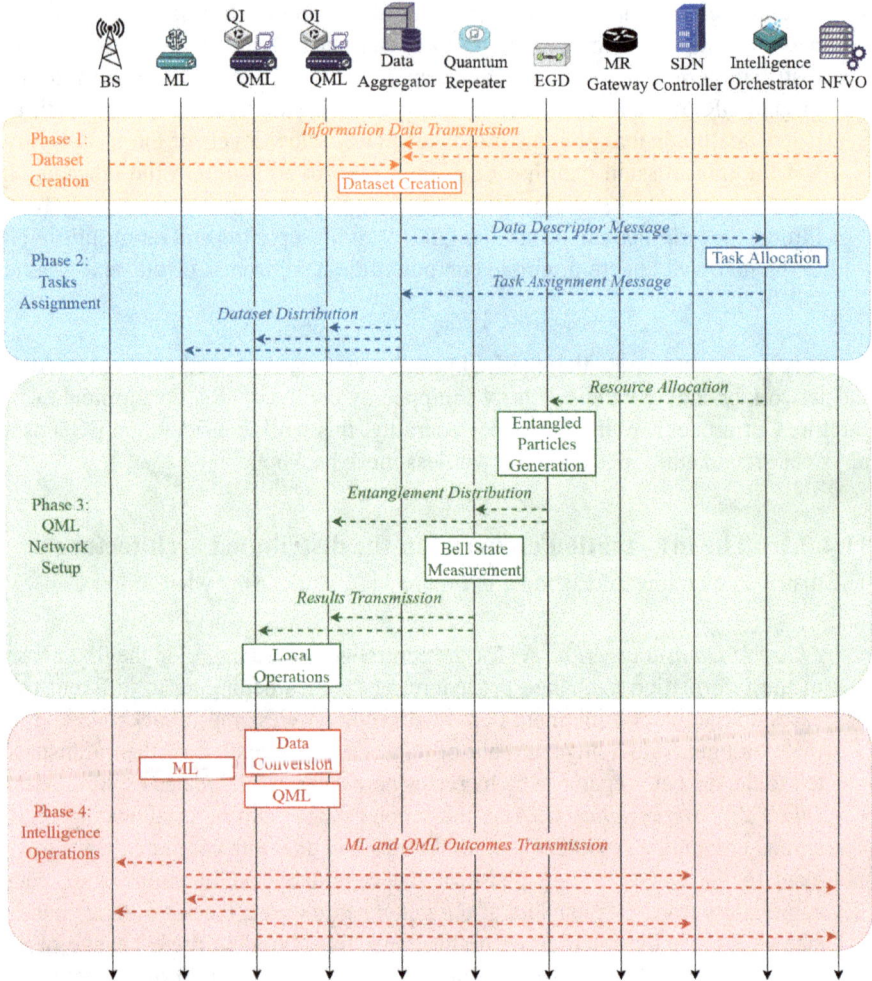

Figure 11.7   Message sequence chart of the distributed architecture

involved quantum nodes. They are, then, able to establish a long-distance end-to-end entanglement through the entanglement swapping procedure.

- *Phase 4: Intelligence operations.* Again, when quantum computers are involved in the computing operation, the received dataset is converted by the QI devices before executing QML algorithms. The outcomes of ML and QML operations are, finally, transmitted to the nodes of the network for different purposes (e.g., the SDN controllers for optimal routing procedures, the NFVO for allocating virtual resources, BSs for optimal resource scheduling).

11.4.2.2 Benefits and research challenges

The issues pointed out for the centralized architecture in WMNs can be easily overcome by employing the proposed distributed architecture. In fact, the deployment of multiple quantum computers into the wireless mesh backbone mitigates possible network congestion episodes, improves scalability, and reduces communication latency. On the other side, since quantum Internet is in its fancy, the entanglement distribution and heterogeneity of qubit may represent a first hindrance for the physical implementation of the distributed architecture [39]. In fact, quantum information will be initially transferred to quantum computers belonging to the same cluster and equipped with homogeneous qubits. Then, according to the scientific community long-term vision, the proposed architecture will be practicable when hardware heterogeneity of different quantum computers will be taken into account and entangled particles will be distributed between distant quantum computers. Another issue is related to the simplicity of involved quantum computers in terms of the number of qubits. In this case, complex QML tasks require a high cluster size that drastically increases the number of transferred quantum states and, in turn, the delays and error rates due to the quantum teleportation procedure/t of congestion for the load balancing of computing services. Here, since many quantum computers are involved, it is important to efficiently distribute quantum algorithms among them by jointly minimizing the size of the cluster and the user-quantum computer distance. Another research challenge to take into account is the optimal allocation of quantum computers where more computational tasks are expected.

11.5 Conclusions

The need of combining QC and ML to fulfill the requirements of future WMNs envisages the definition of proper network architectures with the introduction of new logical entities. After describing the role of different ML algorithms for WMNs and the main properties of QC, this work introduced the application of quantum ML for WMNs. Specifically, it proposed a centralized and a distributed architecture where quantum ML capabilities are placed in the cloud or at the edge, respectively. Design principles and information exchange procedures are deeply discussed for both these architectures, also highlighting their advantages, disadvantages, and possible future research directions.

References

[1] Hossain E., Leung K.K. *Wireless Mesh Networks: Architectures and Protocols*. Springer; 2007.

[2] Akyildiz I.F., Wang X., Wang W. 'Wireless mesh networks: a survey'. *Computer Networks*. 2005;47(4):445–87.

[3] Akyildiz I.F., Wang X. 'A survey on wireless mesh networks'. *IEEE Communications Magazine*. 2005;43(9):S23–30.

[4]   Bano M., Qayyum A., Bin Rais R.N., Gilani S.S.A. 'Soft-mesh: a robust rout-
      ing architecture for hybrid SDN and wireless mesh networks'. *IEEE Access.*
      2021;9:87715–30.

[5]   Kusumoto H., Okada H., Kobayashi K., Katayama M. 'Performance compari-
      son between single-user MIMO and multi-user MIMO in wireless mesh net-
      works'. The 15th International Symposium on Wireless Personal Multimedia
      Communications; IEEE; 2012. pp. 202–6.

[6]   Jaafar W., Ajib W., Tabbane S. 'The capacity of MIMO-based wireless mesh
      networks'. 2007 15th IEEE International Conference on Networks; IEEE;
      2007. pp. 259–64.

[7]   Kumar R., Venkanna U., Tiwari V. 'A time granular analysis of software de-
      fined wireless mesh based IoT (SDWM-IoT) network traffic using supervised
      learning'. *Wireless Personal Communications.* 2021;116(3):2083–109.

[8]   Elzain H., Yang W. 'QoS-aware topology discovery in decentralized software
      defined wireless mesh network (D-SDWMN) architecture'. Proceedings of
      the 2018 2nd international conference on computer science and artificial intel-
      ligence; 2018. pp. 158–62.

[9]   Al-Jarrah M.A., Alsusa E., Al-Dweik A., Alouini M.-S. 'Performance analy-
      sis of wireless mesh backhauling using intelligent reflecting surfaces'. *IEEE
      Transactions on Wireless Communications.* 2021;20(6):3597–610.

[10] Karunaratne S., Gacanin H. 'An overview of machine learning ap-
      proaches in wireless mesh networks'. *IEEE Communications Magazine.*
      2019;57(4):102–8.

[11] Zornoza J., Mujica G., Portilla J., Riesgo T. 'Merging smart wearable de-
      vices and wireless mesh networks for collaborative sensing'. 2017 32nd
      Conference on Design of Circuits and Integrated Systems (DCIS); IEEE;
      2017. pp. 1–6.

[12] Nam J., Kim S.M., Min S.G. 'Extended wireless mesh network for VANET
      with geographical routing protocol'. 11th International Conference on Wireless
      Communications, Networking and Mobile Computing; 2015. pp. 1–6.

[13] Lee H.-C., Ke K.-H, . 'Monitoring of large-area IoT sensors using a LoRa
      wireless mesh network system: design and evaluation'. *IEEE Transactions on
      Instrumentation and Measurement.* 2018;67(9):2177–87.

[14] Nawaz S.J., Sharma S.K., Wyne S., Patwary M.N., Asaduzzaman M.
      'Quantum machine learning for 6G communication networks: state-of-the-art
      and vision for the future'. *IEEE Access.* 2019;7:46317–50.

[15] Khan T.M., Robles-Kelly A. 'Machine learning: quantum vs classical'. *IEEE
      Access.* 2020;8:219275–94.

[16] Zhang C., Patras P., Haddadi H. 'Deep learning in mobile and wire-
      less networking: a survey'. *IEEE Communications Surveys & Tutorials.*
      2019;21(3):2224–87.

[17] Wang X., Wong J.S., Stanley F., Basu S. 'Cross-layer based anomaly detec-
      tion in wireless mesh networks'. 2009 ninth annual international symposium
      on applications and the internet; IEEE; 2009. pp. 9–15.

[18] Shams E.A., Rizaner A. 'A novel support vector machine based intrusion detection system for mobile ad hoc networks'. *Wireless Networks*. 2018;24(5):1821–9.

[19] Das B., Roy A.K., Khan A.K., Roy S. 'A new approach for gateway-level load balancing of wmns through k-means clustering'. 2014 International Conference on Computational Intelligence and Communication Networks; IEEE; 2014. pp. 515–9.

[20] Zaidi Z.R., Hakami S., Landfeldt B., Moors T. 'Real-time detection of traffic anomalies in wireless mesh networks'. *Wireless Networks*. 2010;16(6):1675–89.

[21] Boushaba M., Hafid A., Belbekkouche A., Gendreau M. 'Reinforcement learning based routing in wireless mesh networks'. *Wireless Networks*. 2013;19(8):2079–91.

[22] Yin M., Chen J., Duan X., Jiao B., Lei Y. 'Qebr: Q-learning based routing protocol for energy balance in wireless mesh networks'. 2018 IEEE 4th International Conference on Computer and Communications (ICCC); IEEE; 2018. pp. 280–4.

[23] Tang F., Mao B., Fadlullah Z.M., *et al.* 'On removing routing protocol from future wireless networks: a real-time deep learning approach for intelligent traffic control'. *IEEE Wireless Communications*. 2017;25(1):154–60.

[24] Nie L., Jiang D., Yu S., Song H. 'Network traffic prediction based on deep belief network in wireless mesh backbone networks'. 2017 IEEE Wireless Communications and Networking Conference (WCNC); IEEE; 2017. pp. 1–5.

[25] Mnih V., Kavukcuoglu K., Silver D., *et al.* 'Human-level control through deep reinforcement learning'. *Nature*. 2015;518(7540):529–33.

[26] Liu Q., Cheng L., Jia A.L., Liu C. 'Deep reinforcement learning for communication flow control in wireless mesh networks'. *IEEE Network*. 2021;35(2):112–9.

[27] Yin C., Yang R., Zou X., Zhu W. 'Research on topology planning for wireless mesh networks based on deep reinforcement learning'. 2020 2nd International Conference on Computer Communication and the Internet (ICCCI); IEEE; 2020. pp. 6–11.

[28] Xu Y., Yu J., Headley W.C., Buehrer R.M. 'Deep reinforcement learning for dynamic spectrum access in wireless networks'. MILCOM 2018-2018 IEEE Military Communications Conference (MILCOM); IEEE; 2018. pp. 207–12.

[29] Gyongyosi L., Imre S. 'A survey on quantum computing technology'. *Computer Science Review*. 2019;31(9):51–71.

[30] Nielsen M.A., Chuang I.L. *Quantum Computation and Quantum Information: 10th Anniversary Edition*. New York: Cambridge University Press; 2010.

[31] Cacciapuoti A.S., Caleffi M., Van Meter R., Hanzo L. 'When entanglement meets classical communications: quantum teleportation for the quantum Internet'. *IEEE Transactions on Communications*. 2020;68(6):3808–33.

[32] Sergioli G. 'Quantum and quantum-like machine learning: a note on differences and similarities'. *Soft Computing*. 2020;24(14):10247–55.

[33]   Schuld M., Petruccione F. *Supervised Learning with Quantum Computers*. Vol. 17. Berlin: Springer; 2018.

[34]   Sierra-Sosa D., Telahun M., Elmaghraby A. 'TensorFlow quantum: impacts of quantum state preparation on quantum machine learning performance'. *IEEE Access*. 2020;8:215246–55.

[35]   O'Quinn W., Mao S. 'Quantum machine learning: recent advances and outlook'. *IEEE Wireless Communications*. 2020;27(3):126–31.

[36]   Phillipson F. Quantum machine learning: benefits and practical examples. QAnswer; 2020. pp. 51–6.

[37]   Bhat J.R., Alqahtani S.A. '6G ecosystem: current status and future perspective'. *IEEE Access*. 2021;9:43134–67.

[38]   Fastovets D., Bogdanov Y.I., Bantysh B., Lukichev V.F. 'Machine learning methods in quantum computing theory'. International Conference on Micro- and Nano-Electronics 2018; International Society for Optics and Photonics; 2019. p. 110222S.

[39]   Caleffi M., Chandra D., Cuomo D., Hassanpour S., Cacciapuoti A.S. 'The rise of the quantum Internet'. *Computer*. 2020;53(6):67–72.

[40]   Yin J., Cao Y., Li Y.-H., *et al.* 'Satellite-based entanglement distribution over 1200 kilometers'. *Science*. 2017;356(6343):1140–4.

[41]   Liu H.-Y., Tian X.-H., Gu C., *et al.* 'Drone-based entanglement distribution towards mobile quantum networks'. *National Science Review*. 2020;7(5):921–8.

Chapter 12

Game theoretical-based task allocation in malicious cognitive Internet of Things

Marco Martalò[1], Virginia Pilloni[1], Talha Faizur Rahman[2], and Luigi Atzori[1]

In this chapter we consider a heterogeneous Internet of Things (IoT) scenario in which sensor nodes belong to different platforms and form different clusters managed by their own cluster head (CH). This configuration is exploited to foster nodes collaboration in sensing activities with coordinated resource usage. We assume that the considered nodes have cognitive radio (CR) and exploit device-to-device (D2D) communications. The nodes collaboratively sense the spectrum, through standard energy detection, to find spectrum holes to be opportunistically exploited for task allocation by the CH. Some of the nodes may act as malicious nodes (MNs) trying to disrupt this process by providing tampered data, trying to lead to a higher overall probability of error of the spectrum sensing. At this point, task allocation is performed by means of a game theoretical-based approach considering two elements: the gain that is won for its contribution to sensing and for the execution of the task (in case, it wins the competition), and the cost in terms of energy to be consumed in case the task is executed. In particular, we investigate the impact of tampered data in the former aspect so that the MNs try to gain as much as possible control of the allocated task, thus performing a Denial of Service (DoS) attack. Extensive simulations are performed to evaluate the impact of the main system parameters on the overall performance and provide guidelines for future work.

12.1   Introduction

Wireless mesh network (WMN) is a well-established scheme in which nodes can wirelessly communicate with each other through a dynamic and non-hierarchical topology and with the external Internet through proper gateways [1].

[1]Networks for Humans (Net4U) Lab, Department of Electrical and Electronic Engineering, University of Cagliari, and also the National Telecommunication Inter University Consortium, Research Unit of Cagliari, Italy

[2]Department of Electrical and Computer Engineering, Mississippi State University, Starkville, MS, USA

Such a communication technology can be a benefit in the IoT scenarios where resource-constrained objects, such as sensors, personal electronics, and smart vehicles [2], cooperate to perform some tasks of applications of interest, for example, communication of data sensed from the physical world, computation of quantities of interest, data storage, and so on [2]. In fact, many IoT applications involve the pervasive aggregation of data from devices in order to manage the physical world by means of D2D short-range communications among the involved devices.

As in WMNs, the IoT network scenario can foresee the use of CHs to maintain a hierarchy among the participants in the communication infrastructure. CHs should perform extra functionalities (extraction of information from data sent by other nodes, authentication and authorization of nodes, etc.) and have Internet connectivity to perform as a gateway to forward the information to the application server. Therefore, the CH may be in charge of allocating the task to be performed to a suitable executor among several available providers, i.e., nodes in the IoT networks.

Inefficient use of these resources for IoT task performance may be a limiting performance factor because it can lead to an untimely depletion of available resources. To this end, cooperative resource allocation is needed to guarantee correct task execution, while prolonging the network lifetime as much as possible. Even if cooperative resource allocation may be designed to optimally achieve the desired goals, some of the nodes in the network may perform data falsification, by tampering with their data so that the task allocation procedure is overall degraded. This type of attack is also called Byzantine attack and the attacking nodes will be denoted in the following as MNs, in contrast with honest nodes (HNs) [3].

In this chapter, we consider a task allocation problem in a cognitive IoT scenario, i.e., a scenario where IoT nodes are equipped with CR capabilities and cooperate to detect spectrum holes in legitimate bands served by Licensed (primary) Users (LUs). The motivation behind the use of CRs is arisen by the fact that in the future IoT traffic will skyrocket, and the allotted spectrum together with ISM (Industrial, Scientific, Medical) bands might not be sufficient to accommodate the traffic while maintaining an adequate level of quality in device communications. Thus, an IoT node with CR capabilities, namely a cognitive node (CN), is capable of using the spectrum without harming the LU transmissions with interference. Once the spectrum hole is detected by the CNs, the CH assigns each requested task to a single CN in the cluster exploiting the available spectrum for service provisioning according to the game theoretical-based approach in Rahman *et al.* [4]. Even if the problems of task allocation in IoT networks and the presence of MNs in cognitive spectrum sensing are separately studied, the main contribution of this chapter is to assess the performance of such IoT networks to vulnerabilities to one or more MNs that intend to disrupt the spectrum sensing process by reporting tampered measurements to the CH in order to gain as much control of the allocated task as possible. This allows the MNs to provide some DoS attacks.

The rest of this chapter is structured as follows. In Section 12.2, we overview the main literature on the considered task allocation and cognitive networks under MN topics. In Section 12.3, the reference scenario is described. In Section 12.4, the considered task allocation strategy in the presence of MNs is presented. In Section

12.5, numerical results are provided. Finally, in Section 12.6, concluding remarks are given.

12.2 Related work

The task allocation issue has been extensively addressed in WSNs. Haghighi [5] studied the application of market-based algorithms for energy management, resource allocation, and task assignment in WSNs, considering the execution of multiple concurrent tasks at the same time. Multi-objective optimization problems in WSNs are investigated by Fei *et al.* [6], where the authors compare a number of approaches focused on finding a trade-off among several different criteria, such as network lifetime, coverage, and packet loss.

IoT is based on WSN, but the IoT scenario is different from most WSN scenarios. This is mainly due to the fact that in the IoT the requester/owner has complete control over objects, for instance, requesters/owners can switch off/ on their objects (mobile phones, tablets, etc.) depending on their personal needs. Furthermore, objects' mobility affects the network topology, which dynamically changes over time. As a consequence, objects' connectivity becomes unreliable [7]. This requires strong coordination among IoT objects, which have to dynamically adapt to network changes (i.e., objects leaving or joining) by updating task allocations accordingly. In this regard, [8] and [9] study the resource allocation for IoT applications where the aim is to find and allocate the resources in a way to optimize service execution among the IoT objects. Task allocation for resource usage optimization in the mobile cloud and edge computing is a crucial issue, as confirmed by numerous works, such as [10–12]. In particular, the cloud offloading policy is discussed with the aim of determining the cases where it is more convenient to use local resources or cloud resources. In the study by Wang *et al.* [13], the task assignment is performed also taking into account objects' trustworthiness. The proposed approach is based on an auctioning mechanism for multi-objective optimization of system performance (more specifically mission reliability, utilization variance, and delay to task completion).

In a recent work [7], the IoT_Prose framework is proposed in order to exploit D2D communications among the nodes to implement IoT applications. A game-theoretical solution is proposed by defining a utility function that allows objects to maximize the utility of individual nodes. Pilloni *et al.* [7] considered a repeated task allocation problem where nodes are executing the *won* tasks repetitively together with the bidding for new ones if the request comes in. In this case, it is obvious that nodes would deplete their energy bank quickly causing cutbacks in network lifetime. Moreover, a static spectrum allocation is considered by the authors.

CR together with D2D is able to utilize resources efficiently and minimize the interference between D2D and cellular users when D2D users are operating in in-band mode [14, 15]. Kakalou and Psannis [16] propose a system architecture that makes use of CR capabilities in order to enhance energy and spectral efficiency together with the provisioning of quality of service to applications via software-defined network flows. More recently, CR techniques are considered for the realization of IoT services. In this regard, Aijaz and Aghvami [17] comprehensively studied the concept of CR technology from the perspective of machine-to-machine (M2M) IoT. Moreover, the authors conclude that cognitive M2M operation requires an energy-efficient, reliable, and Internet-enabled protocol stack with enabled intelligence from the physical layer to the transport layer. In Khan *et al.* [18], the authors emphasize on the fact that IoT objects can be able to exploit spectrum resources effectively in spectrum constrained world.

The problem of malicious spectrum sensing attacks has also been extensively studied in the literature, since wireless networks are, by nature, prone to suffer from various kinds of security threats, for example, jamming, spoofing, wiretap, and so on. A comprehensive survey of the main Byzantine attack and defense techniques is provided in Zhang *et al.* [3]. More recently, in Gao *et al.* [19] the performance limits of cooperative spectrum sensing under Byzantine attacks are analyzed, and a novel scheme with low complexity is proposed to counter Byzantine attacks, by taking into account the statistics of the attack itself. In Wu *et al.* [20], under a generalized soft Byzantine attack model, the authors theoretically analyze the attack strategy to evaluate the security of Byzantine attackers and provide design guidelines for a robust defense strategy.

12.3   Reference scenario

The scenario considered in this chapter is that of an opportunistic CR-based IoT system, where devices perform *sensing tasks* requested by the system on a best-effort basis, whenever conditions are suitable. Accordingly, devices' functionalities can be hired by any other device that is part of the system, even a remote server or external groups of devices, to perform the task (see Figure 12.1). This scenario was first investigated by Rahman *et al.* [4]. Note that in this chapter we only consider sensing tasks, but the model could be straightforwardly extended by considering any task (e.g., processing, storage, and data forwarding). In the scenario depicted in Figure 12.1, devices with similar and replaceable sensing functionalities are organized into clusters, where they can communicate directly through D2D communications. Each cluster is coordinated by a CH, which is selected among the cluster nodes due to its extra functionalities and the capability to establish a reliable direct Internet connection to the application server.

Whenever the CH receives a task request from the application server, it sends a relevant control signal to the clustered CNs over a broadcast control channel to initiate the spectrum sensing. Accordingly, the CNs in the cluster observe the spectrum band chosen by the CH to estimate the received signal energy, if any. The information is then shared with the CH, which estimates the absence of any LU by fusing the received signal energy data. The CNs may be either HNs or MNs, depending on if they report correct or tampered data, respectively.

Upon successfully detecting the *free* spectrum space, the CH triggers the distributed task allocation algorithm in the cluster to decide which CN should be selected

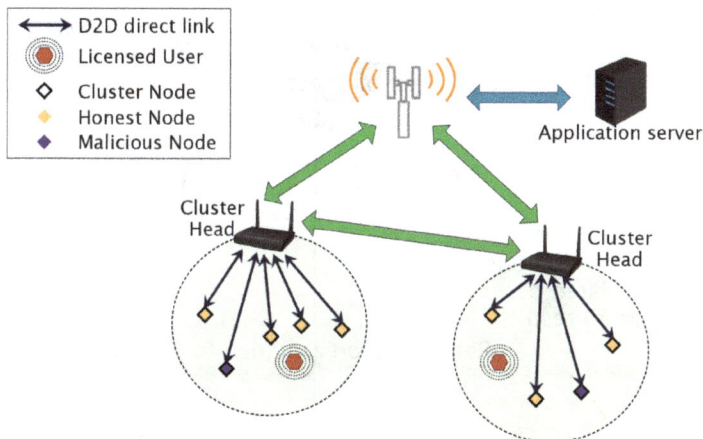

Figure 12.1 *Considered scenario of opportunistic nodes collaboration, where the CH is the only device connected to the application server via an Internet link and coordinates the assignment of tasks to all the nodes*

to perform the task. Such distributed allocation relies on a non-cooperative auction game in which CNs bid to win the task, in an effort to selfishly increase their own utility. Each CN chooses the bid value that maximizes its utility, which is defined as a trade-off between the cost in terms of energy to be consumed by the winner for the task to be executed, and the gain that is won by the CN for its contribution to spectrum sensing and for the execution of the task (in case it wins the competition). The CH assigns the task to the CN that proposes the lowest bid value and, after sending the task results to the CH, the winning CN receives by it a reward that corresponds to the proposed bid. Indeed, the competition is won by the lowest bid so that the CH (and the overall system it represents) can keep the rewards to be given for each task allocation low. The workflow of the proposed scheme is summarized in Figure 12.2.

12.4 The task allocation strategy

As demonstrated by Rahman *et al.* [4], the model described above works well if all the involved nodes are benevolent. In the remainder of this chapter, we will focus on a scenario with MNs attacking the system by simply tampering with the spectrum sensing information sent to the CH. The following paragraphs will focus on analyzing the effects caused by the presence of MNs in the reference scenario and providing guidelines for the design of possible countermeasures that could mitigate them.

12.4.1 Spectrum sensing in malicious cognitive IoT

In the reference system, CRs are employed to detect spectrum white spaces in licensed bands, where intra-cluster communication can be established. If any LU

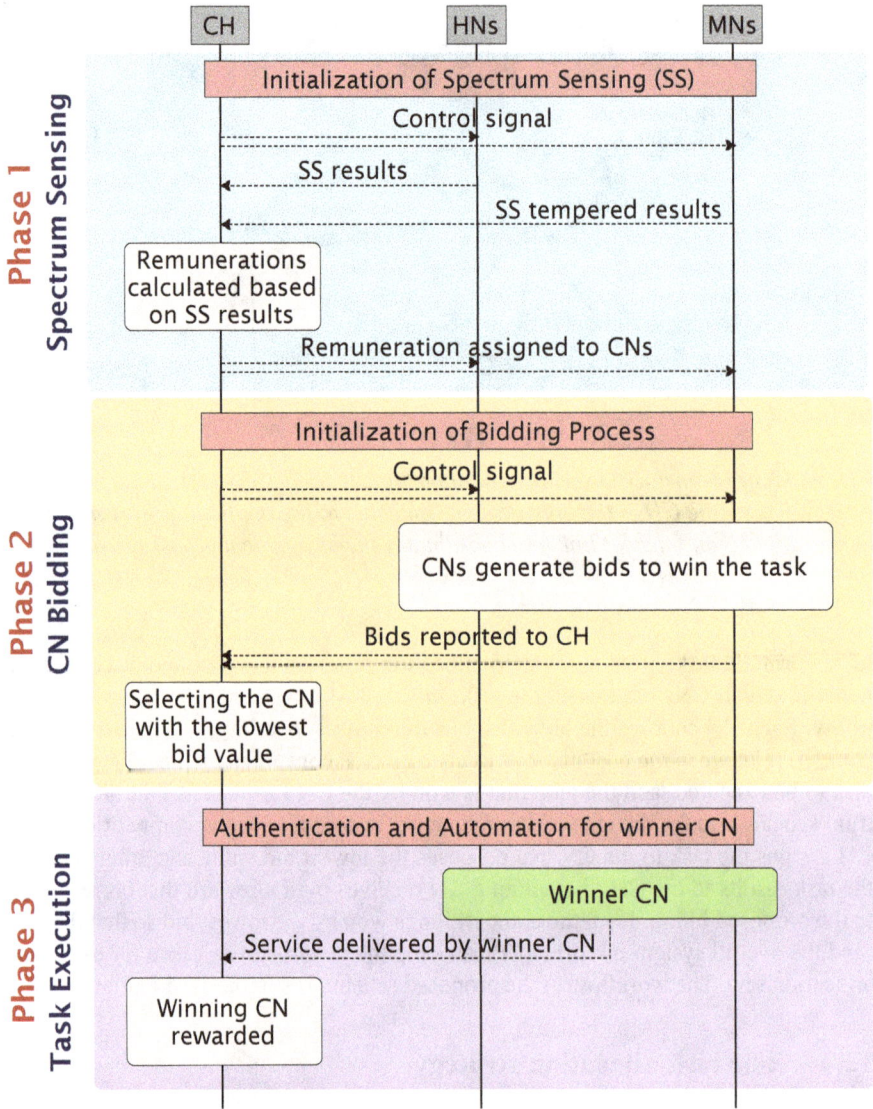

Figure 12.2   Workflow and timeline of messages for the proposed reward-based strategy. Several control signals are necessary between CH and CNs for better coordination and execution of services. Such control signals are characterized as lossless and low-rate signals. Multiple shades are used in order to differentiate different phases of the strategy for optimal task allocation.

activity is detected in the band, CNs either delay the scheduling process or look for other opportunities in other bands. The signal $y_i(t)$ received by CN i at a time t can be expressed as

$$y_i(t) = s_i(t) + n_i(t) \tag{12.1}$$

where $s_i(t)$ is the LU signal at the i-th CN with variance $\sigma_{s,i}^2$, and $n_i(t)$ is the additive white Gaussian noise (AWGN) with zero-mean and variance $\sigma_{n,i}^2$. The signal-to-noise ratio (SNR) γ_i of CN i, computed over N samples in an AWGN-only environment, is

$$\gamma_i = \frac{\sigma_{s,i}^2}{\sigma_{n,i}^2} \tag{12.2}$$

It is evident from this equation that each CN experiences different γ_i values, which is consistent with their diverse positions inside the cluster.

We assume that all the CNs in a cluster collect N samples to have a robust measure of spectrum sensing. Given its low complexity and simple implementation, spectrum sensing based on energy detection is considered here, being it an efficient technique [21]. Such technique relies on the energy of the signal received by each CN, defined as

$$\xi_i = \sum_{t=1}^{N} |y_i(t)|^2 \tag{12.3}$$

Based on the energy calculation ξ, the CH first estimates the CNs' reliability in terms of local probabilities of detection P_d^i and false alarm P_{fa}^i, considering the operating channel environment (i.e., AWGN or fading channel) [21]. More specifically, the P_d^i expresses the probability that, based on the information delivered by CN i, is possible to correctly assume that the spectrum band is occupied by an LU. On the other hand, the P_{fa}^i expresses the probability associated with the CN i that it detects the presence of an LU by mistake. Therefore, a favorable condition would be having high values of P_d^i and low values of P_{fa}^i.

We first assume that CNs are all honest. As defined in [22], P_d^i and P_{fa}^i can be computed as

$$P_d^i = Pr\left(\xi_i \geq \zeta_i | \Lambda_1\right) = Q\left(\frac{\xi_i - N\sigma_{n,i}^2\left(\gamma_i + 1\right)}{\sqrt{N}\sigma_{n,i}^2\left(\gamma_i + 1\right)}\right) \tag{12.4}$$

$$P_{fa}^i = Pr\left(\xi_i \geq \zeta_i | \Lambda_0\right) = Q\left(\frac{\xi_i - N\sigma_{n,i}^2}{\sqrt{N}\sigma_{n,i}^2}\right) \tag{12.5}$$

where, Λ_1 corresponds to the presence of an LU in the spectrum band (i.e., $\sigma_{s,i}^2 > 0$), Λ_0 corresponds to the absence of an LU (i.e., $\sigma_{s,i}^2 \approx 0$), ζ_i is a decision threshold, and $Q(\cdot)$ is the Q-function. By fixing the P_{fa}^i value so that it is low enough, it is possible to set the decision threshold ζ_i by inverting 12.5 as follows:

$$Q^{-1}\left(P_{fa}^i\right) = \left(\frac{\xi_i - N\sigma_{n,i}^2}{\sqrt{N}\sigma_{n,i}^2}\right) \tag{12.6}$$

Energy Detection based Spectrum Sensing

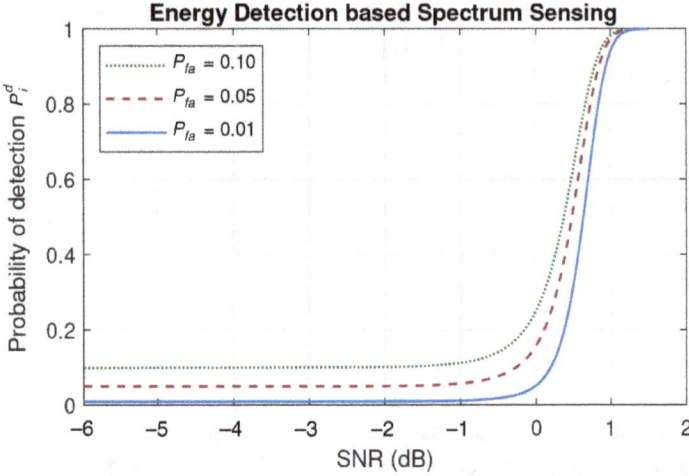

Figure 12.3   *Probability of detection P_d^i values for changing SNR values, with fixed probability of false alarm P_{fa}^i values*

According to (12.4), once the decision threshold ζ_i is set, the performance of the energy detector in correctly detecting an LU increases for high values of γ_i. This behavior is demonstrated by Figure 12.3, where the performance of the considered spectrum sensing technique for different P_{fa}^i values is shown. This figure depicts how P_d^i changes for SNR values ranging between −30 and 0 dB.

Based on the relevant information sent by CNs to the CH, it decides whether an LU is present or not by fusing the soft information sent by CNs. Different fusion rules can be considered. In particular, in this chapter, we assume that the CH checks, once the threshold ζ_i is determined, how many CNs decide for LU absent or present; at this point, one may apply AND (the LU is declared present if all CNs are favorable), OR (the LU is declared present if at least one CNs is favorable), or majority (the LU is declared present if most of the CNs are favorable) fusion rule. Even if several fusion rules exist (see, e.g. [23]), the AND and OR rules are of interest for the considered IoT scenario due to their simplicity. Such decision is complemented by the global probabilities of detection and false alarm, respectively P_D and P_F, computed the corresponding fusion rule. In particular, considering the AND, one as [22]

$$P_D = \prod_{i=1}^{N_c} \left(P_d^i\right), \qquad P_F = \prod_{i=1}^{N_c} \left(P_{fa}^i\right) \qquad (12.7)$$

and OR-rule as

$$P_D = 1 - \prod_{i=1}^{N_c} \left(1 - P_d^i\right), \qquad P_F = 1 - \prod_{i=1}^{N_c} \left(1 - P_{fa}^i\right) \qquad (12.8)$$

with N_c being the number of CNs that operate under a CH. Accordingly, the CH can evaluate the probability P_e of detecting the spectrum erroneously as [24]

$$P_e = P_0 \cdot P_F + P_1 \cdot (1 - P_D) \tag{12.9}$$

where P_0 is the a priori probability that an LU is not present in the spectrum band, otherwise $P_1 = 1 - P_0$. The CH learns about P_0 from the past appearances of the LU in the band.

In the presence of a malicious behavior, each MN tampers the energy sent to the CH according to

$$\tilde{\xi}_i = \begin{cases} \xi_i - \Delta_i & \text{if } \xi_i < \zeta_i \\ \xi_i + \Delta_i & \text{if } \xi_i \geq \zeta_i \end{cases} \tag{12.10}$$

where Δ_i is the attack intensity. In other words, the goal of the MN is to enforce the final decision on the spectrum occupancy in the same direction of the other nodes so that it can take advantage in the following game for task alloca-tion (as will be shown later in this section). A similar approach is considered in Gao *et al.* [19], where the MN tampers data toward the opposite direction: if the LU is present, the MN tries to force the final decision toward LU absent, and vice versa. Similarly to Gao *et al.* [19], the attack intensity is normalized as $\Delta_i = \rho_i \mathbb{E}\{\xi | \Lambda_0\}$, being ρ_i the attack intensity parameter. After rearranging terms and taking into account the node's false alarm and detection probabilities, under Λ_0 the i-th MN sends the following quantity to the CH:

$$\tilde{\xi}_i = \begin{cases} \xi_i - \Delta_i & \text{with probability } 1 - P_{fa}^i \\ \xi_i + \Delta_i & \text{with probability } P_{fa}^i. \end{cases} \tag{12.11}$$

On the other hand, under Λ_1 the following soft information is sent by the i-th MN:

$$\tilde{\xi}_i = \begin{cases} \xi_i - \Delta_i & \text{with probability } 1 - P_d^i \\ \xi_i + \Delta_i & \text{with probability } P_d^i. \end{cases} \tag{12.12}$$

At this point, the final decision at the CH is taken, according to (12.7) and (12.9), as all CNs are honest without trying to detect an MN behavior or counter-acting it (see, e.g. [19] and references therein). This extension goes beyond the scope of this chapter and is a matter of future investigation.

Each CH that provided correct information on the presence or absence of the LU is then awarded a remuneration factor ς_i, which is assessed consider-ing an approach similar to the one in Rajasekharan *et al.* [25]. Accordingly, ς_i is computed as a function of P_d^i, so as to encourage a reduction of the uncertainty about the LU activity

$$\varsigma_i = 1 - \left[-P_d^i \log_2 \left(P_d^i \right) - \left(1 - P_d^i \right) \log_2 \left(1 - P_d^i \right) \right] \tag{12.13}$$

Figure 12.4 shows how ς_i changes for different values of P_d^i. As it is evident from the figure, when $P_d^i < 0.5$ the remuneration is given to the CN only if the CH estimates that the LU is absent. On the other hand, when $P_d^i > 0$, the remuneration is given to the CN only if the CH estimates that the LU is present. It is worth noticing that for $P_d^i = 0.5$ the remuneration is zero, consistently with the fact that the infor-mation on the spectrum sensing provided by that CH is meaningless. Note that if a

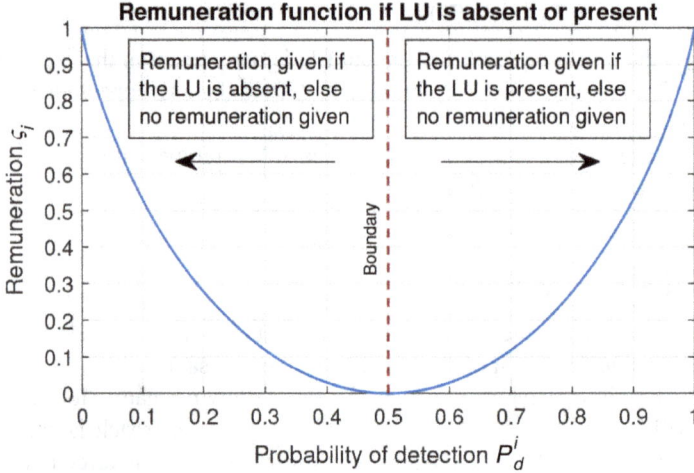

Figure 12.4   Remuneration values, computed according to (12.13), are shown as a function of P_d^i

CN has a malicious behavior, the remuneration is altered according to the SNR (and, correspondingly, the detection probability) experienced by the MN. This means that when the MN provides higher $\tilde{\xi}_i$ values it increases its P_d^i value, whereas when it provides lower $\tilde{\xi}_i$ values its P_d^i value decreases. Therefore, under Λ_0 condition, if $\tilde{\xi}_i = \xi_i + \Delta_i$ the remuneration value ς_i decreases, whereas if $\tilde{\xi}_i = \xi_i - \Delta_i$ the remuneration value increases. Note that, according to (12.11), the latter is the most likely to occur. On the other hand, when the LU is present (Λ_1), if $\tilde{\xi}_i = \xi_i + \Delta_i$ the remuneration value ς_i increases, whereas if $\tilde{\xi}_i = \xi_i - \Delta_i$ the remuneration value decreases. Considering (12.12), the former is the most likely to occur. These last conclusions are in line with the fact that the objective of MNs is to tamper the reported sensing value to increase its remuneration.

12.4.2   Cluster node bidding

Upon detecting the spectrum white spaces, the CH proceeds with the task allocation procedure, based on a first-price sealed-bid auction [26]*. Accordingly, each CN evaluates the bid $b_{k,i}$ that maximizes its own utility function $H_i(b_{k,i})$, expressed as a trade-off between the cost of executing task k and the benefit of having a reward in return. Indeed, the cost is expressed in terms of the energetic cost of CN i to perform task k, which is in inverse proportion with its residual energy after executing the task and transmitting the result to the CH: the lower the residual energy after task k's execution, the higher the cost for CN i. Assuming that the initial energy budget of CN i is E_β^i, the residual energy $\chi_{k,i}$ is given as

---

*Note that the auction described below can be straightforwardly changed to consider multiple winners and/or different auction types (e.g., Discriminatory, Uniform Price, Vickrey Multi-Unit [27]), by adjusting the relevant equations: the results would change, but the process would remain the same.

$$\chi_{k,i} = E_\beta^i - E_{tx}^{k,i} \tag{12.14}$$

where $E_{tx}^{k,i}$ is the energy required by the i-th CN to deliver task k to the CH.

The utility function also accounts for the fact that the higher the proposed bid the lower the probability to win the competition with the other CNs. The CN utility function is then defined as

$$H_i(b_{k,i}) = \Psi(b_{k,i}) \left(\varsigma_i b_{k,i} - \alpha \frac{1}{\chi_{k,i}} \right) \tag{12.15}$$

where $\Psi(b_{k,i})$ is the probability of CN i to win the competition for task k by proposing the bid $b_{k,i}$ as assessed individually by every CN based on past outcomes (refer to [4] for details), ς_i is the remuneration factor (see (12.13)) resulting from the spectrum sensing, α is a weighing factor and $1/\chi_{k,i}$ is the cost associated with the residual energy level if CN i wins the competition to perform task k. Considering a first-price sealed-bid auction, the aim of each CN is to find the lowest bid value that maximizes its own utility function. Such a game can be modeled using a Nash game [28], where the Nash equilibrium point (NEP) is exactly the point where no CN can propose a lower bid value without decreasing its own utility for the considered task. This means that the solution to this non-cooperative game is an NEP that ensures that no object has any incentive to deviate unilaterally from it [26]. The detailed solution of such a game can be found in Rahman *et al.* [4].

It is also possible to prove that the NEP corresponds to the solution that maximizes the system utility. Indeed, the overall system utility is maximized when the selected CN (i.e., the CN that wins the competition) achieves its highest utility with a reward that is the lowest among the CNs in the cluster. Therefore, the expected system utility for task k is defined as

$$E[\mathcal{U}_k] = \sum_{i=1}^{N_C} H_i(b_{k,i}) \cdot x_{k,i}, \qquad \text{with } x_{k,i} = \{0, 1\} \tag{12.16}$$

where $x_{k,i}$ is 1 if task k is assigned to CN i, whereas it is 0 otherwise. Considering also that task k is assigned to CN i when

$$b_{k,i} = b^{min} = \min \{ b_{k,1}, b_{k,2}, \cdots, b_{k,N_C} \} \tag{12.17}$$

the highest expected system utility value for task k corresponds to the solution of this optimization problem

$$\max_{\mathbf{x}} \sum_{i=1}^{N_C} H_i(b) \cdot x_{k,i} \tag{12.18}$$

$$\text{s.t.} \sum_i x_{k,i} = 1 \tag{12.18a}$$

$$H_i(b_{k,i}) \geq H_i(b) \quad \forall i \tag{12.18b}$$

$$b^{min} = \sum_i b_{k,i} \cdot x_{k,i} \tag{12.18c}$$

$$b_{k,i} \geq b^{min} \quad \forall i \tag{12.18d}$$

Table 12.1   Simulation parameters

Parameters	Value
Total nodes (HNs + MNs)	10
Energy cost coefficient, α	−2
E^i_β[mJ]	Uniformly distributed [95,100]
E^i_{tx}[mJ]	5
$\sigma_{s,i}$[dB]	[−10, 0]
MN	1, 2
Probability of occupancy, P_1	0.7
Attack Intensity, ρ	0.1, 0.3, 0.5
MN activity time	0, 0.3, 0.5, 1

where $\mathbf{X} = [[x_{k,i}]^{N_c}, b, [b_{k,i}]^{N_c}, b^{min}]$ is the array of the considered variables. Equations (12.18a)–(12.18d) ensure that the following conditions are satisfied: task k is assigned to exactly one CN (12.18a); the highest value of the expected utility function $H_i(b)$ for CN i corresponds to a bid equal to $b_{k,i}$, i.e., the optimal bid for CN i (12.18b); the value of the winning bid b^{min} is equal to the value of the optimal bid of the winning CN (12.18c); all the optimal bids are equal to or greater than the winning bid b^{min} (12.18d).

Finally, the comprehensive utility of the system, defined as the utility that the system achieves when k tasks are assigned to the optimal CNs and appropriate rewards are provided in return, can be then computed as

$$u = \sum_{i=1}^{N_c} \sum_{k=1}^{K} \left(\varsigma_i b_{k,i} - \alpha \frac{1}{\chi_{k,i}} \right) \cdot x_{k,i} \tag{12.19}$$

12.5   Simulation results

The game theoretical-based task allocation method with the presence of malicious cognitive spectrum sensing presented in Section 12.4 is validated here by means of MATLAB®-based simulations. We consider an illustrative scenario with parameters listed in Table 12.1.[†] The proposed approach follows sequential mechanisms of cooperative spectrum sensing and auctioning (bidding) using game theoretical framework. At first, we evaluate the performance of the game theoretical framework against malicious cooperative spectrum sensing when MNs are deliberately

---

[†]The N samples are acquired at a sampling rate of 125 kbps for a sensing time of 0.24 msec.

tempering their spectrum sensing results. Indeed, the remuneration factor ς_i depends on the type of fusion rule employed in cooperative spectrum sensing. We, therefore, consider the AND and OR fusion rules for cooperative spectrum sensing for different scenarios and throughout the simulations we are keeping the probability of spectrum occupancy by LU as $P_1 = 0.7$. For each task, 1,000 independent Monte-Carlo simulations are run and performance is averaged over these runs.

We first investigate the impact of MN activity on the offered bid. In Figure 12.5, the average bid values offered by each CN are shown, as a function of the number of tasks in the game, with 1 MN with $\rho = 0.5$ and two possible values for the false alarm rates P_{fa}^i, which in the following is referred to with P_f: (1) $P_f = 0.01$ and (2) $P_f = 0.1$. The bold violet line is for a minimum value of the bid for the

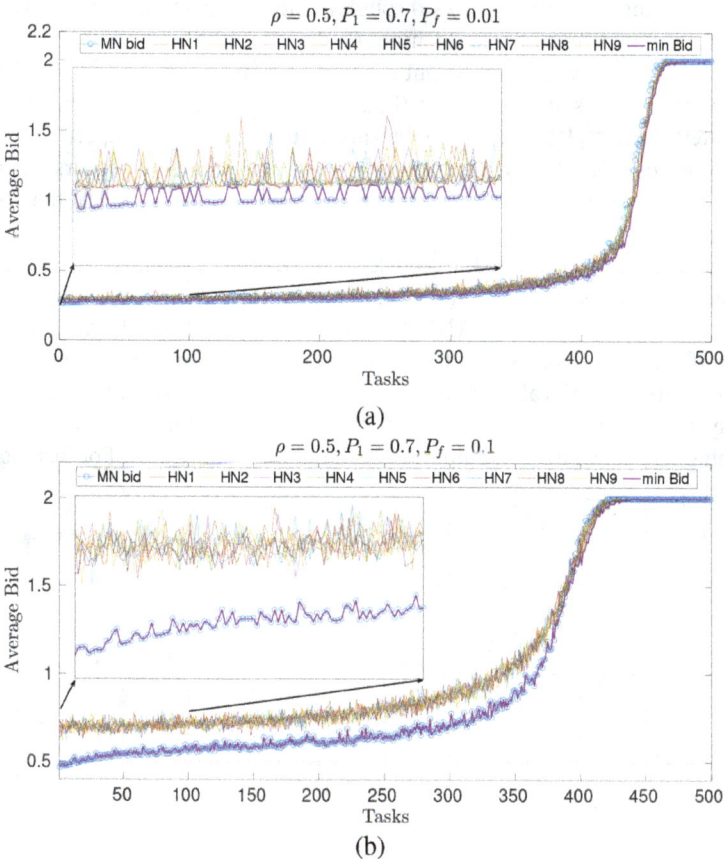

(a)

(b)

Figure 12.5   *Average bid values offered by each CN, as a function of the number of tasks in the game, with 1 MN with $\rho = 0.5$ and two possible false alarm rates: (A) $P_f = 0.01$ and (B) $P_f = 0.1$. The bold violet line is for a minimum value of the bid for the specific task. AND fusion rule is considered.*

specific task. AND fusion rule is considered. The probability of false alarm P_f controls the reliability of spectrum sensing that has an impact on the following bidding process. In fact, P_f controls the probability of detection P_d that eventually drives the remuneration factor ς_i, inversely related to the bid value. This property is exploited by MNs by sending in tempered P_d values in order to attain maximum ς_i among all the nodes in the network. If P_f is relatively low, for example, 0.01 as in case (A), the minimum bid is not very much deviated from the rest of the bids. In fact, with $P_f = 0.01$, the spectrum sensing is regarded as reliable making the competition stiff. On the other hand, when P_f increases, for example, to 0.1, this leads to a huge difference between the winning bid (bold) value and rest of the bids from the other users indicating unreliability in the system which is a consequence of unreliable spectrum sensing. This unreliability in the system is exploited by MNs by winning most of the tasks through malicious spectrum sensing. However, it should be noted that in case of high P_f, i.e., 0.1, the nodes' bid values reach the saturation point with a significant number of tasks still to be auctioned by CH. Note that when the saturation point (bid = 2) is reached the nodes have depleted their battery energy. Beyond this point, nodes are unable to compete for task and the game is considered over. In case of $P_f = 0.01$, the saturation occurs when more than 450 tasks are performed by the nodes, stamping the efficiency of the proposed approach when spectrum sensing is reliable.

This aspect is investigated in Figure 12.6, where the average number of tasks allocated to MNs is shown, as a function of the attack intensity, for various numbers of MNs and false alarm rates. AND fusion rule is considered. Recall that the total number of available tasks is 100. It can be observed that, for increasing attack intensity, the number of tasks allocated to the MNs increases. In particular, multiple MNs in the system attain aggressive policy with higher attack intensities when it comes to winning the tasks through tempering spectrum sensing results. For instance, with

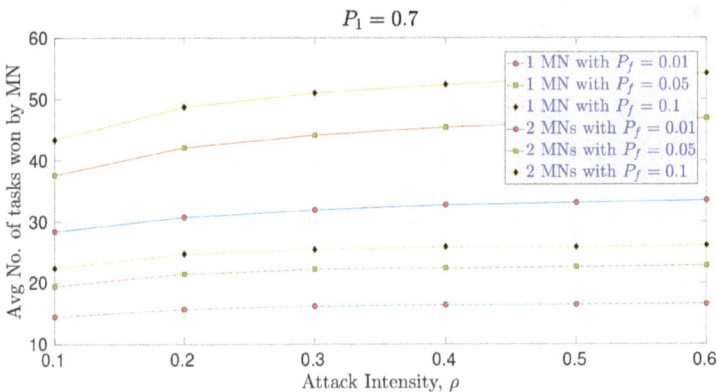

Figure 12.6   *Average number of tasks allocated to MNs, as a function of the attack intensity, for various number of MNs and false alarm rates. AND fusion rule is considered.*

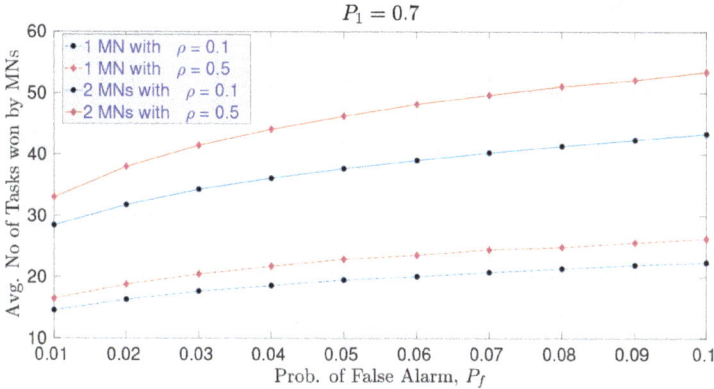

Figure 12.7   *Average number of tasks won by MNs, as a function of the false alarm rate, for a different number of MNs and attack intensity. AND fusion rule is considered.*

2 MNs more than 50 tasks are collectively allocated to them when the false alarm rate is 0.1. The trend appears to be saturated as the attack intensity increases due to depleting energy resources of MNs. The tasks allocated to the MNs reduce with the false alarm rate arriving to 10–20 with just 1 MN and $P_f = 0.01$.

The impact of the false alarm rate is deeper, as investigated in Figure 12.7, where the average number of tasks won by the MNs is shown, as a function of the false alarm rate, for a different number of MNs and attack intensity. AND fusion rule is still considered. Similar considerations to those given in Figure 12.6 are valid in this case as well.

Finally, we investigate the impact of unreliable spectrum sensing resulting in unallocated tasks, i.e., tasks that are not allocated either because (i) an LU is correctly detected, (ii) an LU is detected by mistake (determining a false alarm), and (iii) an LU is not detected even though it is present (determining a missed detection), thus causing a collision in the communication between the nodes and the CH, during or after the task assignment. In Figure 12.8, the average number of unallocated tasks is shown, as a function of the false alarm rate, for different values of the MN activity time. This parameter is representative of the percentage of time that a CN operates as an MN instead of an HN. For instance, MN activity time equal to 0.3 means that a CN acts as malicious for 30% of the time. Both AND and OR fusion rules are considered in cooperative spectrum sensing in order to analyze system performance in terms of unallocated tasks. In Figure 12.8, a bimodal behavior can be observed at a false alarm rate of 0.04. With false alarm rate below 0.04, average unallocated tasks are due to miss-detection of LU that occupies the spectrum 70% of the time. When MN activity time is zero, the unallocated task is approximately equal to 70 and increases as MN activity time increases. As the false alarm rate increases beyond 0.04, the overwhelming effects of false alarm appear to dominate and unallocated tasks increase logarithmically.

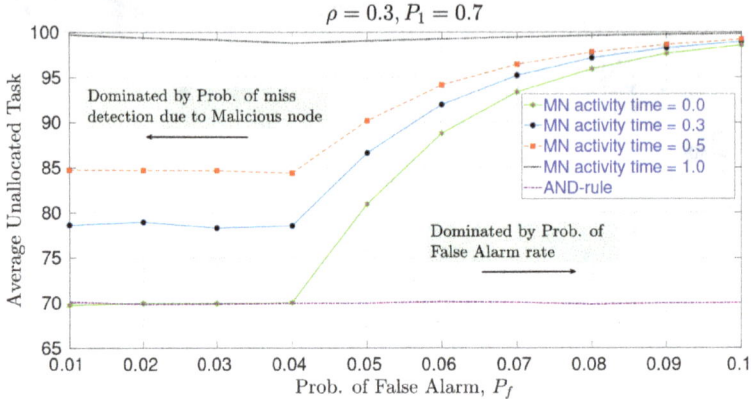

Figure 12.8   Average number of unallocated tasks, as a function of the false alarm rate, for different values of the MN activity time. Both AND and OR fusion rules are considered.

12.6   Concluding remarks

In this chapter, we have considered a cognitive IoT-based scenario, in which sensor nodes belonging to different platforms form different clusters managed by their own CH. The nodes collaboratively sense the spectrum, through standard energy detection, to find spectrum holes to be opportunistically exploited for task allocation by the CH. In particular, a game-theoretical approach for task allocation taken from the literature is considered. We extend this approach to analyze the overall performance in the presence of MNs trying to disrupt this process by providing tampered data, so that they try to gain as much as possible control of the allocated task, thus performing a DoS attack. Our simulation-based results have highlighted the impact of the main system parameters on the overall performance and provide guidelines for future work in the direction of MN activity detection.

References

[1]   Benyamina D., Hafid A., Gendreau M. 'Wireless mesh networks design — a survey'. *IEEE Communications Surveys & Tutorials*. 2012, second quarter;14(2):299–310.

[2]   Atzori L., Iera A., Morabito G. 'Understanding the Internet of things: definition, potentials, and societal role of a fast evolving paradigm'. *Ad Hoc Networks*. 2017;56(6):122–40.

[3]   Zhang L., Ding G., Wu Q., Zou Y., Han Z., Wang J. 'Byzantine attack and defense in cognitive radio networks: a survey'. *IEEE Communications Surveys & Tutorials*. 2015, third quarter;17(3):1342–63.

[4]   Rahman T.F., Pilloni V., Atzori L. 'Application task allocation in cognitive IoT: a reward-driven game theoretical approach'. *IEEE Transactions on Wireless Communications*. 2019;18(12):5571–83.

[5] Haghighi M. 'Market-based resource allocation for energy-efficient execution of multiple concurrent applications in wireless sensor networks' in Park J., Adeli H., Park N., Woungang I. (eds.). *Mobile, Ubiquitous, and Intelligent Computing. Lecture Notes in Electrical Engineering*. 274. Berlin, Heidelberg: Springer; 2014. pp. 173–8. Available from https://doi.org/10.1007/978-3-642-40675-1_27.

[6] Fei Z., Li B., Yang S., Xing C., Chen H., Hanzo L. 'A survey of multiobjective optimization in wireless sensor networks: metrics, algorithms, and open problems'. *IEEE Communications Surveys & Tutorials*. 2017, first quarter;19(1):550–86.

[7] Pilloni V., Abd-Elrahman E., Hadji M., Atzori L., Afifi H. 'IoT_ProSe: exploiting 3GPP services for task allocation in the Internet of things'. *Ad Hoc Networks*. 2017;66(5):26–39.

[8] Guinard D., Trifa V., Mattern F., Wilde E. in Uckelmann D., Harrison M., Michahelles F. (eds.) *From the Internet of Things to the Web of Things: Resource-Oriented Architecture and Best Practices*. Berlin, Heidelberg: Springer; 2011. pp. 97–129.

[9] Silverajan B., Harju J. 'Developing network software and communications protocols towards the internet of things'. Proceedings of the Fourth International ICST Conference on COMmunication System softWAre and middlewaRE. COMSWARE '09; New York: ACM; 2009. pp. 9:1–9:8.

[10] Kwak J., Kim Y., Lee J., Chong S., *et al.* 'Dream: dynamic resource and task allocation for energy minimization in mobile cloud systems'. *IEEE Journal on Selected Areas in Communications*. 2015;33(12):2510–23.

[11] Dinh T.Q., Tang J., QD L., *et al.* 'Offloading in mobile edge computing: task allocation and computational frequency scaling'. *IEEE Transactions on Communications*. 2017;65(8):3571–84.

[12] Wang P., Yao C., Zheng Z., Sun G., Song L. 'Joint task assignment, transmission, and computing resource allocation in multilayer mobile edge computing systems'. *IEEE Internet of Things Journal*. 2019;6(2):2872–84.

[13] Wang Y., Chen I.-R., Cho J.-H., Tsai J.J.P. 'Trust-based task assignment with multiobjective optimization in service-oriented adhoc networks'. *IEEE Transactions on Network and Service Management*. 2017;14(1):217–32.

[14] Sakr A.H., Tabassum H., Hossain E., Kim D.I. 'Cognitive spectrum access in device-to-device-enabled cellular networks'. *IEEE Communications Magazine*. 2015;53(7):126–33.

[15] Rahman T.F., Sacchi C. 'Opportunistic radio access techniques for emergency communications: preliminary analysis and results'. 2012 IEEE First AESS European Conference on Satellite Telecommunications (ESTEL); 2012. pp. 1–7.

[16] Kakalou I., Psannis K.E. 'Sustainable and efficient data collection in cognitive radio sensor networks'. *IEEE Transactions on Sustainable Computing*. 2019-March;4(1):29–38.

[17] Aijaz A., Aghvami A.H. 'Cognitive machine-to-machine communications for internet-of-things: a protocol stack perspective'. *IEEE Internet of Things Journal*. 2015;2(2):103–12.

[18] Khan A.A., Rehmani M.H., Rachedi A. 'Cognitive-radio-based internet of things: applications, architectures, spectrum related functionalities, and future research directions'. *IEEE Wireless Communications*. 2017;24(3):17–25.

[19] Gao R., Zhang Z., Zhang M., Yang J., Qi P. 'A cooperative spectrum sensing scheme in malicious cognitive radio networks'. Proceedings of IEEE Global Telecommunications Conference (GLOBECOM); 2019. pp. 1–5.

[20] Wu J., Li P., Chen Y., *et al.* 'Analysis of Byzantine attack strategy for cooperative spectrum sensing'. *IEEE Communications Letters*. 2020;24(8):1631–5.

[21] Digham F.F., Alouini M.-S., Simon M.K. 'On the energy detection of unknown signals over fading channels'. *IEEE Transactions on Communications*. 2007;55(1):21–4.

[22] Li T., Yuan J., Torlak M. 'Network throughput optimization for random access narrowband cognitive radio Internet of things (NB-CR-IoT'. *IEEE Internet of Things Journal*. 2018;5(3):1436–48.

[23] Yang T., Wu Y., Li L., Xu W., Tan W. 'Fusion rule based on dynamic grouping for cooperative spectrum sensing in cognitive radio'. *IEEE Access*. 2019;7:51630–9.

[24] Althunibat S., Granelli F. 'Novel energy-efficient reporting scheme for spectrum sensing results in cognitive radio'. 2013 IEEE International Conference on Communications (ICC); 2013. pp. 2438–42.

[25] Rajasekharan J., Eriksson J., Koivunen V. 'Cooperative game theory and auctioning for spectrum allocation in cognitive radios'. IEEE International Symposium on Personal, Indoor and Mobile Radio Communications; Toronto, Canada; 2011. pp. 656–60.

[26] Kelechi A.H., Abdullah N.F., Nordin R., Ismail M. 'Smart: coordinated double-sided seal bid multiunit first price auction mechanism for cloud-based TVWS secondary spectrum market'. *IEEE Access*. 2017;5:25958–71.

[27] Markakis E., Telelis O. 'Uniform price auctions: equilibria and efficiency'. *Theory of Computing Systems*. 2015;57(3):549–75.

[28] Maschler M., Solan E., Zamir S. *Game Theory*. New York, NY, USA: Cambridge University Press; 2013.

Chapter 13

Conclusions and future perspectives

Luca Davoli[1] and Gianluigi Ferrari[1]

The aim of the chapters contained in this book has been to introduce, overview, and discuss the concept of wireless mesh network (WMN) and its adoption in different heterogeneous contexts and scenarios. We have highlighted how this network topology has (and will have in the future) several implications in the architecture's modeling and requires to take decisions and actions at different layers.

First of all, the choice of the communication protocol (at low layers) at the basis of the WMN is crucial. As discussed in the book, Bluetooth Low Energy (BLE) and IEEE 802.11 seem to be attractive candidates to implement WMNs, yet requiring (i) a careful analysis of the constraints relevant for each network protocol proper of each network protocol and (ii) the definition and adoption of proper routing protocols to be able to forward the data from source to destination. Routing is fundamental in WMNs, since WMNs will be more and more heterogeneous (in terms of nodes' capabilities and architectures' characteristics) and, thus, will require to save even more energy in the presence of constrained devices. At the same time, reliability will also play a key role, with data having to flow from one side of the network to the other. In this regard, WMNs will also require appropriate decisions on network nodes' addressing policies. In fact, a large number of connected devices may require the adoption of (i) flooding-like strategies, in this case, the backbone portion of a WMN can support this mechanism (e.g., a BLE-oriented network would fit to this approach) or (ii) strategies which allow to address and target specific subsets of devices (e.g., *unicast*, *multicast*, and *broadcast*).

The choices described in the previous paragraph have impact on all scenarios, including the case of body sensor networks (BSNs), which will represent the next generation of wearable networks for monitoring people remotely, thus targeting the so-called Health 4.0 paradigm. To this end, *intra-*, *inter-*, and *beyond-*BSNs will face different requirements and challenges associated with each layer of their specific communication stacks. Moreover, BSNs will need to take into account emerging security aspects and threats that these particular WMNs will face: for example, wearable devices may represent gateways for the BSN and, in turn, should be

[1]Internet of Things (IoT) Laboratory, Department of Engineering and Architecture, University of Parma, Parco Area delle Scienze, Parma, Italy

carefully protected. Finally, there will be a need to carefully manage aspects such as energy efficiency, reliability, delay optimization, and bio-compatibility of the BSN nodes, since these will play a crucial role even on the communication side for the signal propagation and data formatting (e.g., customizable protocols will be needed to support emerging BSN applications).

The same considerations will apply even in the case of WMNs to be deployed in the presence of mobile nodes, such as flying drones used to monitor large areas on the ground and conveying data to remote processing entities through multi-hop communications (e.g., exploiting other Unmanned Aerial Vehicle (UAV) as relay nodes). Therefore, these mobile scenarios will require carefully modeling the interactions among the drones composing the storm and the backbone network (but possibly involving also other types of mobile nodes), as well as analyzing and defining alternative routing solutions in case certain routes disappear (i.e., their performance will become unacceptable). In this case, "mixed" networks solutions based on the parallelization of multiple heterogeneous network technologies, to be used as "backup" solutions in these situations, will represent a key decision to avoid connectivity lags and lacks.

Taking into account these heterogeneous contexts and scenarios, it is clearer how mesh-oriented architectures are suitable and useful to extend the network coverage in both *indoor* and *outdoor* contexts, from smart agriculture and rural-area-monitoring scenarios to smart industry and smart city-oriented applications. In several cases, in fact, either because of geographical position or because of the presence of obstacles, it is not always possible to rely on cellular networks or standard IEEE 802.11 connectivity. Therefore, all the data generated inside these WMNs— likely *big data*—not only will bring new challenges, in terms of data processing and management requirements but will also play a fundamental role, representing the real "treasure" which developers and entrepreneurs should exploit, especially in terms of information analysis and processing, to provide enhanced and intelligent services, especially through the application of artificial intelligence (AI)-based mechanisms and techniques.

Luca Davoli
Gianluigi Ferrari

Index